TRENDS IN

Molecular Biology
and
Biotechnology

TRENDS IN

Molecular Biology
and
Biotechnology

Editors

SHEELA SRIVASTAVA
Department of Genetics
South Campus, Delhi University, New Delhi 110021

P.S. SRIVASTAVA
Department of Environmental Botany
Hamdard University, New Delhi 110062

B.N. TIWARY
Department of Botany
Patna University, Patna 800005

CBS Publishers & Distributors Pvt. Ltd.

New Delhi • Bengaluru • Chennai • Kochi • Kolkata • Mumbai
Hyderabad • Nagpur • Patna • Pune • Vijayawada

ISBN: 81-239-0469-X

First Edition: 1996
Reprint: 2000, 2007, 2012, 2019

Published by **Satish Kumar Jain** and produced by **Varun Jain** for
CBS Publishers & Distributors Pvt. Ltd.,
4819/XI Prahlad Street, 24 Ansari Road, Daryaganj, New Delhi - 110002
delhi@cbspd.com, cbspubs@airtelmail.in • www.cbspd.com
Ph.: 23289259, 23266861, 23266867 • Fax: 011-23243014

Corporate Office: 204 FIE, Industrial Area, Patparganj, Delhi - 110 092
Ph: 49344934 • Fax: 011-49344935
E-mail: publishing@cbspd.com • publicity@cbspd.com

Branches:
• *Bengaluru:* 2975, 17th Cross, K.R. Road, Bansankari 2nd Stage,
  Bengaluru - 70 • Ph: +91-80-26771678/79 • Fax: +91-80-26771680
  E-mail: cbsbng@gmail.com, bangalore@cbspd.com
• *Chennai:* No. 7, Subbaraya Street, Shenoy Nagar, Chennai - 600030
  Ph: +91-44-26681266, 26680620 • Fax: +91-44-42032115
  E-mail: chennai@cbspd.com
• *Kochi:* Ashana House, 39/1904, A.M. Thomas Road, Valanjambalam,
  Ernakulum, Kochi • Ph: +91-484-4059061-65
  Fax: +91-484-4059065 • E-mail: cochin@cbspd.com
• *Kolkata:* 6-B, Ground Floor, Rameshwar Shaw Road, Kolkata - 700014
  Ph: +91-33-22891126/7/8 • E-mail: kolkata@cbspd.com
• *Mumbai:* 83-C, Dr. E. Moses Road, Worli, Mumbai - 400018
  Ph: +91-9833017933, 022-24902340/41 • E-mail: mumbai@cbspd.com

Representatives:

• Hyderabad: 0-9885175004	• Nagpur: 0-9021734563
• Patna: 0-9334159340	• Pune: 0-9623451994
• Jharkhand: 0-9811541605	• Uttarakhand: 0-9716462459

Printed at:
J.S. Offset Printers, Delhi (India)

Late Professor Umakant Sinha
(1938-1990)

Foreword

I am highly privileged to write this foreword for the book **"Trend in Molecular Biology and Biotechnology"** that is being brought out in the memory of the late Professor Umakant Sinha.

Umakant Sinha was a research student in the Department of Genetics at Glasgow University when work on the genetics of *Aspergillus nidulans* was creating the basis of what came to be called somatic cell genetics. In the three years during which he collaborated with me I had the privilege of getting to know him as a person apart from recognizing his ability as a researcher. Details of his work at that time on the biochemical genetics of aromatic aminoacid biosynthesis and its reference to p-fluorophenylalanine resistance are discussed later in this book. Here I shall limit myself to what I learned of Umakant's personality.

Exceptional friendliness and tact towards other members of the Department, stimulating, yet reserved, participation in discussion and readiness to share information still stand out in my memory from those distant days. These qualities, and thoroughness in research, no doubt stood out in his later activities. They struck me again when many years later I was fortunate to be his guest in the group which he had so successfully created at Patna University. There, and later in New Delhi, Umakant and his wife were hosts of outstanding kindness and sensitivity. The activity and the excellent atmosphere of his group in Delhi and Patna leave no doubt that he did very well for genetics in India.

It is a matter of great satisfaction that Drs. P.S. Srivastava, Sheela Srivastava, and B.N. Tiwary, his associate and students decided to bring out this volume to perpetuate the memory of Professor Sinha. Their efforts are to be commended in not only being successful in their endeavour but also in getting an illustrious panel of contributors to write on highly varied topics.

I am hopeful that this book will serve the purpose of providing latest information on the exciting topics that were so dear to Umakant Sinha.

His friends and students sadly miss him on his tragic departure.

G. PONTECORVO

LONDON

Prefatory Note

This book owes its genesis to the fond memory of late Prof. Umakant Sinha, whose sad and untimely demise has left a void in the field of biological sciences more so in molecular genetics.

Born on February 5, 1938, Late Prof. Sinha had an illustrious academic career. Recipient of the University Gold Medal for obtaining first position in B.Sc. and M.Sc. (Botany), in 1964 he joined Prof. G. Pontecorvo, FRS, an international authority in *Aspergillus* genetics, at the University of Glasgow, U.K. on a Commonwealth Scholarship. He successfully completed the doctoral thesis on the genetics of aromatic amino acid biosynthesis in *A. nidulans* in 1967. His subsequent collaboration with Prof. G.S. Stent and Prof. J.M. Ashworth provided him the necessary training in the field of Physiology and Genetics of bacteriophages and *Dictyostelium*. In 1969, he joined Department of Botany, Delhi University and was responsible for establishing and strengthening the teaching and research in Microbial and Molecular Genetics. He took up the Professorship at the U.G.C. Centre of Special Assistance, Department of Botany, Patna University in 1981 which he subsequently headed till he breathed his last.

Prof. Umakant Sinha's major interest remained in the genetics of *Aspergillus nidulans* where he used the analogue (p-fluorophenylalanine=FPA) resistance as a model to understand the regulation of aromatic amino acid biosynthesis. The contribution by him and his students on FPA resistance has been highly acclaimed. He showed keen interest in fungal differentiation and haploidizing action of antimitotic chemicals like FPA, chloral hydrate and griseofulvin in microbial systems that was further extended to plants too. In later years he also investigated upon the biology of thermophilic bacterium *Thermoactinomyces*.

Prof. Sinha was a dedicated researcher and an accomplished teacher testified by more than 25 students who got their doctoral training under his able supervision. He was a prolific writer that is borne out by over 125 original technical papers in journals of national and international repute. He edited several volumes and co-authored two books, one on genetics and the other on bacteria for both Honours and Post-graduate students. He realized the importance of the subject and organized several symposia, seminars and workshops in emerging areas of modern biology and supervised a number of research projects.

Late Prof. Sinha was an extremely warm-hearted, pleasant personality. His forthrightness and helpful nature made him extremely popular amongst his associates and friends. For his contribution in the field of Botany he was awarded the prestigious Prof. Panchanan Maheshwari Medal in 1989. He was elected a member of the Council of the Indian Science Congress

Association (1988-89) and was appointed the first Recorder of the newly introduced section of Bio- chemistry, Biophysics, and Molecular Biology of the 78th session of Indian Science Congress, a job that remained unfulfilled.

After his premature demise in a road accident, it was strongly felt by us that his interest in the field of Genetics/Molecular Biology should be commemorated by bringing out a volume. For this, we contacted a large number of scientists who were associated with Prof. Sinha both in the country and abroad, inviting them to contribute articles for the present volume entitled "Trends in Molecular Biology and Biotechnology". The overwhelming response that was received speaks about the regard and affection earned by Prof. Sinha in the scientific community. Although many of his associates could not contribute to this volume due to prior commitments, they showed keen interest in the project and sent in their good wishes. The editors also feel most fortunate in getting Prof. G. Pontecorvo, FRS to write the Foreword for the book.

Molecular biology and biotechnology have emerged as the thrust areas in biological sciences. The tremendous impact generated by the discovery of gene structure, isolation and unravelling its function has led to researches to solve the enigma of life processes. The surge of literature is making it increasingly difficult to compile all the new information together. The rapid development coupled with successes make it imperative to bring together the available data for the benefit of interested researchers. The present book is an attempt in this direction. The volume comprises 16 chapters spread over 250 pages. These chapters relate to various aspects of molecular biology that would prove informative to the readers. The last two articles although do not conform to the basic theme of molecular biology, deal with the applications of the new technology that can be realized.

The editors are extremely thankful to all the contributors for their support in this endeavour that we wish to place on record. At this juncture, we also wish to express our gratitude to Dr (Ms) Sunita Sinha, wife of Late Professor Umakant Sinha, who has not only taken keen interest in this project but also been a constant source of inspiration. While preparing this book we have received unflinching support from our colleagues and students. Dr (Ms) Madhumati Purohit, Mr Deepak Gupta, Ms Deepshikha Pande and Ms Nisha Gupta helped us in various ways and deserve special praise. As our humble tribute to the illustrious personality of Late Prof. Umakant Sinha, we dedicate this book to his memory.

Finally, we wish to thank CBS publications for their cooperation and enthusiasm in bringing out this book.

EDITORS

New Delhi, April 1, 1996

Contents

Contributors

Shafique Alam
Department of Botany
Forbesganj College
Forbesganj
Bihar
INDIA

Bimal Kumar Bachhawat
Department of Biochemistry
University of Delhi
South Campus, Benito Jaurez Road
New Delhi 110021
INDIA

S.N. Bagchi
Department of Biological Sciences
R.D. University
Jabalpur 482001
INDIA

A. Banik
Molecular Biology Laboratory
Botany Department
Calcutta University
Calcutta 700019
INDIA

J.K. Bhattacharjee
Department of Microbiology
Miami University
Oxford, OH 45056
U.S.A.

I. Chaudhuri
Molecular Biology Laboratory
Botany Department
Calcutta University
Calcutta 700019
INDIA

R.K. Chaudhuri
Molecular Biology Laboratory
Botany Department
Calcutta University
Calcutta 700019
INDIA

P.K. Chitnis
Division of Biochemical Sciences
National Chemical Laboratory
Pune 411008
INDIA

A. John Clutterbuck
Genetics Department
University of Glasgow
Glasgow G11 5JS
Scotland
U.K.

K. Dharmalingam
Genetic Engineering Research Unit
Department of Biotechnology
School of Biological Sciences
Madurai Kamraj University
Madurai 625021
INDIA

Seymour Fogel
Department of Plant Biology
111 GPBB, University of California
Berkley, California 94720
U.S.A.

Ujjal K. Ghosh
Microbial & Molecular Genetics Lab
Department of Botany
Patna University
Patna 800005

V.S. Gupta
Division of Biochemical Sciences
National Chemical Laboratory
Pune 411008
INDIA

Habib Haque
Department of Biochemistry
Rajendra Agricultural University
Pusa 848125, Bihar
INDIA

Seyed E. Hasnain
National Institute of Immunology
Shahid Jeet Singh Marg
New Delhi 110067
INDIA

S.K. Jain
Department of Biochemistry
Hamdard University
Hamdard Nagar
New Delhi 110062
INDIA

Ayyamperumal Jeyaprakash
Department of Plant Biology
111 GPBB, University of California
Berkley, California 49720
U.S.A.

Prakash K. Jha
National Institute of Immunology
Shahid Jeet Singh Marg
New Delhi 110067
INDIA

Walter Klingmüller
Department of Genetics
University of Bayreuth
W 08580
Bayreuth
GERMANY

Santosh Kumar
Department of Botany
L.S. College
University of Bihar
Muzaffarpur 842001
INDIA

M.D. Lagu
Division of Biochemical Sciences
National Chemical Laboratory
Pune 411008
INDIA

P. Mallik
Molecular Biology Laboratory
Botany Department
Calcutta University
Calcutta 700019
INDIA

Mona Moonis
•Department of Biochemistry
University of Delhi, South Campus
New Delhi 110021
INDIA

D.K. Mukhopadhyay
Molecular Biology Laboratory
Botany Department, Calcutta University
Calcutta 700019
INDIA

V.V. Pethe
Division of Biochemical Sciences
National Chemical Laboratory
Pune 411008
INDIA

P.K. Ranjekar
Division of Biochemical Sciences
National Chemical Laboratory
Pune 411008
INDIA

Ajai K. Sharan
P.G. Centre of Botany
Maharaja College, Arrah
U.K.S. University
Arrah
INDIA

Alik N. Singh
P.G. Centre of Botany
Maharaja College, Arrah
U.K.S. University
Arrah
INDIA

Navin K. Sinha
Waksman Institute of Microbiology
Rutgers—The State University of
New Jersey
Piscataway, NJ 08855

Padma Sridhar
National Institute of Immunology
Shahid Jeet Singh Marg
New Delhi 110067
INDIA

Ashwani K. Srivastava
Faculty of Basic Sciences & Humanities
Rajendra Agricultural University
Pusa 848125
INDIA

Sheela Srivastava
Department of Genetics
University of Delhi
South Campus, Benito Jaurez Road
New Delhi 110021
INDIA

M.S. Swaminathan
Centre for Research on Sustainable
Agricultural and Rural Development
Madras
INDIA

Bhupendra N. Tiwary
Microbial and Molecular Genetics Lab
Department of Botany
Patna University
Patna 800005
INDIA

Kailash C. Upadhyaya
Waksman Institute of Microbiology
Rutgers—The State University of
New Jersey
Piscataway, NJ 8855

Jack C. Vaughn
Department of Zoology
Miami University
Oxford, OH 45056
U.S.A.

1

DNA mismatch repair mutants in *Saccharomyces cerevisiae:* Isolation and characterization

Seymour Fogel* and Ayyamperumal Jeyaprakash

Department of Plant Biology, 111 GPBB,
University of California Berkeley, California, 94720, USA

ABSTRACT

Mutations in three non-allelic yeast *PMS* genes (*pms1, pms2* and *pms3*) were implicated in specific DNA mismatch correctional repair during mitotic and meiotic divisions. Bacterial mutant strains with defective mismatch repair systems share certain similarities with the yeast *pms* mutants. The *PMS* gene products function during DNA replication and recombination and hence regulate the fidelity of DNA sequences. The *pms* mutants displayed increases in the frequencies of aberrant 5+:3- and 3+:5- postmeiotic segregations (PMS) at several heterozygous reporter loci occupied by mutant alleles attributable to simple DNA base substitutions, single base insertion frameshift mutations and small deletions/insertions.

INTRODUCTION

The ascomycetes comprise a group of organisms uniquely adapted to the study of recombination from both the viewpoints of genetic analysis and physical or molecular characterization via recombinant DNA techniques. A highly tractable yeast species, *Saccharomyces cerevisiae,* can be grown in unlimited quantities on chemically defined media. The discrete vegetative yeast cells exhibit a characteristic budding cycle, and appear either elliptical or spherical depending on whether they are polyploid, diploid or haploid. Two mating types, *MATa and MATα* are known. They can agglutinate each other and each excretes a pheromone that arrests the opposite mating type in the G1 growth phase. These cells may fuse, generating zygotes which can be

---

* Corresponding Author

readily induced to sporulate giving rise to a four-spored zygotic ascus. Alternatively, the zygote may undergo repeated vegetative budding cycles and the resultant diploid population may be sporulated by shifting early stationary phase, well-nourished cells to a nitrogen deficient medium containing 2% potassium acetate *in lieu* of a fermentable carbon source. The resulting biological unit, the ascus or tetrad, represents the total meiotic output of a single diploid cell that can be subjected to enzymatic digestion to remove the ascal sac. The subsequent micromanipulation in a perfectly controlled, orderly manner yields four intimately related ascosporal colonies. Each heterozygous site present in the parental diploid is expected to segregate in a mandelizing 2+: 2-pattern, where + and – denote the wild-type and mutant alleles of a single genetic locus. Unlike other eukaryotes, yeast hybrids simultaneously segregating for as many as 20 miscellaneous heterozygous sites, both linked and unlinked, are fully fertile and their tetrads can be routinely and unambiguously scored, either by replica plating or by means of various multiple inoculation devices. These tactical strategies are well known and the requisite simple equipment is readily available (see Guthrie & Fink, 1991).

Protocols relating to the isolation of DNA, preparation of genomic libraries, construction of plasmid shuttle vectors, transformation, molecular cloning, DNA sequencing etc., are equally well known and readily executed (Sherman *et al.*, 1986; Guthrie & Fink, 1991).

Given these mutually complementary approaches, our most recent findings concerning some specialized aspects of general recombination as revealed by a study of gene conversion, postmeiotic segregation (PMS), and genes that specifically affect the processing of the presumptive meiotic recombination intermediate, heteroduplex DNA or h: DNA have been described.

RESULTS

Gene conversions and postmeiotic segregations in yeast

Wild-type (+) and mutant (-) alleles present as heterozygous sites in diploids segregate meiotically in a normal mendelian pattern and are detected after ascus dissection as tetrads with 2+:2- spore colonies (or they may be formally considered as octads consisting of four spore pairs with 4+:4- spore colonies). Aberrant segregation patterns such as gene conversions and postmeiotic segregations (PMS) were known to exist in several related ascomycetous fungi. Zickler (1934) first reported 6+:2- and 2+:6- gene conversion events from *Bombardia lunata* and Olive (1956) reported the 5+:3- and 3+:5- postmeiotic segregations from *Sordaria fimicola*. Both gene conversions and PMS events were observed in every tested ascomycete including *Ascobolus* (Leblon, 1972 a,b), *Neurospora* (Case & Giles, 1964), *Saccharomyces cerevisiae* (Esposito, 1971; Fogel et al., 1979) and *Schizosaccharomyces pombe* (Gutz, 1971; Goldman, 1974). The significance of aberrant 4+:4- segregations has been emphasized elsewhere (see Fogel et al., 1979). In the four spored *S. cerevisiae* the normal mendelian segregation pattern was detected as 2+:2- tetrads, gene conversions as 3+:1- or 1+:3- asci and postmeiotic segregations as 2+:1-:1+/- or 1+:2-:1+/- o respectively, 5+:3- and 3+:5- in a tetrad array considered as an octad. The presence of a sectored (+/-) spore colony with half the cell population displaying wild-type and the other half mutant phenotypes from a given ascosporal colony from a tetrad array is the signal characteristic of postmeiotic segregation (PMS). The two sides of the sector must be identical as regards all other segregating genes. To maintain uniform nomenclature, the gene conversions (GC) are designated as 6+:2- or 2+:6- and the postmeiotic segregations (PMS) as 5+:3- or 3+:5-, where the first and second integers respectively denote the number of wild-type and mutant spores. Other wider ratio segregations such as aberrant (ab) 4+:4-, 7+:1-, 1+:7, and 8+:0- or 0+:8- are known but they are extremely rare.

To account for the underlying molecular events that eventuate as gene conversions or postmeiotic segregations, detailed classical genetic analysis of meiotic recombination in *Saccharomyces cerevisiae* and other ascomycetous fungi were carried out (see Fogel *et al.*, 1979; Esposito & Klapholz, 1981; Orr-Weaver & Szostak, 1985; Petes *et al.*, 1992). Detailed molecular models have been developed, notably the single strand initiation model of Meselson and Radding (1975) and the double strand break model of Szostak *et al.* (1983). The former model invokes mismatch correction or failure of correction in heteroduplex DNA recombination intermediates to account for the related phenomena such as postmeiotic segregations, parity polarity, fidelity, co-conversion, conversion associated crossing over and the position of the associated reciprocal exchange, as well as fine structure map expansion (Fogel & Hurst, 1967; Fogel & Mortimer, 1969, 1970; Fincham & Holliday, 1970; Stahl, 1979; Fogel *et al.*, 1981).

At the molecular level 6+:2- and 2+:6- gene conversions may also arise via double strand gap repair, a mechanism that need not involve repair of mismatched nucleotides in heteroduplex DNA intermediates at the converted locus (Szostak *et al.*, 1983). The double strand break presumably occurs due to an endonuclease activity. The gap is enlarged and then followed by replacement of an allele with a DNA segment bearing wild-type or mutant copy resulting in gene conversion (Resnick & Martin, 1976; Szostak *et al.*, 1983). The 5+:3 and 3+:5- postm eiotic segregations presumably arise due to generation of rather limited heteroduplex DNA mismatches localized to the terminals of the gap and a subsequent correction failure at the included mutant sites.

Ultimately, the mismatch delivered into the ascospore is resolved replicationally during the first mitotic division after germination of the ascospore.

The 5+:3- or 3+:5- postmeiotic segregation as well as the comparatively rare aberrant (ab) 4+:4- segregations are encountered as tetrads containing one or two sectored ascosporal colonies. These comprise the *prima facie* genetic evidence for meiotically generated but uncorrected heteroduplex DNA containing mismatched base-pairs or heterologies that are resolved replicationally during the first mitotic replication cycle of the ascospore.

Recent advances in recombinant DNA techniques have made it possible to investigate certain aspects of recombination at the molecular level (see Petes *et al.*, 1992). The study reported here is based on the premise that heteroduplex DNA is the major molecular intermediate that gives rise to both, gene conversion and post-meiotic segregation. Thus, when the largely asymmetrical heteroduplex DNA segment or tract embraces the heterozygous mutant site, repair may generate either a gene conversion or a restitution, while the absence of repair would generate a postmeiotic segregation. Accordingly, if both genetic outcomes are heteroduplex DNA dependent, then mutants defective in correctional repair are expected to manifest a diminution of gene convesions and a corresponding increase in postmeiotic segregations. If conversion and restitution are equally probable then the rise in PMS should be equal to twice the decrease in gene conversion frequency.

A set of yeast mutants displaying increase postmeiotic segregation frequencies were generated in this laboratory (Williamson *et al.*, 1985). These mutants exhibit two principal phenotypes. First, they function as mutators in mitotic cells and, additionally, they increase PMS frequencies of several different loci at the expenses of gene conversions among unselected tetrads. Hence, they are unequivocally deficient in mismatch corrections. This was conclusively demonstrated by transforming mitotic cells with plasmids containing heteroduplex DNA bearing well-defined base pair mismatches (Bishop *et al.*, 1987; Kramer *et al.*, 1989). Taken collectively, these advances facilitate the molecular cloning of genes affecting heteroduplex DNA repair via positive complementation responses of the mitotic mutator phenotype after transformation.

Isolation of mutants affecting postmeiotic segregations (PMS)

Yeast mutants with effects on meiotic (Roth & Fogel, 1971; Fogel & Roth, 1974) and mitotic (Rodarte-Ramon & Mortimer, 1972; Maloney & Fogel, 1980; Esposito & Bruschi, 1982; Esposito *et al.*, 1986) gene conversions were successfully isolated using disomic aneuploid strains. However, genetic analysis of the hypo-rec mutants isolated in these studies proved to be

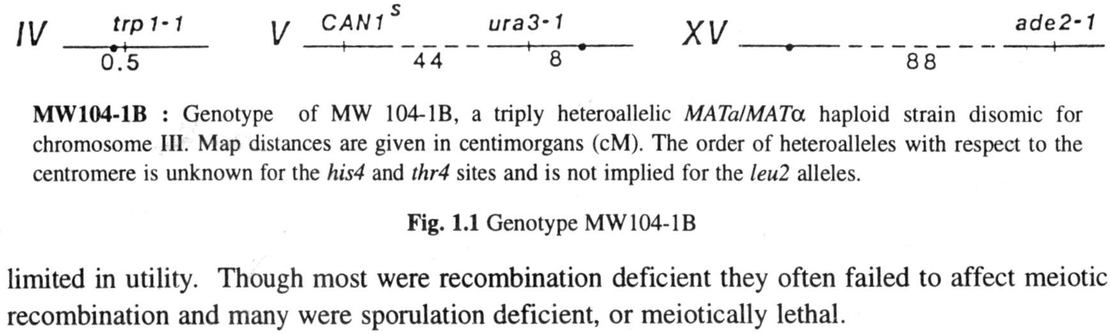

MW104-1B : Genotype of MW 104-1B, a triply heteroallelic *MATa/MATα* haploid strain disomic for chromosome III. Map distances are given in centimorgans (cM). The order of heteroalleles with respect to the centromere is unknown for the *his4* and *thr4* sites and is not implied for the *leu2* alleles.

Fig. 1.1 Genotype MW104-1B

limited in utility. Though most were recombination deficient they often failed to affect meiotic recombination and many were sporulation deficient, or meiotically lethal.

Mutants yielding enhanced postmeiotic segregation frequencies in unselected tetrads and additionally a mitotic mutator effect were isolated by exploiting the *MATa/MATα* chromosome III disomic (N+1) strain MW104-1B (Fig. 1.1). The strain carried heteroalleles at *his4, leu2* and *thr4* loci. The *his4-4°* and *Leu2-1°* ochre-suppressible alleles are revertible simple base substitution mutants. In contrast, *his4-519* is single base insertion frameshift mutation, while *thr4-1, thr4-16 and leu2-27* can generate simple mismatches since they probably originated as base substitutions or small insertion/deletion mutations. A stationary phase culture of the disomic chromosome III strain MW104-1B was mutagenized with ethylmethane sulfonate (EMS) and about 6000 surviving colonies were tested individually for enhanced prototroph frequencies at the three heteroallelic loci by replica-printing single surviving colonies first to KAC plates and then subsequently to synthetic complete media respectively deficient for histidine, leucine or threonine. Mutations recessive in heteroduplex DNA corrections-repair genes would be manifest as isolates that displayed increased potassium acetate or incipient meiosis-induced prototroph frequencies at the single loci compared to normal controls. Such recessive mutants are expected to arise, if single site conversions were increased at the expense of normal co-conversions or if overall repair efficiency were diminished. Five putative mutant isolates that consistently displayed elevated prototroph frequencies were identified (Table 1.1). The mutant isolates also showed elevated mitotic forward mutation frequencies as $Can1^s$ to $can1^r$ and *his4* to wild-type. Mutant rates differed from wild-type controls by several orders of magnitude. Often, the most conspicuous heteroallelic meiotic prototroph increase was observed at *his4*. The putative mutants were chosen for further study based on the rationale that mitotic mutators with elevated meiotic prototroph frequencies were likely to be affected in correcting DNA replication errors or base repair mismatches present in meiotic recombination intermediates or h:DNA. In addition, the mutator phenotype is directly observable in vegetative colonies, providing a techni-

Table 1.1 Phenotypic attributes of disomic haploid putative mutants

Allele	Spontaneous mutation frequency to *canl^r*	Meiotic His$^+_4$	Meiotic Thr$^+_4$	Meiotic Leu$^+_2$	Mitotic His$^+_4$
pms1-1	+++	++++	+++	0	+
pms1-2	+++	++++	+++	0	++
pms2-1	++	++++	+++	0	+
pms2-2	+++	++++	++	-	+
pms3-1	++	+++	+++	0	+++
wild-type	0	0	0	0	0

Mutant phenotypes assessed at 30°C. The + indicates respective relative frequencies higher than wild-type; 0 indicates a wild-type response; + = 25 fold increase in frequency over the wild-type response and correspondingly ++, +++ and ++++ indicate 50, 75 and 100 fold increases respectively.

cal advantage in genetic manipulation and molecular cloning.

Determining the number of PMS genes

The five putative mutant disomic isolates designated *pms1-1, pms1-2 pms2-1, pms2-2, pms3-1*, were screened for the appearance of mating proficient. These were crossed to isogenic strains wild-type for the PMS genes and the resulting tetrads analysed for ascosporal haploid colonies displaying increases in meiotic prototroph frequencies at the heteroallelic loci. In sum, the data collected indicate that each of the five haploid mutant isolates carried a single, recessive and unique genetic alteration that displayed a conventional and expected 2+:2- mendelian segregation pattern. The haploid derivatives of the five mutator isolates were crossed in all pair-wise combinations and tested for allelism by complementation. Tetrad data from the crosses established that the five mutator isolates carry alterations in three nonallelic genes (*pms1, pms2 and pms3*). The *pms* mutants are designated as such by virtue of the enhanced postmeiotic segregation frequency that their homozygotes and double mutants (see Table 1.2) generate on miscellaneous reporter loci distributed over the genome. Additionally, all mutants display a mitotic mutator phenotype. It is likely that other PMS loci remain unidentified. In fact, a *pms4* mutant has been isolated but the strain was lost. Of the fifteen reporter loci scored for meiotic gene conversion and PMS, only eight markers displayed distinctive, statistically significant enhancement compared to the wild-type. These are *trp1-1, ura3-1, leu2-1, thr4-1, lys1-1, met13, his2* and *thr1*. Some loci were entirely unaffected: *Mat, Mal2, Cup1^r, ade8-18*, while *hom3-10, his4-4* and *Can1^s* were affected in some mutants, e.g., *pms1-1, pms2-1* and *pms3*.

Gene conversion and postmeiotic segregation (PMS) profiles among *pms* mutants

In this investigation the effects of pms mutations on many heterozygous sites distributed over the yeast genome were monitored by tetrad analysis for correction or noncorrection of DNA mismatches, i.e. frequencies of gene conversions (6+:2- or 2+:6-) and postmeiotic segregations (5+:3- or 3+:5-). Heterozygous, and homozygous single, double mutant *pms* diploids were constructed and subjected to tetrad analysis (see Table 1.2). Conspicuous in this investigation is the occurrence of PMS events in a substantial fraction of the aberrant segregations at certain heterozygous reporter sites. Eight of the fifteen tested sites were affected. Commonly, 50% or more of aberrant segregations display PMS at many affected sites in mutant homozygotes (Table 1.2). Even higher PMS frequencies are observed among unselected tetrads from double mutant homozygotes. Heterozygous sites that are known to constitute simple base substitution mismatches (e.g. sites heterozygous for suppressible ochre alleles *leu2-1°, lys1-1°, his4-4°* and the amber mutant *trp1-1^a*) display increased PMS frequencies among all mutants. The enhanced PMS frequency was also observed at sites heterozygous for *his4-519* and *hom3-10*, single base insertion frameshift mutations, in all mutants except *pms3-1*. The similar mutant postmeiotic segregation profiles observed at *ura3-1, thr4-1, thr1, met13, his2* and *leu2-1* may indicate that heterozygosities at these sites reflect simple mismatches formed by base substitution or small

Table 1.2 Comparison of aberrant segregations in homozygous *pms* single mutants, double mutant and wild-type control diploids[2]

Locus	Pms+/Pms+ (1441 tetrads)		pms1-1/pms11 (356 tetrads)		pms2-1/pms21 (169 tetrads)		pms3-1/pms31 (213 tetrads)		pms1-1/pms31 (40 tetrads)		pms21/pms31 (105 tetrads)	
	% GC	% PMS	% GC	% PMS	% GC	% PMS	% GC	% PMS	% GC	% PMS	% GC	% PMS
trp1-1	0.6	0.1	0.4	0.4	0	1.8	0.5	1.1	-	-	0	0
ura3-1	1.5	0.1	1.1	2.9	0	0.6	1.9	0.8	0	5.3	3.9	3.9
leu2-1	3.5	0.2	1.8	1.8	1.8	1.8	0.8	3.9	0	4.8	1.4	4.9
thr4-1	4.2	0.8	1.2	1.8	1.8	4.9	4.0	9.6	5	27.5	1.2	17.9
thr1	5.9	0.7	5.2	4.5	4.2	4.2	2.4	8.4	5.4	5.4	2.1	12.5
lys1-1	8.1	0.2	2.3	9.3	2.9	0.6	1.4	5.5	2.4	9.5	2.1	6.2
met13	9.7	0.1	7.6	1.8	6.4	4.1	7.4	12.1	7.1	9.5	3.9	16.5
his2	13.0	0	9.3	11.4	8.5	1.2	12.6	5.1	-	-	7.5	11.0
Average	5.8	0.3	3.6	4.2	3.2	2.4	3.9	5.8	3.3	10.3	2.8	9.1
%GC or %PMS[1]	95%	5%	46%	54%	57%	43%	40%	60%	24%	76%	24%	76%

% GC, gene conversion, or% PMS defined as:

$$\frac{\text{total gene conversion or PMS events}}{\text{total aberrant segregation events}} \times 100$$

[2] Unpublished data from M. Williamson & S. Fogel

insertion/deletion mutations.

It may be noted that while a given *pms* allele may function as a potent mutator for forward mutation at *Can1^s*, the same mutant is not an obligatory strong mutator for his4-4° reversion. Among the mutants, *pms1-1* and *pms2-1* have the highest reversion/mutation rates at *hom3-10* and *Can1^s*, while the *pms3-1* strain has the highest *his4-4* reversion rate. These results imply that the genes identified by the *pms* mutations exert different functional specificities. Thus, allele-specific or sequence-specific effects cannot be ruled out.

Yeast mutants that share certain phenotypic similarities with the *pms* mutants include mutator mutants that are radiation insensitive, e.g. *MIC9* and *MIC12* (Maloney & Fogel, 1980), *mut1* and *rem1-2* (Golin & Esposito, 1977; Malone & Hoekstra, 1984), *mut1* and *MUT6* (Gottleib & von Borstel, 1976; Hastings *et al.*, 1976; Nasim & Brychy, 1979) and possibly some *rec* mutants (Esposito & Bruschi, 1982; Esposito, 1984). These mutants were studied primarily for their mitotic effects. None showed increase in meiotic prototroph frequencies and in the absence of direct complementation tests it is reasonable to presume, at least tentatively, that the PMS genes may be nonallelic to these loci.

The PMS genes also participate in mitotic gene conversion. Spontaneous and UV induced prototroph frequencies are slightly elevated in the *pms* mutants. These increases, based on qualitative visual screening, are of a lower order in magnitude, but generally follow the pattern observed for meiotic prototroph frequency increases. The PMS functions are probably not essential for radiation repair, since none of the *pms* mutant strains are either UV or X-ray sensitive. This result accords with the observation that meiotic gene conversion and reciprocal recombination are unaffected in several excision-defective (UV-sensitive) yeast *rad* mutants (Dowling *et al.*, 1985).

Bacterial mutator strains (*mut*) with defective mismatch repair systems share certain similarities with the yeast *pms* mutants. In *E. coli*, four *mut* genes identified by Glickman and Radman (1980) comprise a single mismatch correction epistasis group that is primarily independent of UV excision repair (Glickman & Radman, 1980; Radman *et al.*, 1980; Claverys & Lacks, 1986; Radman & Wagner, 1986; Modrich, 1987). Mismatch correction mediated by the four *mut* genes acts preferentially with some mismatch specificity on base substitution and frameshift mutations (Wagner et al., 1984; Dohet *et al.*, 1985, 1986). Based on results obtained from lambda virus crosses, Glickman and Radman (1980) proposed that mismatch correction is absent during recombination in *mutL* and *mutS*, while correction tracts may be shorter in *mutH*.

Streptococcus pneumoniae hexA mutants represent another example of prokaryotic mismatch repair defective mutators relevant to this study (Tiraby & Fox, 1973; Claverys *et al.*, 1983). In *hex$^+$* strains, the transformation efficiency of given sequences containing different base substitutions varies considerably, while these same sequences transform with equally high efficiency in *hex$^-$* strains, presumably because mismatches are not removed from heteroduplex DNA (Claverys *et al.*, 1980, 1983).

Heteroduplex DNA correction profiles among pms mutants

Our rationale proposed for the meiotic *pms* mutant phenotype was that mismatches in heteroduplex DNA intermediates formed during meiotic recombination were corrected either with diminished efficiency or that the average correction tract length was materially shortened and thereby increasing single site conversions that eventuate as prototrophs at the expense of normal, longer tract coconversions which are persistently auxotrophic. Similarly, the mitotic mutator phenotype of the *pms* mutants could also be accounted for by a defect in DNA mismatch repair if replication errors were not removed by editorial postreplicative mismatch correction. Bishop *et al.* (1987) demonstrated by transformation of yeast cells with *in vitro* constructed heteroduplex DNAs containing self-complementary loop structures that the *pms1-1* mutation affects the repair of such structures. Furthermore, Kramer *et al.* (1989) constructed a variety of plasmids in this laboratory with defined DNA mismatches localized in the cloned *ADE8* locus and estimated the correction efficiency by transforming the *pms* mutants with single base substituted *ade8* alleles. The repair of the base/base mismatches G/T, A/C, G/G, A/G, G/A, A/A, T/T, T/C and C/T were severely reduced in all *pms* mutants and yielded a high proportion of sectored colonies (66% to 87%). The *pms3* mutation, however, conferred a different phenotype. The repair of the single nucleotide loop in the *pms3* mutant was not altered (A loop) or only slightly reduced (T loop). Thus, +1 frameshift mutations may be corrected via a pathway which requires *PMS1* and *PMS2* proteins, but not the *PMS3* protein. Alternatively, the *PMS3* gene product may be involved in the general mismatch recognition process. The mismatch C/C and the 38-nucleotide loop of *ade8-18* were corrected with low efficiency in both mutant and wild-type strains. This substrate specificity pattern resembles those found in *Escherichia coli* and *Streptococcus pneumoniae*, suggesting an evolutionary relationship of DNA mismatch repairs in prokaryotes and eukaryotes (Claverys & Lacks, 1986; Radman & Wagner, 1986; Modrich, 1991).

Map locations of the PMS genes

The *pms1* gene was mapped close to the *met4* locus on chromosome *XIV*. Unselected tetrads were analyzed from a strain heterozygous for *pms1-1* and *met4*. The data 33PD:0NPD:6TT, clearly demonstrate linkage between these two loci with map distance estimated to be 7.7 cM. The *pms2* gene was mapped close to the *pet8* locus on the same chromosome and the data, 22PD:2NPD:17TT, indicate a distance of 35.4 cM between the two genes. Preliminary results indicate linkage of the *pms3* mutation to the *leu1* locus on chromosome *VII*, though further tetrad data are required to confirm this location.

The precise function of the various PMS loci cannot be fully ascertained from the classical genetic behaviour of the various mutants. A more precise characterization of the various genes, singly and in all possible combinations, is best obtained from a study of the effect of specific deletions and their combinations. Such results are attainable by molecular cloning.

Cloning and characterization of the PMS genes

A genomic library from a wild-type or *Pms+* strain was constructed on plasmid YCp50. This was used to transform a *pms1-1* mutant strain. A 7.9 Kb genomic DNA fragment that complemented the recessive *pms1* mutation was cloned and sequenced in this laboratory (Kramer *et al.*, 1989). The transformants harboring the cloned DNA plasmid pWBK3 displayed mitotic forward mutation rates at the *Can1^s* locus or reversion rates at the *hom3-10* locus comparable to isogenic strains carrying the wild-type *Pms1+* gene. The cloned *Pms1+* DNA contained a 2712 base-pair open reading frame and encoded a 103 kilodalton protein with significant homologies to the *E. coli* MutL protein (15%, 99 identical amino acids of 658) and *S. pneumoniae HexB* protein (13%, 87 identical amino acids of 677). At the region of marked homologies between the genes, the identity to the *MutL* and *HexB* proteins were 32% and 33%, respectively. The homology between the *Pms1, MutL,* and *HexB,* proteins implies that DNA mismatch repair emerged early in evolution, at least before the divergence of prokaryotes and eukaryotes.

Mismatch rectification pathways in *E. coli* implicate *MutS, MutL, MutH* and other cofactors (Modrich 1991). This view also predicts the existence of similar protein counterparts in yeast and repair genes other than *PMS2* and *PMS3* as potential counterparts. No specific functions could be attributed to *PMS1, MutL* and *HexB* proteins and the exact molecular mechanism is not yet fully understood. Replacement of the *Pms1+* allele with homozygosity for a deleted copy (*pms1*) did not eliminate gene conversion among unselected tetrads from a well marked diploid (see Kramer et al., 1989) carrying nine miscellaneous reporter genes. In all, 97 tetrads from the deletion homozygote were compared to 175 controls. Some 17 conversions were identified in the experimental group. In this connection it might be supposed that the observed 6+:2- and 2+:6- conversions might actually represent postmeiotic segregations accompanied by an obligate lethal sectoring. Patently, this is not the case, as established by extensive pedigree analysis or double dissection of numerous unselected tetrads.

A DNA mismatch repair model

The phenotypic characteristics shared by the yeast *pms* mutants and *E. coli mut* genes, as well as the sequence homologies between the *PMS1* and *MutL* genes, strongly suggest the presence of similar DNA mismatch repair pathway that comes from studies with *E. coli* and this appears to resemble the yeast pathway. Initiation of heteroduplex repair is dependent on a mismatch, and four additional agents, i.e.: *MutS, MutL, MutH* and ATP. The multifunctional 97 kilodalton product of the *MutS* gene probably binds to the DNA mismatch, and ATP and promotes formation of a loop shaped structure (Su, 1987). The *MutL* protein binds to this tripartite complex followed by the addition of the *MutH* product. The 25 kilodaltons *MutH* product is responsible for the d(GATC) site recognition and a Mg^{2+} dependent endonuclease activity that incises the DNA 5' to the G of d(GATC) (Welsh *et al.*, 1987). DNA mismatch correction efficiency depends on the nature of the mismatch and is influenced by the d(GATC) sequences in the DNA environment in which the mismatch resides. The efficiency of DNA mismatch repair increases in the proximity of d(GATC) sequences (Bruni *et al.*, 1988). In contrast to *MutS* and *MutH*, no simple function can be attributed to *MutL*. Conceivably, the 70 kilodalton MutL protein might function as a protein-protein interface between *Mut*S and *Mut*H (Modrich, 1991). Other proteins such as DNA helicases, single strand binding protein (SSB), exonucleases, polymerases and ligases are undoubtedly involved in subsequent steps that specify the incision, remove the DNA strand containing the mismatch and replace the strand with a corrected copy.

CONCLUSIONS

Mutations in three non-allelic yeast *PMS genes* (*pms*1, *pms*2 and *pms*3) were implicated in specific DNA mismatch correctional repair during mitotic and meiotic divisions. Bacterial mutant strains with defective mismatch repair systems share certain similarities with the yeast *pms* mutants. The *PMS* gene products function during DNA replication and recombination and hence regulate the fidelity of DNA sequences. The pms mutants displayed increases in the frequencies of aberrant 5+:3- and 3+:5- postmeiotic segregations (PMS) at several heterozygous reporter loci occupied by mutant alleles attributable to simple DNA base substitutions, single base insertion frameshift mutations and small deletions/insertions.

Clearly, the pms mutants are defective in correction of specific DNA mismatches. The *Pms*1+ gene sequence shares homologies to the *E. coli MutL* and *S. pneumoniae HexB* DNA repair genes, indicating common repair pathways in both prokaryotes and eukaryotes. It is to be expected that other unrecognized *PMS* genes in yeast remain to be discovered and isolated. Identification, cloning and deletion of the *PMS* genes (*pms*) *from their chromosomal loci will continue to define and illuminate the DNA repair pathways in yeast. In conclusion, we may emphasize that the pms*1, *pms*2 and *pms*3 functions do not alter the gene conversion or *PMS* profile at the heterozygous *ade8-18* site, a known 38 base pair deletion. This finding augers for the existence of still another recombinational repair pathway.

Acknowledgements

We thank our colleague Marsha S. Williamson for valuable unpublished data and Madhav Goyal for providing excellent technical assistance throughout. This work was supported by a NIH grant GM17317 to Seymour Fogel.

The author are pleased to dedicate this paper to Professor Umakant Sinha who provided singular impetus towards developing microbial and molecular genetics in India.

References

Bishop, D.K., Williamson, M.S., Fogel, S. & Kolodner, R.D. (1987). The role of heteroduplex correction in gene conversion in *Saccharomyces cerevisiae*. Nature (London) **328**, 362-364.

Bruni, R., Martin, D. & Jiricny, J. (1988). d(GATC) sequences influence *Escherichia coli* mismatch repair in a distance dependent manner from both upstream and downstream of the mismatch. Nucleic Acids Research **16**, 4875-4890.

Case, M.E. & Giles, N.H. (1964). Allelic recombination in *Neurospora:* tetrad analysis of a three-point cross within the *pan-2* locus. Genetics **49**, 529-540.

Claverys, J.P. & Lacks, S. (1986). Heteroduplex deoxyribonucleic acid base mismatch repair in bacteria. Microbiol. Rev. **50**, 133-165.

Claverys, J.P., Mejean, V., Gasc, A.M. & Sicard, A.M. (1983). Mismatch repair in *Streptococcus pneumoniae:* relationship between base mismatches and transformation efficiencies. Proceeding National Academy of Sciences, USA **80**, 5956-5960.

Claverys, J.P., Roger, M. & Sicard, A.M. (1980). Excision and repair of mismatched base pairs in transformation of *Streptococcus pneumoniae.* Molec. Gen. Genet. **178**, 191-201.

Dohet, C., Wagner, R. & Radman, M. (1985). Repair of defined single base-pair mismatches in *Escherichia coli.* Proceedings National Academy of Sciences, USA **82,** 503-505.

Dohet, C., Wagner, R. & Radman, M. (1986). Methyl-directed repairs of frameshift mutations in heteroduplex DNA. Proceedings National Academy of Sciences, USA **83**, 3395-3397.

Dowling, E.L., Maloney, D.H. & Fogel, S. (1985). Meiotic recombination and sporulation in repair-deficient strains of yeast. Genetics **109**, 283-302.

Esposito, M.S. (1971). Post-meiotic segregation in *Saccharomyces.* Molecular and General Genetics **111**, 297-299.

Esposito, M.S. (1984). Molecular mechanisms of recombination in *Saccharomyces cerevisiae:* testing mitotic and meiotic models by analysis of hypo-rec and hyper-rec mutations. Symposium Society of Experimental Biology **38**, 123-159.

Esposito, M.S. & Bruschi, C.V. (1982). Molecular mechanisms of DNA recombination: testing mitotic and meiotic models. In The Berkeley Workshop on Recent Advances in Yeast Molecular Biology, Edited by M.S. Esposito. University of California, Berkeley. Vol. 1, pp 242-253.

Esposito, M.S., Maleas, D.T., Bjornstad, K.A. & Holbrook, L.L. (1986). The *REC46* gene of *Saccharomyces cerevisiae* controls mitotic chromosome stability, recombination and sporulation: cell-type and life cycle stage-specific expression of the *rec46-1* mutation. Curr. Genet. **10**, 425-433.

Esposito, R.E. & Klapholz, S. (1981). Meiosis and ascospore development. In The Molecular Biology of the Yeast *Saccharomyces*: Life Cycle and Inheritance, Edited by J.N. Strathern, E.W. Jones and J.R. Boach. Cold Spring Harbor Laboratory, Cold Spring Harbor, New York, pp 211-287.

Fincham, J.R.S. & Holliday, R. (1970). An explanation of fine structure map expansion in terms of excision repair. Molec. Gen. Genet. **109**, 309-322.

Fogel, S. & Hurst, D.D. (1967). Meiotic gene conversion in yeast tetrads and the theory of recombination. Genetics **57**, 455-480.

Fogel, S. & Mortimer, R.K. (1969). Informational transfer in meiotic gene conversion. Proceedings National Academy of Sciences, USA **62**, 96-103.

Fogel, S. & Mortimer, R.K. (1970). Fidelity of meiotic gene conversion in yeast. Molec. Gen. Genet. **109**, 177-185.

Fogel, S. Mortimer, R.K. & Lusnak, K. (1981). Mechanisms of meiotic gene conversion, or "wanderings on a foreign strand". In The Molecular Biology of the Yeast *Saccharomyces:* Life Cycle and Inheritance, Edited by J.N. Strathern, E.W. Jones and J.R. Broach. Cold Spring Harbor Laboratory, Cold Spring Harbor, New York, pp 289-339.

Fogel, S., Mortimer, R.K., Lusnak, K. & Travares, F. (1979). Meiotic gene conversion: a signal of the basic recombination event in yeast. Cold Spring Harbor Laboratory Symposium on Quantitative Biology **43**, 1325-1341.

Fogel, S. & Roth, R. (1974). Mutations affecting meiotic gene conversion in yeast. Genetics **130**, 189-201.

Glickman, B.W. & Radman, M. (1980). *Escherichia coli* mutator mutants deficient in methylation-instructed DNA mismatch correction. Proceedings National Academy of Sciences, USA **77**, 1063-1067.

Goldman, S.L. (1974). Studies on the mechanism of the induction of site specific recombination on the ade-6 locus of *Schizosaccharomyces pombe*. Molec. Gen. Genet. **132**, 347-361.

Golin, J. & Esposito, M.S. (1977). Evidence for joint genic control of spontaneous mutation and genetic recombination during mitosis in *Saccharomyces*. Molec. Gen. Genet. **150**, 127-135.

Gottleib, D.J.C. & von Borstel, R.C. (1976). Mutators in *Saccharomyces cerevisiae, mut1-1, mut1-2* and *mut2-1*. Genetics **83**, 655-666.

Guthrie, C. & Fink, G.R. (1991). Guide to Yeast *Genetics and Molecular Biology*. Academic Press, Inc. San Diego, California. p 933.

Gutz, H. (1971). Site specific induction of gene conversion in *Schizosaccharomyces pombe*. Genetics **69**, 317-337.

Hastings, P.J., Quah, S.K. & von Borstel, R.C. (1976). Spontaneous mutation by mutagenic repair of spontaneous lesions in DNA. Nature (London) **264**, 719-722.

Kramer, B., Kramer, W., Williamson, M.S. & Fogel, S. (1989). Heteroduplex DNA correction in *Saccharomyces cerevisiae* is mismatch specific and requires functional *PMS* genes. Molec. Cell. Biol. **9**, 4432-4440.

Kramer, W., Kramer, B. Williamson, M.S. & Fogel, S. (1989). Cloning and nucleotide sequence of DNA mismatch repair gene PMS1 from *Saccharomyces cerevisiae:* homology of *PMS1* to prokaryotic *MutL* and *HexB*. J. Bacteriol. **171**, 5339-5346.

Leblon, G. (1972a). Mechanism of gene conversion in *Ascobolus immersus*. I. Existence of a correlation between the original of mutations induced by different mutagens and their conversion spectrum. Molec. Gen. Genet. **115**, 36-48.

Leblon, G. (1972b). Mechanism of gene conversion in *Ascobolus immersus*. II. The relationship between the genetic alterations in *b1* and *b2* mutants and their conversion spectrum. Molec. Gen. Genet. **116**, 322-335.

Malone, R.E. & Hoekstra, M.F. (1984). Relationship between a hyper-rec mutation (*REM1*) and other recombination and repair genes in yeast. Genetics **107**, 33-48.

Maloney, D. & Fogel, S. (1980). Mitotic recombination in yeast: isolation and characterization of mutants with enhanced mitotic gene conversion rates. Genetics **94,** 825-829.

Meselson, M. & Radding, C. (1975). A general model for genetic recombination. Proceeding of National Academy of Sciences, USA **72**, 358-361.

Modrich, P. (1987). DNA mismatch correction. Ann. Rev. Biochem. **56**, 435-466.

Modrich, P. (1991). Mechanisms and biological effects of mismatch repair. Ann. Rev. Genet. **25**, 229-253.

Nasim, A. & Brychy, T. (1979). Cross sensitivity of mutator strains to physical and chemical mutagens. Can. J. Genet. Cytol. **21**, 129-137.

Olive, L.S. (1956). Genetics of *Sordaria fimicola*. I. Ascospore color mutants. Am. J. Bot. **43**, 93-107.

Orr-Weaver, T.L. & Szostak, J.W. (1985). Fungal recombination. Microbiol. Rev. **49**, 33-58.

Petes, T.D., Malone, R.E. & Symington, L.S. (1992). Recombination in yeast. In The Molecular and Cellular Biology of the Yeast *Saccharomyces:* Genome Dynamics, Protein Synthesis and Energetics, Edited by J.N. Strathern, E.W. Jones and J.R. Broach. Cold Spring Harbor Laboratory Press, Cold Spring Harbor, New York, pp 407-521.

Radman, M. & Wagner, R. (1986). Mismatch repair in *Escherichia coli*. Ann. Rev. Genet. **20**, 523-538.

Radman, M., Wagner, R., Glickman, B.W. & Meselson, M. (1980). DNA methylation mismatch correction and genetic stability. In Progress in Environmental Mutagenesis, Edited by M. Alcevic. Elsevier/North Holland Biomedical Press, Amsterdam, pp 121-130.

Resnick, M.A. & Martin, P. (1976). The repair of double-stranded breaks in the nuclear DNA of *Saccharomyces cerevisiae* and its genetic control. Molec. Gen. Genet. **143,** 119.

Rodarte-Ramon, S.U. & Mortimer, R.K. (1972). Radiation induced recombination in *Saccharomyces:* isolation and genetic study of recombination deficient mutants. Radi. Res. **49,** 133-147.

Roth, R. & Fogel, S. (1971). A system selective for yeast mutants deficient in meiotic recombination. Molec. Gen. Genet. **112**, 295-305.

Sherman, F., Fink, G.R. & Hicks, J.B. (1986). Laboratory Course Manual for Methods in Yeast Genetics. Cold Spring Harbor Laboratory Press, Cold Spring Harbor, New York, p 186.

Stahl, F.W. (1979). Genetic Recombination: Thinking about it in Phage and Fungi. W.H. Freeman and Company, San Francisco.

Su, S-S. (1987). Methyl-directed DNA mismatch repair. Ph.D. thesis. Duke University, Durham, North Carolina.

Szostak, J.W., Orr-Weaver, T.L., Rothstein, R.J. & Stahl, F.W. (1983). The double strand break repair model for recombination. Cell **33**, 25-35.

Tiraby, J.C. & Fox, M.S. (1973). Marker discrimination in transformation and mutation of *Pneumococcus*. Proceedings National Academy of Sciences, USA, **40**, 3541-3545.

Wagner, R., Dohet, C., Jones, M., Doutriaux, M.-P. & Radman, M. (1984). Involvement of *Escherichia coli mismatch* repair in DNA replication and recombination. Cold Spring Harbor Symposium on Quantitative Biology **49**, 611-615.

Welsh, K.M., Lu, A.L., Clark, S. & Modrich, P. (1987). Isolation and characterization of the *Escherichia coli mut*H gene product. J. Biol. Chem. **262**, 15624-15629.

Williamson, M.S., Game, J.C. & Fogel, S. (1985). Meiotic gene conversion mutants in *Saccharomyces cerevisiae*: I. Isolation and characterization of *pms1-1* and *pms1-2*. Genetics **110**, 609-646.

Zickler, H. (1934). Genetishe Untersuchungen an einen heterothallischen Askomyzeten (*Bombardia lunata* nov. spec.). Planta **22**, 573-613.

2

Potential of plant repeated DNA sequences species-specific markers, DNA fingerprint markers and transposable elements

P.K. Ranjekar, V.V. Pethe, P.K. Chitnis, M.D. Lagu
and V.S. Gupta
Division of Biochemical Sciences, National Chemical Laboratory
Pune 411 008, INDIA

ABSTRACT

Multi-loci probes (the human minisatellites 33.1 and 33.6 and the M13 phage DNA) are indeed efficient in cultivar identification, paternity analysis and detection of individual specific DNA fingerprints in wide range of plant species. This potentially efficient tool can be used to produce markers for patent protection of breeder's rights and aid the breeding programmes.

INTRODUCTION

Reiterated or repeated DNA sequences exist in both animal and plant nuclear genomes and their proportion ranges from 35-90% of total DNA (Flavell, 1980; Deshpande & Ranjekar, 1980; Jelinek & Schmid, 1982; Ranjekar, 1982; Sorenson, 1984; Walbot & Cullis, 1985; Flavell, 1986). Based on their genomic organization, two main types of repetitive sequences have generally been described: (i) SINES and LINES (ii) Satellite sequences. Sequences of the former type are inter-spersed with each other and with the so called 'unique' or 'coding' sequences. They themselves are thus scattered and dispersed throughout the genome (Jelinek & Schmid, 1982). Dispersed repeats include transposons as well as other sequences. In addition to the inter-spersed repetitive DNA elements, many plants possess highly repetitive DNA sequences arranged in tandem arrays. These elements, often referred to 'satellite', consist of repetitions of a monomeric unit or basic repeat which can vary in length from a few to several thousand nucleotides. Several such tandemly repeated DNA elements have been cloned, characterized, and sequenced (Kato *et al.*, 1984; Grellet, *et al.*, 1986; Simeons *et al.*, 1988).

Examples of the tandemly repeated DNA sequences include repeats coding for 25 S and 18 S rRNA or 5S rRNA (Vedel & Delseny, 1987). Recently, minisatellite probes which consist of tandem repeats of the 'core' sequence have proven to be very useful in generating individual specific patterns in microbial, plant, and animal genomes (Dallas, 1988; Ryskov *et al.*, 1988). Batteries of repeated DNA sequences (tandem and dispersed) have been identified, isolated, cloned and characterized from a variety of plant genera belonging to different families. Examples of some of these are cited below:

Gramineae: Rice (Gupta *et al.*, 1981; Wu & Wu, 1987; Dhar *et al.*, 1988; Zhao *et al.*, 1989; Dhar *et al.*, 1990; Gupta *et al.*, 1990), wheat (Metzlaff *et al.*, 1986; McIntyre *et al.*, 1988), barley and wheat (Ranjekar *et al.*, 1976), rye (Guidet *et al.*, 1991; Gupta *et al.*, 1990), oats (Fabijanski *et al.*, 1990), Aegilops (Rayburn & Gill, 1986, 1987), and millets (Gupta & Ranjekar, 1982; Sivaraman *et al.*, 1984a,b, 1985, 1986).

Leguminoseae: *Phaseolus* (Tamhankar *et al.*, 1990), *Lupinus* (Sakowicz *et al.*, 1986), *Vicia* (Kato *et al.*, 1984; Lehmann *et al.*, 1990), *Vigna* (Roy *et al.*, 1988), *Cajanus* (Dabak *et al.*, 1988).

Cruciferae: *Brassica* (Iwabuchi *et al.*, 1991; Sibson *et al.*, 1991; Hallden *et al.*, 1987; Lakshmikumaran & Ranade, 1990; Benslimane *et al.*, 1986), *Arabidopsis* (Martinez-Zapater *et al.*, 1986), *Raphanus* (Grellet *et al.*, 1986).

Solanaceae: *Lycopersicon* (Ganal *et al.*, 1988; Schweizer *et al.*, 1988) *Nicotiana tabacum* (Koukalova *et al.*, 1989, 1990; Kuhrova *et al.*, 1991).

Liliaceae: *Lilium* (Sentry & Smith, 1985).

Although several possible roles such as structural (chromatin condensation, centromere stabilization), regulatory, specific effects in meiotic pairing and in general recombination have been suggested (John & Miklos, 1979; Bostock *et al.*, 1985), the precise function of these sequences remains largely under speculation. Some of the repetitive sequences, especially those with long period and short period interspersion patterns, are either transcriptional units for discretely sized RNAs themselves and/or they are extensively homologous to discretely sized RNAs that are transcribed elsewhere (Jelinek & Schmid, 1982). Since these sequences occur in large proportion in the genome, significant research effort has been made world-wide on their identification and characterization in different eukaryotic genomes. In fact, in recent years, these sequences are proving themselves to be very promising tools and lucrative targets in the areas of molecular biology and plant genetics. Lately, repeat regions of DNA have attracted interest among population geneticists because they are often highly polymorphic within populations (Condit & Hubbel, 1991). In the present article, an attempt is made to discuss the potentials of these sequences with respect to a few important aspects such as species-specific character, DNA fingerprint markers and transposable elements.

RESULTS

Repeated sequences as species-specific markers

Species-specific repeated DNA sequences have been identified, isolated and characterised from cereals like wheat (Metzlaff *et al.*, 1986), rye (Guidet *et al.*, 1991), barley (Sonina *et al.*, 1989), rice (Aswidinnoor *et al.*, 1990; Reddy *et al.*, 1990), and sorghum. These sequences are of interest as species-specific molecular markers for selection and taxonomy of plant species. Investigation of their structure and representation in genome is important for elucidation of plant DNA organization and evolution in general. DNA-DNA hybridization studies between wheat, rye, barley, and oats have shown that 16, 22, 28 and 58% of the DNAs of each species, respectively, are species-specific repeated DNA sequences which have probably arisen by amplification of single copy DNA since species divergence (Flavell *et al.*, 1977). Such estimates of amount of species-specific sequences can be used to evaluate the evolutionary genetic distance. For example, 26% of the repetitive DNA sequences of rye hybridizes to all 4 species, 26% only to barley, wheat and rye, 18% only to wheat and rye and 30% only to rye (Rogowsky *et al.*, 1991). The percentages of shared repetitive sequences reflect well at the close phylogenetic relationship between rye and wheat and the more distant relationship between barley and oats. Apart from cereals, dispersed and tandem repeats have been used to assess relatedness and phylogenetic studies in Petunia (Shepherd *et al.*, 1990). Three dispersed repeats ROH15, ROH25, and CAS13 can distinguish species with 14 chromosomes from those with 18 chromosomes in Petunia as well as white-flowered species from pink-flowered species. One sequence CAS73, an A-T rich sequence (75%), characteristic of other highly repeated tandem arrays in plants, is detected only in Petunia species with 14 chromosomes.

In addition to the evolutionary and phylogenetic aspects, species-specific DNA probes have practical importance because they can be used very effectively to mark chromosomes in in situ hybridization and can serve as valuable tools in cytogenetic analysis. Species-specific sequences have a great potential in monitoring the introgression of alien chromatin into wheat (Clarke *et al.*, 1989; Koebner *et al.*, 1986) as well as in varietal analysis (Xin & Appels, 1988; Gupta *et al.*, 1990). A species-specific repeat sequence Dialect-1 from Hordeum vulgare has been isolated and assessed by Southern blot and in situ hybridization (Sonina *et al.*, 1989). Dialect-1 is a moderately repeated sequence (5000 copies per genome) and is dispersed through all the barley genome. It is absent in genomes of several wild barley species as well as in genome of wheat, rye, oat and maize (Sonina *et al.*, 1989). The search for more species-specific sequences from barley is in progress in other labs as well (Chakrabati & Subrahmanyan, 1985; Salina *et al.*, 1986). In rye, the 350-480bp, 5.3 Histone 3 and R173 families are considered to be rye specific, i.e., they do not hybridize to wheat or barley (Bedbrook *et al.*, 1980; Appels *et al.*, 1986; Guidet *et al.*, 1991). Though cereals and millets belong to the same family very little is known about repeat families in millets.

Recently, 2 repeat families, namely 1.3 kbp EcoRI and 1.4 kbp XbaI have been reported which have been found to be specific to great millet. The two families are highly repetitive, largely homogeneous and are dispersed in the genome with typical "clustered and scrambled" organization. When used as probe against the digests of great millet and five other millets namely, barn yard millet, little millet, fox tail millet, finger millet and pearl millet, hybridization is seen only to great millet. Species-specific highly repeated sequences have been implicated to play an important role in chromosome pairing and in affecting the meiotic stability of hybrids (Bennett, 1973) and a similar role has been proposed for the two families in great millet.

Species-specific repetitive DNA sequences which serve as markers in rice backcross breeding programmes have been isolated from the genomes of *Oryza. minuta* (BBCC), *O.* punctata (BB) and *O. officinalis* (CC) (Aswidinnoor *et al.,* 1990). Their genomic organization, chromosome distributions and taxonomic occurrence is under investigation. Recently, genomic DNA sequences specific to wild species of rice, *O. minuta* have been isolated and their role in analysis of alien chromatin introgression in the progeny of interspecific cross of *O. minuta* × *O. sativa* is investigated (Aswidinnoor *et al.*, 1990).

Role as DNA Fingerprint markers

Analysis of minisatellite DNA sequences yielding DNA 'fingerprints' has proven useful in studies of extent and maintenance of genetic diversity and paternity analysis of several different populations (Vassart *et al.*, 1987; Dallas 1988; Rogstad *et al.*, 1988; Ryskov *et al.*, 1988; Nybom *et al.*, 1989). A complex and highly individual-specific pattern obtained after digesting the DNA with restriction enzymes and hybridizing to a radioactively labelled probe is called a DNA fingerprint DNA profile or DNA type. (Nybom *et al.*, 1990). Most of the fragments in this complex pattern are present in the heterozygous state and are inherited as Mendelian traits. Using minisatellite probes (also called multilocus probes), several alleles at many loci can be scanned simultaneously thus revealing band patterns which are individually specific to the degree that justifies the name "DNA fingerprints" (Jeffreys *et al.*, 1985 a,b). DNA fingerprinting is one technique that is based on variation in repeat regions (Jeffreys *et al.*, 1985 b). Eukaryotic genomes have been found to contain highly polymorphic minisatellite loci consisting of variable number of tandem repeats of relatively short sequences (Jeffreys *et al.*, 1985 a,b). These variable numbers are possibly as a result of errors occurring during DNA replication or unequal crossing over (Jeffreys *et al.*, 1985a; Jeffreys, 1987). Another probable reason for deciphering of DNA fingerprints with use of repeat DNA sequence as probes may be due to their high turnover rate which may lead to accumulation of changes in these sequences, thereby making them specific to each individual. Initially, such fingerprints were utilized mainly for paternity testing in human and some animal species (Jeffreys *et al.*, 1985a,b; Burke & Bruford, 1987; Wetton *et al.*, 1987; Burke *et al.*, 1989). Recently, it has been applied in plants to distinguish cultivars and estimate the amount and distribution of genetic variation (Dallas, 1988; Nybom *et al.*, 1989, 1990), paternity analysis (Nybom & Schaal, 1990), as well as to assess gene introgression in breeding programmes (Hillel *et al.*, 1990).

The DNA fingerprinting approach briefly entails digestion of DNA with suitable restriction enzyme, Southern blotting and hybridization with a radioactively labelled probe. The patterns are then analysed taking into consideration presence/absence or differences in the mobility of bands among the accessions under study. Thus, a distinct hybridization pattern specific to a cultivar reveals its own fingerprint.

Different kinds of probes/DNA markers have been used for generation of DNA fingerprints. One of the most commonly used probes for this purpose are the human minisatellite sequences referred to as 33.1 and 33.6 (Jeffreys *et al.*, 1985 a,b). These authors generated probes isolated from the human minisatellite in the myoglobin locus which had a core sequence of 33 nucleotides. A subset of the human minisatellite share a common 10-15bp long consensus sequence. They have gained immense importance due to their applicability in disputed paternity studies of parentage in humans and birds, and forensic investigations (Burke & Bruford, 1987; Burke *et al.*, 1989). These probes have detected minisatellite variation also in plants (Dallas, 1988; Striem *et al.*, 1990). Thus, in cases where the level of RFLPs detected by single copy DNA sequences is low, hypervariable regions that are dispersed throughout the genome provide more information on RFLP markers (Dallas, 1988). JF Dallas (1988) has used these probes to detect RFLPs in rice. When these probes were challenged against the digests of *Orzya sativa* and *O. glaberrima* cultivars, the patterns were highly cultivar-specific; and largely remained unchanged after regeneration of plants from tissue culture (Dallas, 1988). Similarly this probe was used to obtain DNA fingerprint patterns in grapes by digesting *Vitis vinifera genomic* DNA either with *Hinfl* or *HaeIII* (Striem *et al.*, 1990). In horticultural species, the major problem is the difficulty in early identification of the genotype as the life cycle is very long. RFLPs offer an advantage in the identification of the genotype virtually from any plant part at any developmental stage. This is because no variations in RFLP patterns is expected during the course of development. This provides an excellent tool in the hands of an horticulturist.

Another probe in the minisatellite category is DNA fragment derived from M13 phage (Vassart *et al.*, 1987). The probe is prepared by digesting the M13 phage DNA with ClaI and BsmI to generate a 780bp fragment. In fact, this probe has been used as a universal marker since it shows homology to microbial, animal, and plant genomes (Ryskov *et al.*, 1988). Todate, fingerprints using M13 probe have been generated in Raspberries and Blackberries (Nybom et al., 1989, 1990), Acer (Nybom & Rogstad, 1990) and grapes (Striem et al., 1990) as well as in other angiosperms and gymnosperms (Rogstad et al., 1988).

Another class of probes namely oligonucleotides complementary to simple repeated sequences serve as valuable tools to obtain highly informative DNA fingerprints. These regions have again attracted interest among population geneticists because of their highly polymorphic nature. In fact, these have several advantages over the cloned minisatellite probes as they are chemically synthesized, and so pose no problem of stability; hybridization can be done directly on dried gels, and exposure time is much shorter; minute amounts of DNA can be analysed successfully; and can be effectively labelled by non-radioactive methods, for example, biotinylated oligos reveal signals just after 2 hours (Schafer et al., 1988).

Tandem DNA repeats of 2bp or family of simple repeat DNA consisting of simple quadruplet repeats (sqr) fall into this category. Condit & Hubbel (1991) examined the potential of 2bp repeats, namely oligonucleotides poly (GT) and poly (AG) as markers in tropical tree genomes. They report that these 2bp repeats are abundant in plant genomes and could provide informative markers for studies involving plant population genetics.

A family of simple quadruplet repeats (sqr) (GATA/GACA), originally from female-specific snake satellite are promising probes for DNA fingerprinting (Schafer *et al.*, 1988). The first oligonucleotide probes investigated for this purpose were (GACA)4, (GATA)4 and (GATA)2 GACA (GATA)2 (Ali *et al.*, 1986).

Role of repeats as transposable elements

Transposable elements (transposons, insertion sequences, mobile genetic elements) are low copy number repeated sequences and are part of the genomes of a wide range of organisms including bacteria, insects, nematodes, and mammals (Berg & Howe, 1989; Balcells *et al.*, 1991). These elements have been shown to belong to dispersed repetitive sequences, and it is now assumed that the latter sequences have spread in the genome by transposition-like events. (Freeling, 1984; Vedel & Delseny, 1987). In fact, transposition has been described as a major factor in the evolution of plant nuclear genomes. They were first reported in 1948 by Barbara McClintock in maize, based purely on cytogenetical analysis. These are structurally and genetically discrete segments of DNA, able to move from one site to the other either on the same chromosome or a different one (Vedel & Delseny, 1987).

In plants, transposable elements have been studied most extensively only in two plant species, namely maize and snapdragon (Gierl *et al.*, 1989; Coen *et al.*, 1988). They have also been isolated and characterized from a few other plants: for example Tgm1 (Le) element from soybean (Goldberg *et al.*, 1983; Voddkin *et al.*, 1983) and Mu element from rice (Wu & Xie, 1990).

Transposable elements possess characteristic structural features (Vedel & Delseny, 1987). They show presence of terminal inverted repeats, which are supposed to be the substrates for transposase activity, flanked by direct duplication of target site. One of the open reading frame is assumed to encode the transposase.

Recently, transposons are proving to be very useful tools in molecular biology, particularly they serve as tags to isolate genes (Shepherd, 1987; Wienand & Saedler, 1987). An insertional mutagenesis method called transposons tagging has been used to isolate several genes (including regulatory, organellar and other enzyme synthesis genes) from maize and snapdragon. (Balcells *et al.*, 1991). This is due to the peculiar phenomenon of mobile elements which generate mutation or chromosomal rearrangements and thus affect gene expression. For example, in maize the colour of kernel is normally purple due to synthesis of anthocyanin in the cells visible even on the surface of the kernel. However, if one of the proteins required for normal anthocyanin synthesis is not expressed in some of the cells, then the kernel appears colourless in a specific area, leaving the other areas coloured. The mosaic pattern indicates the switching-on and-off of the anthocyanin synthesis in localized areas. This serves as an indirect proof for the presence of a transposon. Thus, insertion of a transposon in-to the gene often inactivates the gene and induces a mutant phenotype. Such insertional events create molecular polymorphisms which can be evidenced by restriction endonuclease analysis, use of appropriate cDNA probes and nucleotide sequencing (Johns *et al.*, 1983).

References

Ali, S., Muller, C.R. & Epplen, J.T. (1986). DNA fingerprinting by oligonucleotide probes specific for simple repeats. Human Genetics. **74**, 239-243.

Appels, R., Moroan, L.B. & Gustafson, J.P. (1986). Rye heterochromation I: Studies on clusters of the major repeating sequence and the identification of a new dispersed repetitive sequence element. Canadian Journal of Genetics and Cytology **28**, 645-657.

Aswidinnoor, H., Dallas, J.F., McIntyre, C.L. & Gustafson, J.P. (1990). Species specific repetitive DNA sequences as markers in rice backcross breeding programme. Abstr. Second International Rice Genetics Symposium, May 14-18, Manila, Philippines.

Balcells, L., Swinburne, J. & Coupland, G. (1991). Transposons a tool for the isolation of plant geness. Trends in Biotechnology **9**, 31-36.

Bedbrook, J.R., Jones, J., O'Dell, M., Thompson, R.D. & Flavell, R.B. (1980). A molecular description of telomeric heterochromatin in *Secale* species. Cell **19**, 545-560.

Bennett, M.D. (1973). Meiotic, gametophytic and early endosperm development in Triticale. In Triticale, Edited by Mac Intyre, E. & Campbell, M. International Development Research Centre, Ottawa, pp 137-148.

Benslimane, A.A., Dron, M., Hartmann, C. & Rode, A. (1986). Small tandemly repeated DNA sequences of higher plants likely to originate from tRNA gene ancestor. Nucleic Acid Reserach **14**, 8111-8119.

Berg, D. & Howe, M.M. (1989). Mobile DNA, American Society for Microbiology.

Bhave, M., Gupta, V. & Ranjekar, P.K. (1986). Arrangement and size distribution of repeat and single copy DNA sequences in four species of Cucurbitaceae. Plant Systematics and Evolution **152**, 133-151.

Bostock, C. (1985). A function for satellite DNA? Trends in Biochemical Sciences **5**, 117-119.

Burke, T. & Bruford, M.W. (1987). DNA fingerprinting in birds. Nature (London) **327**, 147-149.

Burke, T., Davies, N.B., Bruford, M.W. & Hatchwell, B.J. (1989). Parental care and mating behaviour of polyandrous dunnocks *Prunella modularis* related to paternity by DNA fingerprinting. Nature (London) **338**, 249-251.

Chakrabati, T. & Subrahmanyan, N.C. (1985). Analysis of DNA from related and diverse species of barley. Plant Science **42**, 183-190.

Clarke, B.C., Moran, L.B. & Appels, R. (1989). DNA analysis in wheat breeding. Genome **32**, 334-339.

Coen, E.S., Robbins, T.P., Almeida, J., Hudson, A. & Carpenter, R. (1988). In Mobile DNA Edited by Berg, D. & Howe, M.M., Am. Soc. Microbiol. pp 411-434.

Condit, R. & Hubbel, S.P. (1991). Abundance and DNA sequence of 2 base repeat regions in tropical tree genomes. Genome **34**, 66-71.

Dabak, M.M., Ranade, S.A., Dhar, M.S., Gupta, V.S. & Ranjekar, P.K. (1988). Molecular characterization of pigeonpea genome . Indian J. Biochem. Biophys. **25**, 230-236.

Dallas, J.F. (1988). Detection of DNA "fingerprints" of cultivated rice by hybridization with a human minisatellite DNA probe. Proc. Natl. Acad. Sci., USA **85**, 6831-6835.

Deshpande, V.G. & Ranjekar, P.K. (1980). Repetitive DNA in three Gramineae species with low DNA content. Hoppe-Seyler's Z. Physiol. Chem. **361**, 1223-1233.

Dhar, M.S., Dabak, M.M., Gupta, V.S. & Ranjekar, P.K. (1988). Organization and properties of repeated DNA sequences in rice. Plant Sci. **55**, 43-52.

Dhar, M.S., Pethe, V.V., Gupta, V.S. & Ranjekar, P.K. (1990). Predominance and tissue specificity of adenine methylation in rice. Theor. Appl. Genet. **80**, 402-408.

Fabijanski, S., Fedak, G., Armstrong, K. & Altosaar, T. (1990). A repeated sequence probe for the C genome in Avena (oats). Theor. Appl. Genet. **79**, 1-7.

Flavell, R.B. (1980). The molecular characterization and organization of plant chromosomal DNA sequence. Ann. Rev. Plant Physiol. **31**, 569-596.

Flavell, R.B. (1986). Repetitive DNA and chromosome evolution in plants. Philos. Trans. R. Soc. Lond. Ser. B **312**, 227-242.

Flavell, R.B., Rimpau, J. & Smith, D.B. (1977). Repeated sequence DNA relationships in four cereal genomes. Chromosoma **63**, 205-222.

Freeling, M. (1984). Plant transposable elements and insertion sequences. Ann. Rev. Plant. Physiol. **35**, 277-298.

Friedemann, P. & Peterson, P.A. (1982). Mol. Gen. Genet. **187**, 19-29.

Ganal, M. & Hemleben, V. (1986). Different AT rich satellite DNAs in *Cucurbita pepo* and *Cucurbita maxima*. Theor. Appl. Genet. **73**, 129-135.

Ganal, M.W., Lapitan, N.L.V. & Tanksley, S.D. (1988). A molecular and cytogenetic survey of major repeated DNA sequences in tomato (*Lycopersicon esculentum*). Mol. Gen. Genet. **213**, 262-268.

Gierl, A., Saedler, H. & Peterson, P.A. (1989). Annu. Rev. Genet. **23**, 71-85.

Goldberg, R.B., Hoschek, G. & Vodkin, L.O. (1983). Cell **33**, 465-475.

Grellet, F., Delcasso, D., Panabieres, F. & Delseny, M. (1986). Organization and evolution of a higher plant haploid-like satellite DNA sequences. J. Mol. Biol. **187**, 495-507.

Guidet, F., Rogowsky, P., Taylor, C., Song, W. & Langridge, P. (1991). Cloning and characterization of a new rye-specific repeated sequence. Genome **34**, 88-95.

Gupta, V., Gadre, S.R. & Ranjekar, P.K. (1981). Novel DNA sequence organization in rice genome. Biochem. Biophys. Acta **656**, 147-154.

Gupta, V.S., Dhar, M.S., Patil, B.G., Narvekar, G.S., Rawat, S.R. & Ranjekar, P.K. (1990). Molecular cloning and restriction enzyme analysis of long repetitive DNA sequences in rice. J. Biosci. **15**, 261-269.

Gupta, V.S. & Ranjekar, P.K. (1982). Genome organization in pearl millet. Ind. J. Biochem. Biophys. **19**, 167-170.

Hallden, C., Bryngelsson, T., Sall, T. & Gustafesson, M. (1987). Distribution and evolution of tandemly repeated DNA sequence in the family *Brassicaceae*. J. Mol. Evol. **25**, 318-323.

Hillel, J., Schaap, T., Haberfeld, A., Jeffreys, A.J., Plotzky, Y., Cahaner A. & Lavi, U. (1990). DNA fingerprints applied to gene introgression in breeding programs. Genetics **124**, 783-789.

Iwabuchi, M., Itoh, K. & Shimamoto, K. (1991). Molecular and cytological characterization of repetitive DNA sequences in *Brassica*. Theor. Appl. Genet. **81**, 349-355.

Jeffreys, A.J. (1987). Highly variable minisatellites and DNA fingerprints. Biochem. Soc. Trans. **15**, 309-317.

Jeffreys, A.J., Wilson, V. & Thein, S.L. (1985a). Hypervariable "minisatellite" regions in human DNA. Nature **314**, 67-73.

Jeffreys, A.J., Wilson, V. & Thein, S.L. (1985b). Hypervariable "minisatellite" regions in human DNA. Nature **316**, 76-79.

Jelinek, W.R. & Schmid, C.W. (1982). Repetitive sequences in eukaryotic DNA and their expression. Ann. Rev. Biochem. **51**, 813-844.

John, B. & Miklos, G.L.G. (1979). Functional aspects of satellite DNA and heterochromatin. Int. Rev. Cytol. **58**, 1-114.

Johns, M.A., Strommer, J.N. & Freeling, M. (1983). Exceptionally high levels of restriction site polymorphism in DNA near the maize Adh 1 gene. Genetics **105**, 733-743.

Kato, A., Yakura, K. & Tanifuji, S. (1984). Sequence analysis of *Vicia faba* repeated DNA, the Folk I repeat element. Nucleic Acids Res. **13**, 6415-6426.

Koebner, R.M.D., Appels, R. & Shepherd, K.W. (1986). Rye heterochromatin II. Characterization of a derivative from chromosome 1DS. 1RL with reduced amount of the major repeating sequence. Can. J. Genet. Cytol. **28,** 658-664.

Koukalova, B., Reich, J. & Bezdek, M. (1990). A *BamHI* family of tobacco highly repeated DNA - a study about its species specificity. Biol. Plant. **32**, 445-449.

Koukalova, B., Reich, J., Matyasek, R., Kuhrova, V. & Bezdek, M. (1989). A *BamHI* family of highly repeated DNA sequences *Nicotiana tabacum*. Theor. Appl. Genet. **78**, 77-80.

Kuhrova, V., Bezdek, M., Vyskot, B., Koukalova & Fajkus, J. (1991). Isolation and characterization of two middle repetitive DNA sequences of nuclear tobacco genome. Theor. Appl. Genet. **81**, 740-744.

Lagu, M., Bhave, M., Gupta, V.S. & Ranjekar, P.K. (1986). Molecular analysis of Cucurbitaceae genomes: Part IV Homologies among repeated DNA sequences. Ind. J. Biochem. Biophys. **23**, 167-170.

Lagu, M., Ranjekar, P.K. & Pillay, D.T.N. (1986). The genome of cucumber (*Cucumis sativus*): Absence of interspersion of repeated and single copy DNA sequences at a DNA fragment length of 5 kbp. Cell Biol. Intl. Rep. **10**, 869-874.

Lakshmikumaran, M. & Ranade, S.A. (1990). Isolation and characterization of a highly repetitive DNA of *Brassica campestris*. Plant. Mol. Biol. **14**, 447-448.

Lalitha S. Kumar, Gupta, V.S. & Ranjekar, P.K. (1990). Identification and partial characterization of 2 species-specific repeated families in great millet (*Sorghum vulgare*). Plant Syst. Evol. **171**, 249-257.

Leclerc, R.F. & Siegel, A. (1987). Characterization of repetitive elements in several *Cucurbita* species. Plant Mol. Biol. **8**, 497-507.

Lehmann, P. & Stephen, B. (1990). Preliminary characterization of the repeated DNA sequences from *Vicia sativa*. Acta. Biochemica. Polonica **37(1)**, 21-30.

Martinez-Zapater, J.M., Estelle, M.A. & Somerville, C.R. (1986). A highly repeated DNA sequence in *Arabidopsis thaliana*. Mol. Gen. Genet. **204**, 417-423.

McClintock, B. (1948). Mutable loci in maize. Carnegie. Inst. Washington Yearbook 47, 155-169.

McIntyre, C.L., Clarke, B.C. & Appels, R. (1988). Amplification and dispersion of repeated DNA sequences in the Triticeae. Plant Syst. Evol. **160**, 39-59.

Metzlaff, M., Troebner, W., Baldauf, F., Schlegel, R. & Cullum, J. (1986). Wheat specific repetitive DNA sequences- construction and characterization of 4 different genomic clones. Theor. Appl. Genet. **72**, 207-210.

Nybom, H. & Rogstad, S.H. (1990). DNA fingerprints detect genetic variation in *Acer negundo*. Plant Syst. Evol. **173**, 49-56.

Nybom, H., Rogstad, S.H. & Schaal, B.A. (1990). Genetic variation detected by use of the M13 "DNA fingerprint" probe in Malus, Prunus and Rubus (Rosaceae). Theor. Appl. Genet. **79**, 153-156.

Nybom, H. & Schaal, B.A. (1990). DNA "fingerprints" applied to paternity analysis in apples. Theor. Appl. Genet. **79**, 763-768.

Nybom, H., Schaal, B.A. & Rogstad, S.H. (1989). DNA "fingerprints" can distinguish cultivars of black berries and raspberries. Acta. Hortic. 262, 305-310.

Ranade, S.A., Lagu, M.D., Patankar, S.M., Dabak, M.M., Dhar, M.S., Gupta, V.S. & Ranjekar, P.K. (1988). Identification of a dispersed MboIf repeat family in five higher plant genomes. Bioscience Reports **8**, 435-441.

Ranjekar, P.K. (1982). Analysis of plant genomes: A molecular approach. J. Sci. Indus. Res. **41**, 384-393.

Ranjekar, P.K., Pallota, D. & Lafontaine, J.G. (1976). Analysis of the genome of plants II. Characterization of repeated DNA in barley (*Hordeum vulgare*) and wheat (*Triticum aestivum*). Biochem. Biophys. Acta **425**, 30-40.

Rayburn, A.L. & Gill, B.S. (1986). Isolation of a D-genome specific repeated DNA sequence from *Aegilops squarossa*. Plant Mol. Biol. Rep. **4**, 102-109.

Rayburn, A.L. & Gill, B.S. (1987). Molecular analysis of the D genome of the Triticeae. Theor. Appl. Genet. **73**, 385-388.

Reddy, A.S., Cordesse, F., Kiefer, M.C. & Delseny, M. (1990). Isolation, characterization and use of CC genome specific repeated sequences. Abstr. Fourth Annual Meeting of the RF Foundation's International Program on Rice Biotechnology, May 9-12.

Rogowsky, P.M., Manning, S., Liu, J.Y. & Langridge, P. (1991). The R173 family of rye specific repetitive DNA sequences: a structural analysis. Genome **34**, 88-95.

Rogstad, S.H., Patton II, J.C. & Schaal, B.A. (1988). M13 repeat probe detects DNA minisatellite-like sequences in gymnosperms and angiosperms. Proc. Natl. Acad. Sci, USA **85**, 9176-9178.

Roy, P., Bhattacharyya N. & Biswas, B.B. (1988). Isolation, characterization and sequencing of a novel repetitive DNA from mung bean *Vigna radiata*. Gene 73, 57-66.

Ryskov, A.P., Jincharadze, A.G., Prosnyak, M.I., Ivanov, P.L. & Limborska, S.A. (1988). M13 phage DNA as a universal marker for DNA fingerprinting of animals, plants and microorganisms. FEBS Lett. **233**, 388-392.

Sakowicz, T., Galazka, G., Konarzeuska, A., Kwinkowski, M. & Klysik, J. (1986). An unusually high number of direct repeats detected *EcoRI* family fragment in Lupinus luteus L. Planta **168**, 207-213.

Salina, E.A., Vershinin, A.V., Svitashev, S.K. & Shumnyi, V.K. (1986). Isolation and analysis of highly repetitive DNA clone library. Dokl Akad Nauk SSSR, 478-480.

Schafer, R., Zischler, H., Birsner, U., Becker, A. & Epplen, J.T. (1988). Optimized oligonucleotide probes for DNA fingerprinting. Electrophoresis **9**, 369-374.

Schweizer, G., Ganal, M., Ninnemann, H., & Hemleben, V. (1988). Species specific DNA sequences for identification of somatic hybrids between *Lycopersicon esculentum* and *Solanum acaule*. Theor. Appl. Genet. **75**, 679-684.

Sentry, J.W. & Smith, D.R. (1985). A family of repeated sequence dispersed through the genome of *Lilium henryi*. Chromosoma (Berl) **92,** 149-155.

Shepherd, N.S. (1987). In Plant Molecular Biology - A practical approach. Edited by Shaw G.H. IRL Press.

Shepherd, A.L., Anderson, S. & Smith S.M. (1990). Species specific repeated DNA sequences from Petunia. Plant Sci. **67**, 57-95.

Sibson, D.R., Hughes, S.G., Bryant, J.A. & Fitchett, P.N. (1991). Sequence organization of simple highly repetitive DNA elements in *Brassica* species. J. Exp. Botany **42**, 243-249.

Simeons, C.R., Gielen, J., Van Montagu, M. & Juze, D. (1988). Characterization of highly repetitive sequences of *A. thaliana*. Nucleic Acids Res. **16,** 6753-6766.

Sivaraman, L., Gupta, V.S. & Ranjekar, P.K. (1984a). Novel molecular feature of millet genome. Ind. J. Biochem. Biophys. **21**, 299-303.

Sivaraman, L., Gupta, V.S. & Ranjekar, P.K. (1984b). Molecular organization of great millet (*Sorghum vulgare*). J. Biosci. **6**, 795-809.

Sivaraman, L., Gupta, V.S. & Ranjekar, P.K. (1985). Low homology of repeated DNA sequences in millets. Ind. J. Biochem. Biophys. **22**, 268-273.

Sivaraman, L., Gupta, V.S. & Ranjekar, P.K. (1986). DNA sequence organization in the genome of three related millet plant species. Plant Mol. Biol. **6**, 375-388.

Sonina, N.V., Lushnikova, A.A., Tihonov, A.P. & Ananiev, E.V. (1989). Dialect-I species specific repeated DNA sequence from barley *Hordeum vulgare*. Theor. Appl. Genet. **78**, 589-593.

Sorenson, J.C. (1984). The structure and expression of nuclear genes in higher plants. Adv. Genet. **22**, 109-144.

Striem, M.J., Spiegel-Roy, P., Ben Hayyim, G., Beckmann, J. & Gidoni, D. (1990). Genomic DNA fingerprinting of *Vitis vinifera* by the use of multi-loci probes. Vitis **29**, 223-227.

Tamhankar, S.A., Gupta, V.S., Joshi, K.S. & Ranjekar, P.K. (1990). Occurrence and characterization of a dispersed *Mbo*I repeat family in frenchbean (*Phaseolus vulgaris*) seedling DNA. Plant Sci. **68**, 203-211.

Vassart, G., Georges, M., Monsieur, R., Brocas, H., Lequarre, A.S. & Cristophe, D. (1987). A sequence in M13 phage detects hypervariable minisatellite in human and animal DNA. Science **235**, 683-684.

Vedel F. & Delseny, M. (1987). Repetitivity and variability of higher plant genomes. Plant

Physiol. Biochem. **25**(2), 191-210.

Voddkin, L.O., Rhodes, P.R. & Goldberg, R.B. (1983). A lectin gene insertion has the structural features of a transposable element. Cell **34**, 1023-1031.

Walbot, V. & Cullis, C.A. (1985). Rapid genomic change in higher plants. Annu. Revi. Plant Physiol. **36**, 367-396.

Wetton, J.H., Carter, R.E., Parkin, D.T. & Walters, D. (1987). Demographic study of a wild house sparrow population by DNA fingerprinting. Nature 327, 147-149.

Wienand, U. & Saedler, H. (1987). In Plant DNA infectious agents. Edited by Hohn, T.H. and Schell, J., pp 205-227.

Wu, T. & Wu, R. (1987). A new rice repetitive DNA shows sequence homology to both 5S RNA and tRNA. Nucleic Acids Res. **15**, 5913-5923.

Wu, R. & Xie, Y. (1990). Transposable elements in rice. Abstr. Second International Rice Research Genetics Symposium, May 14-18 at IRRI, Manila, Philippines.

Xin, Z.Y. & Appels, R. (1988). Occurrence of rye (*Secale cereale*) 350 - family DNA sequences in Agropyron and other Triticeae. Plant Syst. Evol. **160**, 65-76.

Zhao, X., Wu, T., Xie, Y. & Wu. R. (1989). Genome specific repetitive sequences in the genus *Oryza*. Theor. Appl. Genet. **80**, 201-209.

3

Analyses of some of the processes involved in baculovirus mediated expression of foreign genes in insect cells and caterpillars

Prakash K. Jha, Padma Sridhar and Seyed E. Hasnain
National Institute of Immunology
Shahid Jeet Singh Marg, New Delhi 110067, India

ABSTRACT

Recombinant DNA techniques have been used judiciously over the past few years to obtain large quantities of proteins of use in human and animal health, agriculture and for basic biological studies. For this, the desired gene is expressed in a suitable heterologous system. Of a number of prokaryotic and eukaryotic expression systems developed so far, the insect baculovirus based system has shown remarkable success and great utility for producing large amounts of foreign proteins both in insect cells and caterpillars. Barring some extremely complex type of post-translational modifications the recombinant proteins are identical in almost all respects to the native proteins.

INTRODUCTION

Inspite of the availability of numerous expression systems, such as bacteria, yeasts, filamentous fungi, and cultured mammalian cells, a helper-independent baculovirus expression vector system (BEVS) has become remarkably popular for the expression of a wide variety of foreign genes in insect cells and caterpillars. The reasons for this wide acceptability lie in the high levels of gene expression and the rapidity with which such levels of expression are achieved by employing a simple cell culture technology. Moreover, the recombinant proteins produced using this expression system are antigenically, immunologically, and functionally identical to the native protein (Luckow & Summers, 1988; Maeda, 1989; Miller, 1989; Wood & Granados, 1991). Baculoviruses by virtue of being non-pathogenic to vertebrates and plants (Blissard &

Rohramann, 1990), additionally provide a safer system from drug regulatory point of view. The insect cells in culture or the larvae host used for BEVS are capable of proper post-translational modifications, processing and transport resulting in the synthesis of biologically active recombinant proteins (Summers & Smith, 1987; Maeda, 1989).

Baculoviruses have been extensively used to express several proteins of biomedical, veterinary, and agricultural importance. The first recombinant HIV envelope protein, approved by the Federal Drug Agency of USA for clinical evaluation as a candidate vaccine for AIDS, was synthesized in insect cells using recombinant baculovirus. The recent interest in live caterpillars as a host for baculoviruses has made it possible to produce recombinant proteins on a cost-effective basis. The widespread use of BEVS for the production of proteins for medical research and biotechnology has led to renewed interest in understanding the biology of baculoviruses.

Baculovirus biology

Viruses infecting insects are classified into seven families of which the most common group is the family Baculoviridae (Maeda, 1989) represented by a single genus Baculovirus (Wood & Granados, 1991). This family consists of DNA viruses, pathogenic predominantly to holometabolous insects (those that undergo complete metamorphosis) and is divided into three sub-groups. One of them includes nuclear polyhedrosis viruses (NPVs) which are currently being employed as expression vectors mainly because cell lines permissive for their replication are readily available. NPVs infect several orders of insects, mainly Lepidoptera, and have the unique property of producing proteinaceous nuclear occlusion bodies termed polyhedra in which progeny virions are embedded.

During baculovirus infection two genetically alike but morphologically, biochemically, serologically, and functionally different forms of viral progenies are produced in insect cells: (i) extracellular virus (ECV) which are also called cell-released-virus or non-occluded or budded virus and (ii) occluded virus (OV) or polyhedra derived virus. ECVs consist of a single enveloped nucleocapsid which bud through the plasma membrane of infected cells, about 10-12 hours post infection (p. i.). They spread infection among cells within an individual insect via the host's hemolymph (secondary infection). OVs are contained in large occlusion bodies called polyhedra and are produced in the nuclei of infected cells in maximum numbers at 48-72 hours p.i. and continues until the infected cells lyse (Summers & Smith, 1987). Each polyhedron contains many enveloped virus particles (virions) embedded in a crystalline protein matrix composed primarily of 29 kDa and a 10 kDa protein termed polyhedrin.

The occluded forms are an important part of the natural life cycle of virus providing the means for horizontal transmission (causing infection among insects). Infected dead larvae have millions of polyhedra left in the decomposing tissue. Polyhedra help in protecting the virus particles, embedded in them, from adverse environmental conditions such as ultraviolet light and soil acidity. When healthy larvae feed on contaminated plant foliage, they ingest polyhedra. In the insect midgut, the alkaline pH (9.5-11) aids in dissolution of the polyhedrin, thereby releasing virions. These virions bring about infection which again show temporal regulation (formation of ECV and OV in biphasic manner during the infection process) of progeny virus as mentioned above.

Studies of pulse-labeled infected cell specific proteins as well as those using inhibitors of protein synthesis has suggested that in the infected cells the expression of viral genes and DNA replication occur in an ordered cascade of events in which each successive phase is dependent of the previous phase. Broadly, baculovirus gene expression is divided into two temporal phases: (i) an early phase which precedes viral DNA replication and (ii) a late phase which occurs with or after viral DNA replication begins. The early phase is subdivided into phases of (a) immediate early genes that can be transcribed by host cells and no viral gene products are required for their expression and (b) delayed early genes which need viral gene products for their transcription (Blissard & Rohrmann, 1990). Late phase expression includes late genes and hyperexpressed very late genes. These two classes are differentiated because mRNAs for late genes decline at a very late time p.i. while mRNA level of hyperexpressed late genes remain high throughout the infection cycle after they have been transcribed (Blissard & Rohrmann, 1990). Two of the known hyperexpressed late genes are polyhedrin and p10.

Useful features of BEVS for gene expression

Research on BEVS, using prototype baculovirus, *Autographa californica* nuclear polyhedrosis virus (AcNPV), was pioneered by Professor Max Summers at Texas A&M University, USA, in early eighties. The cell line used as a host for infection by AcNPV was derived from cultured ovaries of *Spodoptera frugiperda* (Sf9 and Sf21). These cells are routinely cultured in monolayer or suspension culture at room temperature, preferably at 27°C. The cells have a doubling time of 18-24 hours and can be passaged in glass or plastic flasks without having a requirement for any minimum seeding density. They do not require carbon dioxide for growth, unlike mammalian cells, and can easily be dislodged from monolayers (Summers & Smith, 1987). The ECV form of AcNPV is only infectious in insect cell culture. The OV form is not infectious because the pH of the culture medium is not high enough to dissolve the polyhedrin matrix. However, if polyhedra are dissolved by alkali treatment, then released virions can infect cultured cells.

Two of the baculoviruses, AcNPV and Bombyx mori (silkworm) NPV (BmNPV) have been extensively used for baculovirus mediated gene expression. This article focusses primarily on AcNPV which has a rod shaped capsid (40-140 x 250-400 nm) containing double stranded, circular, covalently closed, supercoiled DNA of about 128 kilobases. The genome has been mapped with restriction enzymes and a number of RNA transcripts together with their *in vitro* translation products have been located on the map. Approximately 70% of the AcNPV genome has already been sequenced and characterized (Blissard & Rohrmann, 1990; Possee *et al.*, unpublished). The polyhedrin gene has also been mapped and sequenced. Studies with deletion mutants have conclusively proved that polyhedrin is non-essential either for replication of virus or its infectivity. It is an abundant protein expressed very late in the infection cycle constituting more than 50-70% of the total cellular protein. Such level of expression suggests a very strong promoter of the polyhedrin gene and indeed, the polyhedrin gene promoter is one of the strongest promoters known so far in any system.

There are several features which make AcNPV an excellent vector for the propagation and expression of foreign genes such as the potential of the rod shaped virus to encapsidate viral genomes with as much as 10 kilobases of additional foreign DNA, the ability of BEVS to perform mRNA splicing events and the safety of the recombinant viral vector which is non-pathogenic to vertebrates, plants, and non-arthropod invertebrates. The polyhedrin gene also provides many useful features such as (i) a nonessential region of the AcNPV genome wherein a foreign DNA can be inserted, (ii) a hyperactivated promoter which directs transcription late in the infection when host genes and most viral genes are switched off, and (iii) a visual morphological marker useful for the selection of recombinant virus.

Using engineered AcNPV for the expression of foreign genes

The AcNPV genome is very large and contains multiple restriction enzyme sites which causes difficulties in inserting foreign genes by direct cloning. To circumvent the problem, a plasmid transfer vector is used which contains a unique restriction enzyme site, after the polyhedrin promoter, for inserting the foreign gene, and is flanked by viral sequences to allow homologous recombination *in vivo*. For this, permissive insect cells, such as Sf9, are cotransfected by the recombinant plasmid vector together with the wild type viral DNA. When replication of viral DNA begins, recombination, between the viral sequences flanking the gene of interest contained on the plasmid vector and the homologous sequences in the wild type viral DNA, takes place. Consequently, small fraction of progeny viruses (0.1-5%) have a replacement of wild type polyhedrin sequence with the plasmid sequence, thereby transferring the foreign gene into viral genome. Alternatively, the use of a linearized form of an engineered "wild type" AcNPV which carries a unique site for a very rare restriction enzyme *Bsu*361 has been suggested (Kitts *et al.*, 1990) for enhancing the recovery of recombinant viruses. The recombinant viruses thus obtained are functional and stable for generations.

One of the most simple and direct approaches of screening recombinant viruses is by visualization of the difference in the plaque morphology of wild type virus and the recombinant virus. Plaques of wild type virus appear as refractory centres to the naked eye due to the presence of polyhedra in them (Occ+) while recombinant viruses produce plaques that lack occlusion bodies (Occ−) and, therefore, can be visually identified.

A large number of foreign genes of bacterial, viral, plant, animal, and insect origin have been expressed using recombinant baculoviruses. Yields of proteins ranging from 1 mg to 1.2 g per liter have been obtained depending on the nature of genes being expressed but the expression levels rarely reach that of polyhedrin. Some of the numerous genes expressed using AcNPV and BmNPV baculovirus vectors include *E. coli* β-galactosidase, Hepatitis B virus core and surface antigens, human α- and β-interferon, human interleukins, human c-myc proto-oncogene, human erythropoeitin, HIV *env, gp* 41 and *gp* 120 proteins, human tissue plasminogen activator, α- and β subunits of human chorionic gonadotropin, phaseolin and patatin storage proteins of plants, luciferase from firefly, and several others (Luckow & Summers, 1988; Maeda, 1989; Hasnain & Nakhai, 1990; Jha *et al.*, 1992; Nakhai *et al.*, 1991a,b, 1992). This list is by no means a complete one, and the additions are being made at such a rapid rate that literature is flooded with new reports of expression of various genes using BEVS.

Recombinant proteins have been produced as fusion or non-fusion proteins and both intracellular and extracellular proteins have been synthesized. All the genes that have been expressed at high levels are derived from cDNAs or genomic clones that do not contain introns, but genomic DNA can also be used implying that this system is capable of correct splicing. In general, proteins which are non-secretory are produced in higher quantity than those which enter the secretory pathway. Similarly, nuclear located or nonstructural proteins are most highly expressed while those requiring post translational modifications are least expressed. It has also been shown that temporal nature and not the promoter strength that determines quantitative and qualitative expression of genes encoding secretory and extensively processed glycoproteins (Sridhar *et al.*, 1992; Luckow & Summers, 1988). Factors that are responsible for hypertranscription from the polyhedrin promoter are being investigated in our laboratory and elsewhere.

Recombinant proteins produced in insect cells using BEVS undergo appropriate post-translational modifications to produce recombinant products very similar or identical to the native proteins. Recombinant proteins can be secreted, targeted to nucleus or other specific organelle within the cell or targeted to the cell surface, proteolytically cleaved, phosphorylated, signal peptide processed, C-terminus amidated, and glycosylated etc. One notable difference from other advanced eukaryotic systems lies in the quality of complex type of N-linked glycosylation involving trimming of terminal sugars.

Insect caterpillars as living protein factories

Insect caterpillars are being considered a sophisticated, living protein factory for the large scale production of foreign gene products (Miller, 1989; Medin *et al.*, 1990; Jha *et al.*, 1991) because the efficiency of protein synthesis is extremely high in larvae (as high as 1 mg/larva/day). They also show extraordinarily rapid growth (Maeda, 1989) leading to increasingly more protein synthesis. At the same time, it is easy to maintain larvae in mass culture with minimum cost inputs. Most significantly, a high percentage (~55%) of the recombinant virus - infected larvae does not moult into pupa (Table 3.1), and remain in larval stage till their death due to the arrest of metamorphosis caused by the virus coded *egt* gene (O'Reilly & Miller, 1989). The prolonged life span of infected larvae (4-5 days) as compared to mock-infected larvae (Table 3.1) is of special relevance for harvesting and purification of recombinant proteins because almost all of the foreign protein would be synthesized at this stage only.

Silkworm (*Bombyx mori*) and caterpillars of several pest insects such as *Spodoptera frugiper-*

Table 3.1 Effect of viral infection on moulting of larvae. Batch of 20 larvaewere injectedwith 10 µl vAcβhCG-*luc* each and monitored every day for moulting/death. There was10% injection induced mortality within 24 h p.i. Mock infected larvae were injected with distilled water.

Treatments	Days post infection						
	4	5	6	7	8	9	10
	Developmental stage (%)						
Mock infected larvae							
Larvae	0						
Prepupae	100	50					
Pupae	0	50	100	------------------------ moth			
vAcβhCG-*luc* infected larvae							
Larvae	78	55	55	55	55	33	0
Prepupae	22	11	11	11	11	11	0
Pupae	0	33	33	33	33	33	33----- moth
Mortality	0	0	0	0	0	22	66

da, Trichoplusiani etc. have been used as hosts for the baculovirus mediated foreign gene expression. Recombinant proteins can be produced in caterpillars either by directly injecting an ECV form of recombinant virus in the hemolymph, or by feeding them with plant foliage contaminated with occluded recombinant virus. For generating occluded recombinant viruses, the cell cultures are co-infected with the ECV form of both recombinant and wild type viruses (Miller, 1989; Hasnain, 1991). Occluded virions containing both virion types are generated which can be used to infect larvae through oral route. At present there are only about five centres with demonstrated expertise of producing foreign proteins in caterpillars in large quantities for biochemical or structural studies. A few notable examples are: Hepatitis B surface and core antigens, human interferon, interleukin 3, firefly luciferase, and human chorionic gonadotropin. An offshoot of this strategy is the potentials of using engineered ("crippled") baculoviruses as natural biocide for the control of plant pests because these viruses will not be able to persist in the environment due to the lack of polyhedrin coat.

We have also demonstrated that a recombinant baculovirus derived from a transfer vector containing duplicate copies of the polyhedrin promoter is genetically stable and capable of producing, in caterpillars, two foreign proteins- firefly luciferase and ß subunit of human chorionic gonadotropin, simultaneously (Jha *et al.*, 1992).

Regulation of synthesis and targeting of foreign proteins: use of firefly luciferase as reporter

Reporter genes are very good tools for molecular genetic analysis and understanding of eukaryotic gene regulation (Alam & Cook, 1990). Luciferase (*luc*) gene encoding firefly luciferase which catalyzes light emitting chemical reaction, has been widely used (Gould & Subramani, 1988) and expressed in several eukaryotic expression systems (Ow *et al.*, 1986; Gould & Subramani, 1988; Alam & Cook, 1990; Hasnain & Nakhai, 1990; Jha *et al.*, 1991). It has emerged as the most sensitive reporter gene for studying promoter and/or enhanced activity as well as trans-acting factors by monitoring the synthesis of *luc* (Gould & Subramani, 1988). The observations that luc is sorted at peroxisomes in several eukaryotic cells (such as yeasts, fireflies and plants) has raised interest in the development of luc gene as a model for elucidation of signals required for targeting proteins to peroxisomes (Subramani, 1991).

The most exciting aspect of using luciferase fusions is the ability to monitor gene expression visually (Schaeur, 1988). Very simple assay systems for luc are available. Light emission can be detected and measured even from a single insect cell, infected with a recombinant virus carrying the *luc* gene, by exposure to X-ray film (Hasnain & Nakhai, 1990) or quantitated in a luminometer or scintillation counter. In our laboratory, vAcluc was constructed where *luc* gene was placed under the transcriptional control of the viral polyhedrin promoter, replacing the native polyhedrin gene (Hasnain & Nakhai, 1990). Upon infection with v*Ac*luc, high levels of luc were synthesized in insect cells and larvae. A direct comparison with commercially available *luc* standard (Boehringer Mannheim, Germany) revealed that the recombinant luc reached the levels of approximately 150μg /10^6 cells in infected cells (Hasnain & Nakhai, 1990) and 1.3 mg/larva of *Spodoptera* (Jha *et al.*, 1991). In fact, larvae of *Spodoptera* and of *Trichoplusiani* synthesized very high levels of luc which represented ~28% and ~18%, respectively, of the total coomassie blue stainable proteins. Luc which is a non-secretory protein, was retained intracellularly in the infected Sf9 cells, as confirmed by immunofluorescence studies (Hasnain & Nakhai, 1990). Similarly, in the infected larvae, luc was retained in the body tissue and not secreted into the hemolymph (Jha *et al.*, 1991).

Luc has been known to be targeted to peroxisomes due to the presence of a "peroxisomal sorting domain" located within the last 12 amino acids at the C-terminal end (Subramani, 1991). We have designed experiments to distinguish the so called peroxisomal sorting domain from the signal(s) responsible for peroxisomal retention by constructing recombinant derivatives of *luc* lacking this motif (Jha *et al.*, 1991). Luc has also been exploited for constructing a new generation of baculovirus expression vector which renders the selection of recombinant virus very efficient and simple. In the plasmid, pNEluc, the *luc* gene is placed under the control of the viral polyhedrin promoter while another copy of the polyhedrin promoter in an opposite orientation followed by a *Bam*HI cloning site becoming available for inserting a foreign gene. Following co-transfection, recombinant viruses can be rapidly and efficiently detected and purified by monitoring the expression of *luc* gene. This strategy of dual expression, using *luc* as a reporter, has paved the way for the easy and rapid selection of recombinant viruses as opposed to somewhat tricky, laborious and time-consuming process based on plaque morphology (Occ$^+$ vs Occ$^-$) (Hasnain *et al.*, 1991). However, not much information is available on the virus infectivity, dose and routes of infection, level of gene expression and purification of recombinant products for the known baculovirus-larval systems.

Secretion and processing of an extensively glycosylated hormone using BEVS

A number of secretory glycoproteins have been expressed in BEVS. We have been using the extensively glycosylated human chorionic gonadotropin hormone (hCG) as a reporter to address specific issues of protein processing, folding and glycosylation in BEVS using both cells and larvae as host. hCG belongs to a group of glycoprotein hormones consisting of two non-identical subunits, α and ß, held together by non-covalent bonds that are coded by two separate genes. Both subunits have two N-linked complex type carbohydrate each and, in addition, the ß subunit has four o-linked oligosaccharide chains on its unique 30 amino acid long carboxy terminus. The sugars play an important role in secretion, subunit assembly, bioactivity and stability of hCG. β hCG has in addition six intrachain disulphide bonds which are critical in the proper folding and secondary structure of the molecule. At present, hCG is being extensively used for induction of ovulation in women and for treatment of male infertility and birth control vaccines based on the ß subunit of hCG (Talwar & Singh, 1988).

Recombinant viruses were constructed in which the α or β hCG gene was placed under the polyhedrin transcriptional control (Nakhai *et al.*, 1991 a; Hasnain, 1991). Insect cells, upon infection with recombinant virus v*Ach*α*CG,* synthesized ~11.3 μg 10^6 cells/ml of immunoreactive and bioactive hCG (Nakhai *et al.*, 1991a). Although the level of expression is low for BEVS but is well within the range of those obtained for other secretory proteins. In fact, this level was at least an order of magnitude higher than previously reported using other expression systems. It may be noted that in these constructs, transfer vector pAc373 was used which does not contain some of the sequences around the polyhedrin ATG, now shown to be important for high level expression of foreign genes placed under polyhedrin promoter control. By using improved vectors such as pAcYM1, which have the above mentioned sequences, we expect to have a much higher levels of expression.

SDS-PAGE, western blot and radioimmunoassay studies have clearly indicated that hCG synthesized in insect cells are identical to native protein implying that the hormone made in this system is glycosylated. This result is indeed significant because it is often considered that insect cells are incapable of carrying out proper glycosylation. Further, these cells are believed to lack sialyl transferase-- the enzyme that catalyzes the addition of sialic acid residues during protein processing. However, as the recombinant hCG was bioactive, two broad logical deductions can be made: (i) sialylation is not necessary for the bioactivity of hCG or (ii) sialyl transferase activity is present in insect cells. Analyses of βhCG expressions (Sridhar *et al.*, unpublished) have indeed confirmed the ability of insect cells to sialylate.

Interestingly, ^{35}S methionine labeling of proteins of v*Ac*αhCG infected cells revealed that αhCG is not efficiently secreted by these cells. αhCG was present both in the culture supernatant as well as in the infected cell pellet but these molecules had different molecular size. The secreted αhCG molecules moved with a slightly increased mobility due to the cleavage of the 24 amino acid secretory sequence. Densitometric scanning of the gel revealed that as much as ~61-72% of αhCG synthesized in insect cells remained inside the cells and had a significantly reduced electrophoretic mobility. In addition, a ~12 kDa protein was also detected in vAcαhCG infected cells which were absent in mock infected or wild type virus infected cells and possibly represented the overexpressed, unprocessed αhCG unmodified due to cell death caused by the lytic nature of baculovirus (Nakhai *et al.*, 1991a, b). These observations have suggested the existence of "secretory load" which states that the overproduction of a foreign protein which is extensively glycosylated (such as hCG) is not secreted properly as the secretory pathway of these cells is compromised during the later stages of viral infection (Nakhai *et al.*, 1991c).

There are several ways of testing the secretory load hypothesis. For example, the cells can be given more time for processing the foreign proteins either by using a weaker promoter which is activated earlier in the viral infection cycle or by somehow extending the life span of insect cells expressing the foreign gene. Alternatively, the gene can be expressed in a virus-free system. To sort out this issue, βhCG gene was placed under the control of the AcNPV basic protein promoter (MP1) which is activated earlier than polyhedrin promoter (Hill-Perkins & Possee, 1990). The recombinant virus vAcMPßhCG infected cells synthesized immunoreactive and biologically active ßhCG. The level of ßhCG secreted in this case was significantly higher (11.67 µg/10^6 cells/ml) than that obtained when the gene was placed under the polyhedrin promoter (6.93 µg/10^6 cells/ml) at 48 h p.i. (Hasnain, 1991). It may be mentioned that although the MP1 promoter is transcriptionally less active than the polyhedrin promoter, the enhanced level of secretion of hCG is possibly indicative of the increased time available to the infected cells to process the protein. An obvious difference in the degree of glycosylation between ßhCG expressed under MP1 and polyhedrin promoter control has also been noticed (Sridhar *et al.*, unpublished).

When the recombinant virus vAcβhCG was used for infecting the insect cells, these cells synthesized immunoreactive and bioactive βhCG which showed increased mobility on SDS-PAGE indicating that it is hypoglycosylated. Since the recombinant βhCG was identical to the native βhCG in bioactivity, it implies that the hormone made in this system is glycosylated such that there is no impairment of function. However, the differences in molecular weight of recombinant βhCG as compared to the native protein is probably due to the extent of glycosylation taking place in the molecule. Nevertheless, as the insect cells synthesized significantly high amounts of βhCG, this system can be used for understanding the mechanism of polypeptide chain folding, carbohydrate assembly, intracellular migration, and secretion of the hormone (Hasnain, 1991).

CONCLUSIONS

Within a decade since their discovery, baculovirus vectors have catapulted in the field of molecular biology as the best candidate for the expression of important human, animal, and plant genes. They have become a system of choice for the expression of complex eukaryotic genes, including those which are extensively glycosylated, even though much need to be explored about them at the molecular level. Our understanding of baculovirus diversity and molecular biology is rapidly advancing but there is an urgent need to answer some of the fundamental questions pertaining to their unique biology. There is indeed no doubt that this system holds the promise of producing biologically active proteins in a highly cost-effective manner besides fulfilling research needs.

Acknowledgements

We thank all the present and past members of our laboratory for their contribution reported in this review. This research was supported by Grants BT/TF/03/026/007/88 and BT/MS/01/001/89/ silk/baculo from the Department of Biotechnology, Govt. of India to SEH.

References

Alam, J. & Cook, J.L. (1990). Reporter genes: application to the study of mammalian gene transcription. Anal. Biochem. **188**, 245-254.

Awasthi, A.K., Jha, P.K. & Hasnain, S.E. (1992). Trafficking of virus and recombinant protein, synthesized in *Spodoptera* caterpillars, monitored by a genetically engineered baculovirus carrying gene encoding firefly luciferase (In Preparation).

Blissard, G.W. & Rohrmann, G.F. (1990). Baculovirus diversity and molecular biology. Ann. Rev. Entomol. **35z**, 127-155.

Gould, S.J. & Subramani, S. (1988). Firefly luciferase as a tool in molecular and cellular biology. Anal. Biochem. **175**, 5-13.

Hasnain, S.E. (1991). The baculovirus expression vector system: Use of the human chorionic gonadotropin as a reporter for studying post-translational modifications. In Immunology: Perspectives in reproduction and infection, Edited by Gupta, S.K. Oxford & IBH Publ., New Delhi.

Hasnain, S.E. & Nakhai, B. (1990). Expression of the gene encoding firefly luciferase in insect cells using a baculovirus vector. Gene **91**, 135-138.

Hill-Perkins, M.S. & Possee, R.D. (1990). A baculovirus expression vector derived from the basic protein promoter of *Autographa californica* nuclear polyhedrosis virus. J. Gen. Virol. **71**, 971-976.

Jha, P.K., Pal, R., Nakhai, B., Sridhar, P. & Hasnain, S.E. (1992). Simultaneous synthesis of enzymatically active luciferase and biologically active ß subunit of human chorionic gonadotropin in caterpillars infected with a recombinant baculovirus (Submitted).

Jha, P.K., Ehtesham, N.Z., Dhar, R. & Hasnain, S.E. (1991). Intracellular trafficking of proteins in insect cells and larvae infected with recombinant baculovirus. J. Cell Biochem. **15D**, 197.

Kitts, P.A., Ayres, M.D. & Possee, R.D. (1990). Linearization of baculovirus DNA enhances the recovery of recombinant virus expression vectors. Nucleic. Acids Res. **18**, 5667-5672.

Luckow, V.A. & Summers, M.D. (1988). Trends in the development of baculovirus expression vectors. Biotechnology **6**, 47-55.

Maeda, S. (1989). Expression of foreign genes in insects using baculovirus vectors. Ann. Rev. Entomol. **34**, 351-372.

Medin, J.A., Hunt, L., Gathy, K., Evans, R.K. & Coleman, M.S. (1990). Efficient low-cost protein factories: expression of human adenosin deaminase in baculovirus-infected insect larvae. Proc. Natl. Acad. Sci. USA, **87**, 2760-2764.

Miller,L.K. (1989) Insect baculoviruses: Powerful gene expression vectors. Bioessays 11, 91-95.

Nakhai, B., Pal, R., Sridhar, P., Talwar, G.P. & Hasnain, S.E. (1991a). The subunit of human chorionic gonadotropin hormone synthesized in insect cells using a baculovirus vector is biologically active. FEBS Lett. **283**, 104-108.

Nakhai, B., Sridhar, P., Talwar, G.P. & Hasnain S.E. (1991b). Construction, purification and characterization of a recombinant baculovirus containing the gene for alpha subunit of human chorionic gonadotropin. Ind. J. Biochem. Biophys. **28**, 237-242.

Nakhai, B., Sridhar, P. & Hasnain, S.E. (1991c). Is processing of overexpressed proteins in insect cells infected with recombinant baculovirus impaired due to a "secretory load"? J. Cell Biochem. **15D**, 201.

Nakhai, B., Sridhar, P., Pal, R., Talwar, G.P. & Hasnain, S.E. (1992). Overexpression and characterisation of recombinant beta subunit of human chorionic gonadotropin hormone synthesized in insect cells infected with a genetically engineered baculovirus. Ind. J. Biochem. Biophys. (In Press).

O'Reilly, D.R. & Miller, L.K. (1989). Science **245**, 1110-1112.

Ow, D.W., Wood, K.V., DeLuca, M., deWet, J.R., Helinski, D.R. & Howell, S.H. (1986). Transient and stable expression of firefly luciferase gene in plants. Science **234**, 856-859.

Schauer, A.T. (1988). Visualizing gene expression with luciferase fusions. TIBS **6**, 23-27.

Sridhar, P., Panda, A.K., Pal, R., Talwar, G.P. & Hasnain, S.E. (1992). Temporal nature and not promoter strength determines the qualitative and quantitative expression of an extensively processed protein in the baculovirus system. (Submitted)

Subramani, S. (1991). Peroxisomal targeting signals-the end and the beginning. Curr. Sci. **61**, 28-32.

Summers, M.D. & Smith, G.E. (1987). A manual of methods for baculovirus vectors and insect cell culture procedures. Texas Agric. Expt. Stn. Bull No. 1555, College Station, Texas.

Talwar, G.P. & Singh, O. (1988). In Contraception Research for Today and Nineties. Edited by Talwar, G.P., Springer Verlag, 183-197.

Wood, H.A. & Granados, R.R. (1991). Genetically engineered baculoviruses as agents for pest control. Ann. Rev. Microbiol. **45,** 69-87.

4

Molecular evolution in *Solanum nigrum* species complex at repetitive DNA level

**R.K. Chaudhuri, A. Banik, P. Mallik, I. Chaudhuri and
D.K. Mukhopadhyay**
*Molecular Biology Laboratory, Botany Department
Calcutta University, Calcutta 700019, India*

ABSTRACT

Three different cytotypes of *Solanum nigrum* species complex are found in nature. The relative amount of DNA and the amount of repetitive DNA in the three cytotypes does not correlate with their ploidy level. New chromosome types appear in tetraploid and hexaploid. Based on repetitive DNA sequences, it could be inferred that diploid cytotypes are evolving very rapidly in nature through frequent shuffling of these sequences. While a 1.3 kb *Bam* HI repetitive DNA element in the diploid could generate by amplification a 50 kb *Pst* I fragment, a 300 bp and a 1.08 kb *Hind* III repeat element is distributed in all the cytotypes. These observations point out towards similar evolutionary tendencies in diploids and hexaploids but different in tetraploid.

The ribosomal RNA gene sequences in the three cytotypes exhibit gene length heterogeneity which has possibly arisen through unequal crossing over. The additional support to this inference has come from the sequence of a Hind III repeat that has one ORF with 80 percent homology with *psb* A gene.

INTRODUCTION

There are two causes for biodiversity. First, new genotypes are cropping up constantly in a population through environmental mutation, genetic recombination, and also through immigration of individuals, their gametes or propagules. Second, diversity in population is being eliminated by natural selection. It would be correct to assume that biodiversity starts at the molecular levels, and ultimately is being manifested phenotypically. So, mutation and natural selection could generate genetic and morphological variations within a lineage (Stebbins, 1974). It has been argued by many authors that during speciation the original population could be divided into distinct genetic pools, and then each gene pool might acquire a unique set of characteristics through mutation and natural selection (Stebbins, 1950; Grant, 1981).

Among the first to realize the role of eukaryotic chromosomes to evade selection pressure were Britten and Khone (1968), who argued that during organic evolution eukaryotic chromosomes accumulated hundred thousand copies of identical DNA pieces - the repetitive DNA elements. In the same year, Kimura (1968) provided a population genetic rationale by proposing his 'neutral mutation hypothesis', where he had suggested that most differences in the rate of evolution could be accounted for conserved DNA sequences, particularly the sequences which might influence the phenotypic characters. Repetitiveness of a particular DNA sequence might help to preserve a particular 'code'. So, essentially Kimura's hypothesis and Britten and Khone's observation are the same.

An excess of DNA in higher organisms had puzzled cytogeneticists since the time it was first discovered through cytophotometry. The tendency at that time was to label them as 'junk' or useless DNA-sequences. In recent years, this argument had lost its credibility, when transcription of repetitive DNA sequences was reported from many organisms (Dutta & Chaudhuri, 1975; Long & Dawid, 1980; Walbot & Cullis, 1985; Heinz, 1989). It was found to be present in all eukaryotic organisms except in some fungi (Timberlake, 1978). In plants, their amounts can go upto 80 percent of total DNA. However, in plants like *Arabidopsis thaliana* (Leutwiler et al., 1984) and in *Solanum nigrum* (Bhattacharyya et al., 1986) the amount of repetitive DNA sequence in the nuclei is very low. There could be several reasons for this low amount of repetitive DNAs in *A. thaliana* and *S. nigrum* nuclear genomes, but it would be logical to assume as these plants have small genomes (0.07 pg for the former and 0.3 pg for the latter), they show small percentage of DNA sequence size and its repetitious DNA amount. But, this is not a general rule. Even in *S. nigrum* species complex we will notice that total chromatin matter does not tally with multiplication (increase) of chromosome complement.

In plants, repetitive DNA sequences often occur as long tandem clusters, or they may occur as dispersed elements, interspersing with non-repetitive DNA sequences, within a genome. Often, they could just be the relic that have been conserved for millions of years (Flavell, 1981), or in plant species like *S. nigrum* their evolution seems to be rapid. Perhaps, this rapid genome-sequence change might be responsible for successful colonising character of *S. nigrum*. If we trace the distribution of this species complex, we find them in pantropical areas throughout the world, mainly in waste land. The findings indicate that perhaps the repetitive DNA sequences might serve some useful cell functions. They could be transcribed, and definitely they are not 'junk' or senseless copies of the genome. We know that heterochromatic blocks of eukaryotic chromosomes harbor repetitive DNAs, ribosomal RNA cistrons, tRNA genes etc. If heterochromatic segments could help a plant to counteract adverse situations, then it is logical to infer that DNA sequences (repetitive DNAs) residing inside these segments are helping the plant to adapt in a particular ecological condition.

It has been observed by Lima-de-Faria (1983) that sensor mechanism in plant chromosomes is very sensitive. So, plants are more elastic than animals, in avoiding selection pressure after random genome alterations. While working on molecular organization of annual pantropic weed *S. nigrum* L., we observed different evolutionary rates in repetitive DNA elements in different morphological and cytological variants.

RESULTS AND DISCUSSION

S. nigrum is often been considered as a species complex (Dev, 1979). The plants are found in both old World and new World countries. The species has three cytotypes - diploid, tetraploid, and hexaploid, having a basic chromosome number of x=12. In these cytotypes, variants are also observable. They are found to grow in wastelands and roadsides as seasonal weeds. The polyploid cytotypes do not reveal any multivalent formation, and, therefore, some authors (Bhaduri, 1933; Tandon & Rao, 1974; Dev, 1979; Edmonds, 1979) suspected them as interspecific hybrids. The taxonomy and proper nomenclature of these cytotypes are still controversial, and classical taxonomy could not solve the problem. Bhaduri (1933) observed three cytotypes in the same locality, growing side by side. He was of the opinion that the hexaploid form might be an allotetraploid, arising from an intermixing between the diploid and tetraploid forms in nature. In 1950, Stebbins suggested that the chromosome studies of polyploids cannot solve their origin or nature until they are synthesized artificially. Hence, to trace the origin of polyploid forms of *S. nigrum*, Venkateswarlu and Rao (1972), Tandon and Rao (1974), and Edmonds (1979) actually synthesized hexaploid *S. nigrum* using diploid species like *S. nodiflorum* Jacq. syn. *S. americanum* Miller and tetraploid species like *S. villosum* Miller as parents. The immediate hybrids were sterile as they were triploids. Fertility of triploid hybrids was restored by colchicine treatment and hexaploid forms were created. In morphological characters natural and artificially produced hexaploids were found to be identical. Edmonds (1979) had noticed that protein profiles of two groups were similar. It was argued by Edmonds that in nature similar phenomenon is occurring.

Since Dunal's time no one attempted a worldwide revision of this taxon. It was only much later that Edmonds (1972) and Heiser *et al.* (1979) revised this group with the limited available data. They reported that fertility barrier between allied species of this species-complex is very low. In a sense, *S. nigrum* species-complex is perhaps composed of large number of 'microspecies' or 'semispecies' (Grant, 1971). Heiser *et al.* (1979) argued that in many respects these microspecies mimic true species. Our studies on repetitive DNA sequences of three cytotypes of *S. nigrum* suggest that tetraploid forms could be completely different from diploid and hexaploid counterparts, at least for the sequences we had studied. The allotetraploid nature of tetraploid *S. nigrum* could not be substantiated by our studies, except to the fact that it has a low amount (13 percent) of repetitive DNA sequences in their haploid genome (Bhattacharyya *et al.*, 1986), at 50 Cot level using 500-550 base pair DNA pieces. This low amount of repetitive DNA might be due to the presence of heterogeneous DNA sequences in the same nucleus, or due to integration of two different genomes. It might also be possible that this low repeat content in tetraploid *S. nigrum* genome is reflecting the small genome's characteristics. Earlier, we had reported (Bhattacharyya *et al.*, 1989) *S. nigrum* haploid genome as 0.7 picogram - only four time of *Arabidopsis thaliana* genome (Leutwiler *et al.*, 1984). So, similar to *A. thaliana* it is showing a low amount of repetitive DNA at moderate reassociation condition, a short life cycle (6 to 8 weeks) and small plant size. Whether this low amount of repetitive DNA in *S. nigrum* genome is due to the admixture of the two separate genomes in the same nucleus or due to small genome size is difficult to answer at this stage. Unless a detailed investigation of genomes of all cytotypes is possible, the correct interpretation for our observation cannot be stated. However, from distribution patterns of some repetitive DNA sequences we could infer that diploid cytotypes is rapidly evolving in nature, and that the tetraploid cytotype could be different from diploid and hexaploid cytotypes.

Repetitive DNA sequences are now being considered as important chromosome constituents. As stated earlier in this chapter, their transcripts were also reported by several authors including our group in 1975. It is often assumed that so called 'junk DNA sequences' could play significant roles in eukaryote evolution - as a considerable amount of changes within these sequences could be accommodated without any lethal effect. On the contrary, unique DNA sequences, and also house-keeping genes, would show drastic phenotypic effects if similar duplication and/or changes occur within them by shuffling, deletion or duplication.

As repetitive DNA elements could accommodate major shuffling or changes it is expected many of these elements could be traced in distantly related taxa. Bedbrook & Kolodner (1980) observed conservedness of highly repetitive sequences of rye and in other cereal genomes, though in distantly placed cereal species their amount or sequence divergence was found to be different. Even in rye cytotypes, these characteristics do vary for the repeat DNA elements studied. Yakura *et al.* (1987) observed a *Bam* HI repeat family of *Vicia faba* conserved not only in other legumes but also in rat liver DNA. Rather, the conservedness was found to be more in rat liver DNA than in pea or other legumes. The possible explanation in legumes is that this repeat element is fast evolving. Similar instances could be traced in other plant and animal species, with other repeat types. While working on *S. nigrum*, we had observed that relative DNA amount and repeat amount of three cytotypes did not correlate with their ploidy level, i.e. they are not in arithmetical progression as we observed in case of chromosome number (Banik, 1991). New chromosome types were found to appear in tetraploid and hexaploid cytotypes. The total chromatin matter was found to decrease with increase in ploidy (Table 4.1). Also, we have noticed that there is less conservedness of few repetitive DNA sequences in several diploid cytotypes collected from different localities, rather than their conservedness in tetraploid and hexaploid cytotypes. This observation, if analysed on the findings of Bedbrook & Kolodner (1980) with *Bam* HI element of rye, it can be inferred that diploid cytotypes are evolving very rapidly in nature making frequent shuffling of the repetitive sequences we had studied.

A 1.3 kilobase *Bam* HI repetitive DNA element was found in the diploid *S. nigrum* genome.

Table 4.1 Comparative study of the genomes of different cytotypes of *Solanum nigrum*

Ploidy Level	Total chromosome length (μ)	Total chromosome volume (cu. μm)	Range of chromosome length	Karyotype formula	DNA amount (pg)
2n	46.28	39.42	1.64-2.28	A_6B_6	8.17
4n	81.44	69.72	1.19-2.07	$A_7B_{16}C_1$	12.04
6n	114.18	97.74	1.06-1.91	$A_{16}B_{15}C_2D_3$	14.86

This element was found to constitute 1.65% of total diploid genome (Mallik, 1992). With *Pst* I digestion of *S. nigrum* DNA, a 50 kb fragment could be observed in Southern blots. It implies that *Pst* I restriction enzyme (a hexacutter) has no site within this 50 kb DNA which is unlikely unless we assume that the site was affected by base methylation. It should be uneconomical to assume that all the deoxycytidine positions of this 50 kb DNA molecule had been methylated during evolution. However, if we assume that there is rapid amplification (tandem duplication) of this 1.3 kb repeat (*Bam* HI) element, after methylation, we could expect to get a similar situation. A thirtyeight - fold amplification of original *Bam* HI repeat could generate this 50 kb fragment. In the diploid cytotype as in the other cytotypes, this amplification had taken place as no 50 kb fragment could be identified after *Pst* I digestion. Copy number determination gave us an idea that this 50 kb DNA stretch is actually amplified 1.3 kb original fragment. Its autoradiogram is more than twenty times intense (Fig. 4.1).

We have also studied the distribution of a 300 bp *Hind* III repeat element, isolated from diploid cytotype. It shows different distribution patterns in different diploid cytotypes. These cytotypes were collected from wild localities on the basis of the differences in the morphological characters. On the other hand, the distribution of this repeat element in diploid, tetraploid, and hexaploid cytotypes were found to be more or less similar. It again implies that evolution is more rapid in diploid cytotype for this repetitive DNA sequence. A 1.08 kb *Hind* III repeat element was found to be conserved in all cytotypes (Fig. 4.2). This observation suggests that for this DNA sequence there is conservedness and the same perhaps is restricted to a specific chromosome locus of *S. nigrum*. Similar conservedness was observed in case of centromeric and telomeric DNA sequences. It could also be possible that this might be an ancient repeat element. However, without a detailed investigation this argument can be given little weightage.

Based on these observations, it could be inferred that these repetitive DNA elements have evolved separately in *S. nigrum* genomes. From their distribution it could be hypothesized that evolutionary tendencies are very much alike in diploid and hexaploid cytotypes, but it is different in tetraploid cytotype of *S. nigrum*. Tetraploid form might have evolved separately, or, if it is an allotetraploid, as suspected by Tandon and Rao (1974) and Edmonds (1979), this 300 bp

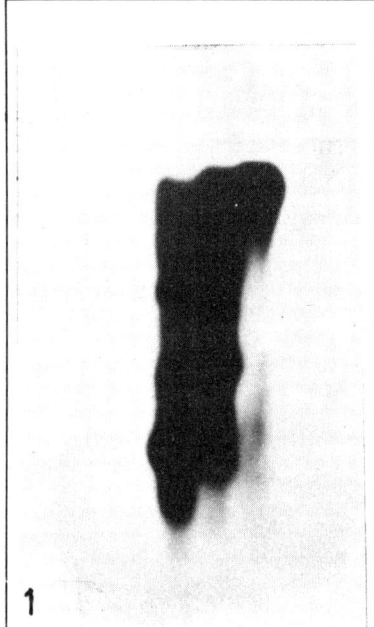

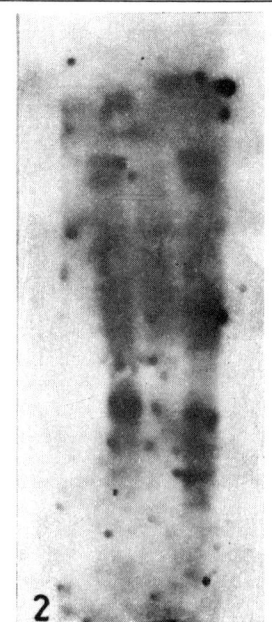

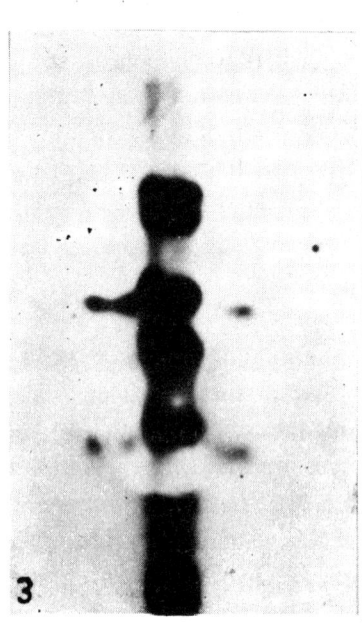

Fig. 4.1 Autoradiogram of the distribution pattern of cloned 1.3 kb Bam HI repeat (pMBDC5) in diploid *S. nigrum* L. genome.
Lane 1 - Genomic DNA digested with Bam HI, Lane 2 - with EcoRV and Lane 3 - with Pst 1.

Fig. 4.2 Autoradiogram shows the hybridization pattern of 1.08 kb Hind III fragment of pMMC9 with S. nigrum L. genomic DNAs digested with Hind III.
Lane 1- Diploid variety, Lane 2 - Tetraploid variety and Lane 3 - Hexaploid variety.

Fig. 4.3 Autoradiogram shows the cDNA distribution pattern of different diploid 'Types' of *S. nigrum* L. *N. crassa* rDNA was used as the probe. All the DNAs were digested with EcoRI.

repeat element was eliminated during hybridization and polyploidization. If we consider diploid and hexaploid as two species (*S. nodiflorum = S. americanum* and *S. nigrum*), they are related at DNA level at least for all the repetitive DNA elements tested so far. Some of these elements were found to be absent in tetraploids. If tetraploid is an allotetraploid, elimination of some sequences could possibly take place. During alloploidy, elimination of genome sequences has been reported by several workers in many plants (Atreya & Subramaniyam, 1989; Mirochnitchenko, 1989).

From distribution patterns, it is apparent that 300 bp *Hind* III repeat element is diploid-specific, and is evolving rapidly in diploid variants. It is also apparent that tetraploid form had evolved long before the diversification of 1.08 kb *Hind* III repeat element, which was found to be conserved in all cytotypes. In distribution patterns of 300 bp *Hind* III repeat element, diploid and hexaploid behave identically indicating a close relationship of the two cytotypes. In morphological characters too the two cytotypes are close.

In the experiments done with ribosomal RNA gene sequences, it was noticed that three cytotypes have rRNA gene length heterogeneity. In *Eco* RI digestion, southern showed two bands in diploid and hexaploid form, but five in morphological variants, so the number of rRNA gene clusters were found to vary from two to four in diploid variants (Fig. 4.3). It confirms our earlier observations that repetitive DNA sequences in diploid cytotype are fast evolving, and the diversification of this gene took place after hexaploid and tetraploid genomes had diverged from progenitors. It is pertinent to mention here that ribosomal RNA gene sequences might change rapidly if there is an unequal crossing over within the gene segment. The sequence divergence of rRNA gene of *S. nigrum* could take place similarly. We have sequenced the *Hind* III repeat, and from computer analysis we noticed that it has a ORF which has over 80 percent homology with *psb*A gene (Mallik, 1992). Walbot and Cullis (1985) had noticed a 'promiscuous DNA' which might be responsible for genetic transfer between nuclear and organelle genomes, at rRNA gene sites. They noticed a 252 base-pair maize DNA sequence, sharing 90 percent homology to the last one-third of *psb*A gene, and are often found to be next to repetitive sequences, active in recombination. Whether that process is taking place in *S. nigrum Hind* III fragment is yet to be determined.

The evolution of repetitive DNA sequences coincides with the organization of eukaryotic chromosomes vis-a-vis evolution of animal and plant species. Lima-de-Faria (1980) proposed a chromosome gradient model for chromosome evolution, where in a chromosome field, larger chromomeres are located near the telomeres and smaller chromomeres near the centromere. Though in mitotic chromosome this gradient is not observable, but in meiotic prophase it is clear. In *Ixeris* it is observable in mitotic prophase too (Lima-de-Faria, 1954). Flavell (1981) was of the opinion that tandem duplication, polyteny, unequal crossing over etc. might generate repetitive DNA elements from unique DNA sequences and mutation in a repetitive element could transform it into a unique sequence. What is the role of environment in these transformations could not be fathomed till date. However, data are sufficient to show that ecology has some role in fixation of genome size. Temperature might play an important role; cold temperature prefers to produce small genomes and warm temperature large genomes. The correlation between large genome sizes and capacity to grow at low temperature, however, was reported in 24 taxa of cereals, grasses, and legumes. Our observations on *S. nigrum* suggest that temperature-escaping and rainfall-sensitive plants like *S. nigrum* have small genomes. Similar to *Arabidopsis thaliana, S. nigrum* also completes its life cycle within couple of weeks and has a small genome (0.3 pg, i.e. 2.7 x 10^8 bp) and also small amount (13 percent) of repetitive DNAs.

References

Atreya, C.D. & Subramaniyam, N.C. (1989). Comparative analysis of repetitive DNA in five *Arachis* (plant) species. Biochem. Syst. Ecol. **17**, 11-14.

Banik, A.S. (1991). Study in alkaloid production in *Solanum* with an aim to its improvement. Ph.D. Thesis, Calcutta University, Calcutta.

Bedbrook, J.R. & Kolodner, R. (1980). The structure of the chloroplast DNA. Ann. Rev. Plant. Physiol. **30**, 593-620.

Bhattacharyya, N., Mukhopadhyay, D.K., Chaudhuri, Ila & Chaudhuri, R.K. (1986). Repetitive DNA amount in *Solanum nigrum* genome. Curr. Sci. **55**, 569-571.

Bhattacharyya, N., Mallik, P., Mallik, D. & Chaudhuri, R.K. (1989). Small angiosperm genome in *Solanum nigrum* Linn. The Nucleus **32**, 80-82.

Bhaduri, P.N. (1933). Chromosome numbers of some Solanaceous plants of Bengal. J. Indian Bot. Soc. **12**, 56-62.

Britten, R.J. & Khone, D.E. (1968). Repeated sequences in DNA. Science **161**, 529-540.

Dev, D.B. (1979). Solanaceae in India. In: The Biology and Taxonomy of Solanaceae. Edited by J.G. Hawkes, R.N. Lester and A.D. Skelding, 87-112, Linn. Soc. Acad. Press, London.

Dutta, S.K. & Chaudhuri, R.K. (1975). Transcription of repetitive DNA in *Neurospora crassa.* Mol. Gen. Genet. **136**, 227-232.

Edmonds, J.M. (1972). A synopsis of the taxonomy of Solanum sect. Solanum (maurella) in South America. Kew Bull. **27**, 95-114.

Edmonds, J.M. (1979). Nomenclature notes on some species of Solanum found in Europe. Bot. J. Linn. Soc. **78**, 213-233.

Flavell, R.B. (1981). Did retroviruses evolve from transposable elements? Nature **289**, 10-11.

Grant, V. (1971). Plant speciation. Columbia Univ. Press, New York, London.

Grant, V. (1981). Plant speciation. Columbia Univ. Press, New York, London.

Heiser, C.B., Donald, J.R., Burton, L. & Schilling, E.E. Jr. (1979). Biosystematic and Taxonomic studies of Solanum nigrum complex in Eastern North America. In The Biology and Taxonomy of Solanaceae, Edited by J.G. Hawkes, R.N. Lester and A.D. Skelding, 513-528, Linn. Soc. Acad. Press, London.

Heinz, H. (1989). Cell Biology, Harper and Row Publisher. New York.

Kimura, M. (1968). Evolutionary rate at the molecular level. Nature **217,** 624-626.

Leutwiler, L.S., Hough-Evans, R.B. & Meyerowitz, R.B. (1984). The DNA of A*rabidopsis thaliana.* Mol. Gen. Genet. **194**, 15-23.

Lima-de-Faria, A. (1954). Chromosome gradient and chromosome field in Agapanthus. Chromosoma **6**, 330-370.

Lima-de-Faria, A. (1980). Classification of genes, rearrangements and chromosomes according to the chromosome field. Hereditas **93**, 1-46.

Lima-de-Faria, A. (1983). Molecular evolution and organization of chromosomes. Elsevier, New York, Oxford.

Long, E.O. & Dawid, I.B. (1980). Repeated genes in eukaryotes. Ann. Rev. Biochem. **49**, 727-764.

Mallik, P. (1992). Cloning of a repeated DNA fragment and distribution of different repeat family in some members of Solanaceae. Ph.D. Thesis, Calcutta University, Calcutta.

Mirochnitchenko, G.P. (1989). Arabidopsis Inf. Serv. No. **26**, 15-28.

Stebbins, G.L. (1950). Variation and Evolution in Plants. Columbia Univ. Press, New York.

Stebbins, G.L. (1974). Flowering plants, evolution above the species level. Harvard Univ. Press, Cambridge Mass.

Tandon, S.L. & Rao, G.R. (1974). In: Evolutionary Studies in World Crops, Edited by J. Hutchinson, 109-117, Cambridge Univ. Press.

Timberlake, W.E. (1978). Low repetitive DNA content in *Aspergillus nidulans*. Science **202**, 973-974.

Venkateswarlu, J. & Rao, M.K. (1972). Breeding system, crossability relationships and isolating mechanisms in the *Solanum nigrum* complex. Cytologia **37**, 317-326.

Walbot, V. & Cullis, C.A. (1985). Rapid genomic changes in higher plants. Ann. Rev. Plant Physiol. **36**, 37-396.

Yakura, K., Kato, A. & Tanifuji, S. (1987). Cytological localization of the highly repeated DNA sequences, the FOK I sequence family and *Bam* HI sequence family in *Vicia faba* chromosomes. Jap. J. Genet. **62**, 325-332.

5

Molecular analysis of the cloned *LYS5* gene of *Saccharomyces cerevisiae*

Sudha Rajnarayan[1], Jack C. Vaughn[2] and J.K. Bhattacharjee[1]
Department of Microbiology[1] and Department of Zoology[2]
Miami University, Oxford, OH 45056, USA

ABSTRACT

Lysine is synthesized via the α -aminoadipate pathway in yeast and other fungi. The *LYS2* and *LYS5* genes of *Saccharomyces cerevisiae* are required for the synthesis of α-aminoadipate reductase α enzyme of this pathway. This enzyme is a large heterodimer. The *LYS2* gene encodes a large polypeptide. The cloned *LYS5* gene has been isolated from a genomic library of *S. cerevisiae* by functional complementation of a LYS5 mutant. This gene has been partially characterized within a 3.2 kb DNA insert of the pSR7 subclone. A *CAL1* gene is also present adjacent to the *Sph*I end of this DNA insert. It is concluded that the *LYS5* gene encodes a small polypeptide of the heterodimeric α-aminoadipate reductase enzyme.

INTRODUCTION

Lysine is an essential amino acid (obtained from diet) for humans (Rose *et al.*, 1955). However, like other amino acids, lysine is synthesized by bacteria, fungi, and plants (Vogel, 1960; Umbarger, 1978). Unlike other amino acids, lysine is synthesized by two distinct pathways in nature. Bacteria, some phycomycetes (lower fungi) and plants use the diaminopimelic acid pathway for the synthesis of lysine (Vogel, 1960), which is a member of the aspartic acid family of amino acids (Gilvarg, 1960; Cohen, 1983). Fungi including certain phycomycetes, yeasts and higher fungi as well as cyanobacteria have evolved a novel α-aminoadipate pathway for the synthesis of lysine (Strassman & Weinhouse, 1953; Vogel, 1960). There are eight enzyme steps in the pathway including homocitrate synthase, the first committed enzyme which catalyzes synthesis of homocitric acid and the first committed intermediate of the pathway (Fig. 5.1). This and the subsequent reactions leading to the synthesis of α-aminoadipic acid are analogous to the reactions of the citric acid cycle involved with the synthesis of glutamic acid (Strassman & Ceci, 1967). Like the citric acid cycle, the enzymes and the intermediates for the synthesis of α-aminoadipic acid (first half of the pathway) are localized within the mitochondria (Betterton *et al.*, 1968). The remaining three enzyme steps for conversion of α-aminoadipate to lysine (second half of the pathway) are cytosolic and all of the enzymes are encoded by nuclear genes (Bhattacharjee, 1983, 1985, 1992). This pathway has been investigated in several fungi including *Neurospora crassa* (Broquist, 1971), *Saccharomyces cerevisiae* (Strassman & Ceci, 1966; Bhattacharjee & Strassman, 1967), *Rhodotorula glutinis* (Kinzel *et al.*, 1983), *Schizosaccharomyces pombe* (Ye & Bhattacharjee, 1988), *Candida maltosa* (Kunze *et al.*, 1987), *Yarwiia lipolytica* (Xuan *et al.*, 1990) and *Penicillium chrysogenum* (Jaklitsch & Kubicek, 1990).

Complementation and recombination analysis of a large number of lysine mutants of *S.*

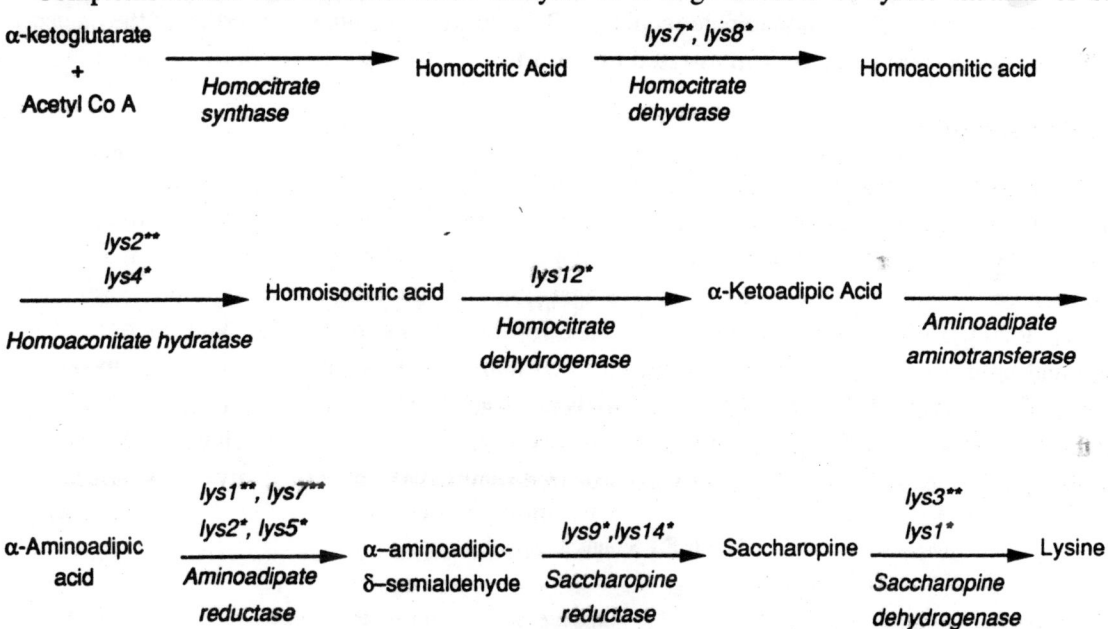

Fig. 5.1 Relationship among the genes, enzymes and intermediates for biosynthesis of lysine in yeast. Mutant loci of *S. cerevisiae* (∗); mutant loci of *S. pombe* (∗∗).

cerevisiae have revealed more than ten loci responsible for the synthesis of lysine (Biswas & Bhattacharjee, 1974; Borell *et al.*, 1984). Lysine loci are not linked to each other, but map to several different linkage groups of *S. cerevisiae* (Hwang *et al.*, 1966, Mortimer & Schild, 1985). Studies of lysine mutants have enabled us to determine the gene, enzyme, and intermediate relationship for lysine biosynthesis (Bhattacharjee, 1985; Fig. 5.1). In the second half of the pathway, *lys2* and *lys5* mutants are blocked in the α-aminoadipate reductase step, *lys9* and *lys14* mutants in the saccharopine reductase step, and a lys1 mutant is blocked in the saccharopine dehydrogenase step for the synthesis of lysine (Broquist, 1971; Bhattacharjee, 1985).

Enzyme purification and mutant studies indicated α-aminoadipate reductase as a hetero-polymeric enzyme and *LYS2* and *LYS5* as the structural genes (Storts & Bhattacharjee, 1989). Saccharopine reductase has been shown to be a homopolymeric enzyme, encoded by the *LYS9* gene. Saccharopine dehydrogenase is a monomeric enzyme, and is encoded by the *LYS1* gene (Fjellstedt & Ogur, 1970; Fujioka & Tanaka, 1981; Bhattacharjee, 1985).

The cloned *LYS1*, *LYS2*, *LYS5*, *LYS9*, and *LYS14* genes of *S. cerevisiae* have been isolated from genomic libraries of *S. cerevisiae* by functional complementation involving transformation of appropriate lysine mutants (Bhattacharjee, 1992). The DNA sequence and transcript size of the *LYS2* gene indicated this gene to encode the large subunit of α-aminoadipate reductase (Barnes & Thorner, 1986; Morris & Jinks-Robertson, 1991). The *LYS5* gene has been cloned in the recombinant plasmid pSC5 (Borell & Bhattacharjee, 1988; Rajnarayan *et al.*, 1992). Molecular studies of the *LYS5* gene are crucial to understand its physical size and precise role in encoding the smaller subunit of α-aminoadipate reductase. Some of the molecular properties of the cloned *LYS5* gene of *S. cerevisiae* are summarized in this article.

Strains and media used

All yeast and bacterial strains used are described in Table 5.1. Yeast strains were maintained on a nutrient agar medium (YEPD) comprising glucose 20g, peptone 10g, yeast extract 20g, and agar 20g per liter of distilled water. Minimal medium containing glucose 10 g, Difco yeast nitrogen base without amino acids 6.7g, and agar 25g with the addition of appropriate supplements, at 50 μg/ml final concentration, were used for the growth of mutants. Nutrient broth and minimal broth media lacked only agar. *E. coli* was maintained on Luria-Bertani medium (LB). Consisting of tryptone 10μg, yeast extract 5g, NaCl 10g per liter of distilled water. Ampicillin at the concentration of 50μg per ml was added to LB for selection. Yeast cultures were grown at 30°C, and *E. coli* at 37°C. Growth was measured in terms of optical density at 560 nm.

Table 5.1 Strains used in this study

Organism	Strain	Genotype	Phenotype	Source
S. cerevisiae				
	RC-1	a, *LYS*	Prototroph	a
	X4004-3A	a, *lys5, trp1, met2, ura3*	Auxotroph	a
	AB9-2	a, *lys5*	Auxotroph	a
S. pombe	lys 7.2	a, *lys7*	Auxotroph	a
E. coli	HB 101	(r B,mB), recA13, amp[5], tet[5]	Sensitive	a

[a] Laboratory collection

Plasmids used

The yeast episomal plasmid YEp24 (vector) is a high copy number cloning vehicle used for the introduction of DNA constructions into *S. cerevisiae*. This plasmid can replicate both in *E. coli* (due to the presence of the pBR322 origin of replication) and in *S. cerevisiae* (due to the presence of the replication determinant of the yeast 2 micron circle plasmid). It carries the ampicillin resistance (Amp) gene of pBR322 for use as a selectable marker in *E. coli* and the URA3 gene of the yeast strain + D4 for use as a selectable marker in yeast. The tetracycline (Tc) gene of pBR322 is present, but separated from its promoter by the URA3 sequence and inactivated by the introduction of *S. cerevisiae* genomic DNA at the *Bam* HI site (Botstein et al., 1979). The *LYS5* gene was isolated by functional complementation of lys5 mutant (strain X4004-3A) of *S. cerevisiae* (Borell & Bhattacharjee, 1988) using a YEp24 plasmid bank constructed by Carlson and Botstein (1982). The cloned *LYS5* gene was contained within a 7.9 kb DNA insert of the recombinant plasmid pSC5. The recombinant plasmid subclone pSR7 contains the *LYS5* gene on a 3.2 kb *Sph* I - *Sau* 3A 1 fragment of the original insert, in the same plasmid vector, YEp24.

Plasmid isolation from *E. coli* and *S. cerevisiae*

In order to screen *E. coli* transformants, crude plasmid preparations were obtained by the rapid alkaline extraction method of Birnboim and Doly (1979) with a few modifications (Maniatis *et al.*, 1982). CsCl purified plasmids were prepared from *E. coli* transformants according to Norgard *et al.* (1979). Recombinant plasmids pSC5 and pSR7 used to transform *S. cerevisiae* were isolated by the method of Lorincz (1985) with a few alterations. Plasmid bearing yeast colonies were individually streaked on selective agar medium to increase the number of cells for the isolation protocol. Several loopfuls of cells were added to 200 µl of the ice cold lysis buffer (100 mM NaCl, 10mM Tris-HCl pH 8.0, 1mM Na_2EDTA and 0.1% lyticase). Sterile glass beads (0.45 mm diameter) were added until just below the level of the buffer. The suspension was vigorously vortexed for 10 min, with incubation at 4°C to prevent over-heating of the sample. The sample was extracted twice, with an equal volume of Tris-buffered phenol, pH 8.0 (Maniatis *et al.*, 1982). This was followed by an extraction with an equal volume of chloroform-isoamyl alcohol, 24:1 (v/v). The plasmid DNA was ethanol precipitated, the pellet resuspended in 50 µl of sterile 0.1X SSC and aliquots were used to transform competent *E. coli* HB101 cells for enrichment and isolation of plasmids.

E. coli and *S. cerevisiae* transformations

An overnight culture of *E. coli* (HB 101) set up in 40 ml of LB broth medium served as pre-inoculum. A 1:100 dilution of cells was obtained by using 0.4 ml of the overnight culture to inoculate 40 ml of fresh medium pre-warmed to 37°C until the culture reached an optical density of 0.6 at A_{600}nm. The cells were pelleted at 500 rpm for 5 min at 4°C in a Sorvall GSA rotor and the supernatant discarded. The cells were resuspended in 20 ml of ice cold 50 mM $CaCl_2$ and incubated in the ice slurry for 20 min. Cells were pelleted as before and resuspended in 2 ml of cold $CaCl_2$ and incubated for 20 min. Aliquots (100μl) of the cell suspension were transferred to sterile 1.5 ml Eppendorf tubes to which 100μl of DNA (0.1 to 0.4μg) was added and the mixture allowed to incubate for 20 min on ice. The cells were subsequently heat shocked at 42°C for exactly 2 min, 1 ml of pre-warmed LB broth medium was added to each tube and incubated at 37°C for an hour. Aliquots of 0.1-0.4 ml were plated on LB-Amp (50 μg/ml ampicillin) plates which were incubated overnight ατ 37°C to yield transformants (Mandel & Higa, 1970). The *LYS5* mutants of *S. cerevisiae* were transformed using published procedures (Ito *et al.*, 1983; Wang *et al.*, 1989).

Aminoadipate reductase assay

Aminoadipate reductase (EC 1.2.1.31) was assayed according to published procedure of Sagisaka and Shimura (1960) and Winston and Bhattacharjee (1987). The reaction mixture consisted of aminoadipate 125 mM, ATP 150 mM, $MgCl_2$ 50 mM, glutathione 10 mM, NADPH 6.5 $MgCl_2$ mM, Tris-HCl pH 8.0, 500 mM, and crude cell extract in a final volume of 1 ml per tube. Control tubes contained all the reagents except the enzyme preparation. The mixture was incubated for 1 hr at 30°C, following which 1.0 ml of a 2% (w/v) solution of p-dimethylamino benzaldehyde (PDAB) was added to the reaction mixture. The tubes were heated in a boiling water bath for 20 min. The boiled suspension was centrifuged for 10 min at 15,000 rpm in a Sorval GSA rotor. The reaction with PDAB produces a yellow coloured adduct of p-dimethyl-amino-benzaldehyde-aminoadipate-semialdehyde, that was measured spectrophotometrically at A_{460}nm.

5'-End labelling and restriction analysis of LYS5 DNA

[32]P-labelling of 5'termini of appropriate DNA fragments with T4 polynuceotide kinase was the method used in most cases (Maniatis *et al.*, 1982). This enzyme catalyzes the transfer of the gamma-phosphate of ATP to a 5'-OH terminus of a DNA molecule. To do this, the 5'-phosphate groups from that fragment were first removed with bacterial alkaline phosphatase (BAP).

Restriction enzyme digestion of plasmids and inserted DNA fragments was done as suggested by the supplier (BRL). Following the digestion, a 1/10 volume of 10X glycerol dye solution [0.25% (w/v) bromophenol blue, 0.25% (w/v) xylene cyanol FF in 50% (v/v) glycerol] was added to the sample. The samples were electrophoresed in 0.8% agarose gels and stained by standard procedure (Maniatis *et al.*, 1982). Fragment sizes were determined by reference to appropriate restriction enzyme digested lambda DNA or pBR322 size standards (Maniatis *et al.*, 1982).

Subcloning of the *LYS5* gene on plasmid pSR7

The LYS5 gene on a 7.9 kb insert in the plasmid pSC5 was subcloned into a smaller 3.2 kb *Sph* I - *Bam* HI fragment in the recombinant plasmid pSR7. This required several controls which included the demonstration of the completeness of digestion by restriction enzymes, the effectiveness of alkaline phosphatase to prevent recircularization of the linearized vector, YEp24 and the ability of the ligase enzyme to ligate non-phosphatase treated linear DNA molecules.

RESULTS

Confirmation of cloning of the *LYS5* gene

Lysine independent (*LYS5+*) transformants were selected through functional complementation of two separate *lys5* mutants, X4004-3A and AB-9. Strain X4004-3A required methionine, tryptophan and uracil as additional supplements in the minimal medium used to select lysine independent transformants. The negative control cells received no pSC5 plasmid DNA, and as expected no colonies grew on the appropriately supplemented minimal agar medium lacking lysine. Cells that received plasmid DNA and plated on medium containing methionine, tryptophan and uracil for *LYS5* selection served as a positive control. The positive control plate and the lysine selection plate showed comparable transformation frequencies, in the range of 105 colonies per 1 to 4 µg of purified plasmid DNA. Transformation frequency was dependent on the concentration of pSC5 DNA (Fig. 5.2). The plasmid pSC5 was then isolated in bulk from *E. coli* HB101 and used for further studies. In a separate experiment, transformation of a lys7 mutant of *S. pombe* (blocked at the α-aminoadipate reductase step) with pSC5 plasmid DNA yielded comparable numbers of transformants (results not shown). In a different experiment, cloned *LYS2* gene of *S. cerevisiae* transformed only lys1 mutant of S. pombe. Results from these experiments indicate that *LYS2* and *LYS5* gene of *S. cerevisiae* are isofunctional to *LYS1* and *LYS7* genes, respectively, of *S. pombe* (Fig. 5. 2).

Yeast mini-preparation

In order to confirm the success of functional complementation of *LYS5* mutants by the recombinant plasmid pSC5, yeast mini preparations were carried out to retrieve plasmid DNA from *LYS5+* transformant colonies, grown on appropriately supplemented minimal agar medium that was selective for lysine prototrophy. Restriction analysis of plasmid (pSC5) DNA retrieved from the yeast transformants, alongwith pSC5 plasmid DNA isolated from *E. coli* HB 101, revealed the identity of the plasmids.

Plasmid loss experiments with transformed strains

Plasmid loss experiments were performed to determine loss of lysine prototrophy of (*LYS5+*) transformed cells as a result of growth on nonselective medium. After initial growth on nonselective YEPD medium, *LYS5+* cells were serially transferred daily to YEPD plates for seven days and from the last set of plates the cells were replica plated to selective **and** nonselective

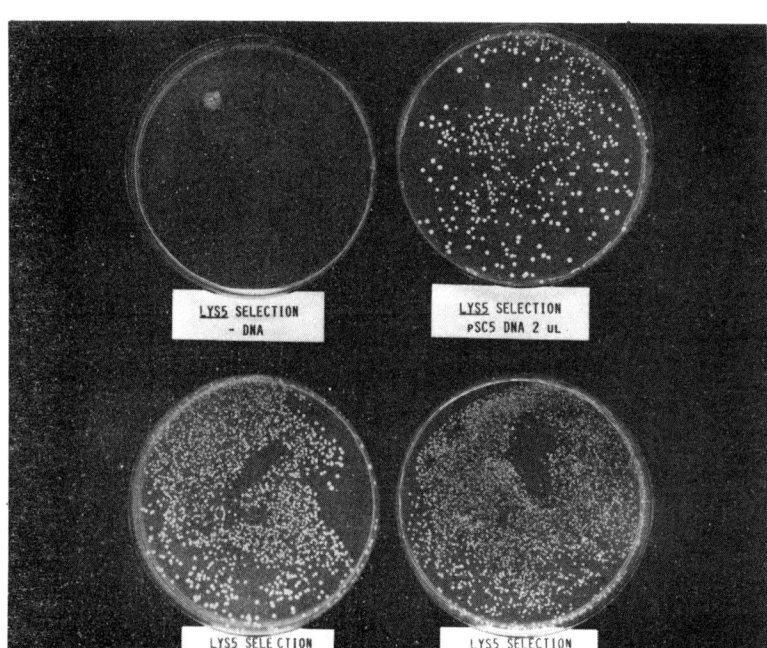

Fig. 5.2 Complementations of *LYS5* mutant with recombinant plasmid pSC5 carrying the *LYS5* gene of *S. cerevisiae*. The *LYS5* mutant X4004-3A was transformed with 2 µl, 5 µl, and 10 µl of purified pSC5 plasmid DNA and equal aliquots of treated cells were plated on appropriately supplemented minimal medium lacking lysine.

media. The *LYS5+* transformed strains showed a 40% loss of the original colonies (results not shown). Controlled *LYS5+* cells serially transferred to minimal medium exhibited no loss of colony forming units.

RESTRICTION ANALYSIS

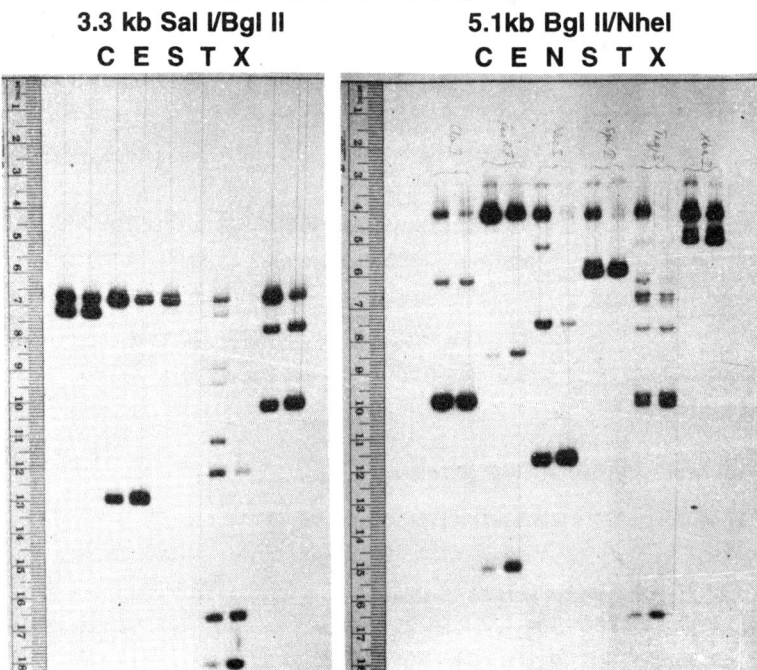

Fig. 5.3 Autoradiograph of the restriction endonuclease digests of the LYS5 gene containing insert DNA in plasmid pSC5. Two singly end-labelled fragments of the original 7.9 kb LYS5 insert were mapped separately. The 3.3 kb *Sal I-Bgl II fragment was mapped using Cla I (C), Eco RV (E), Sph I (S), Taq I (T), and Xba I(X). The 5.1 kb Bgl II-Nhe I* fragment was mapped with all the above enzymes including Nco I (N).

Construction of a detailed restriction map of the *LYS5*- containing DNA insert

The 7.9 kb yeast DNA insert with the gene in the recombinant plasmid pSC5, was mapped with a series of restriction enzymes. Plasmid DNA (pSC5) was digested with *Sal* I and *Nhe* I to yield a 8.4 kb fragment including the *LYS5* DNA with some vector sequence at the ends. The purified fragment was 5'-end labelled with ^{32}P and asymmetrically cut with *Bgl* II. Singly end-labelled 3.3 kb *Sal* I/*Bgl* II and 5.1 kb *Bgl* II/*Nhe* I fragments were then subjected to partial and complete digestion with each enzyme. The partial and limit cuts were run side-by-side on agarose gels, and autoradiographs were made (Fig. 5.3). Fragment sizes were determined from standard curves constructed using λ DNA cut with *Eco* RI + *Bam* HI, and pBR322 DNA cut with Alu I. A detailed restriction map of this insert was constructed with the sites for the enzymes, *Cla* I, *Eco* RV, *Nco* I, *Sph* I, *Taq* I and *Xba* I, indicated by arrow heads (Fig. 5.4). The enzymes for which no sites were found included *Bam* H I, *Kpn* I, *Nhe* I and *Sal* I.

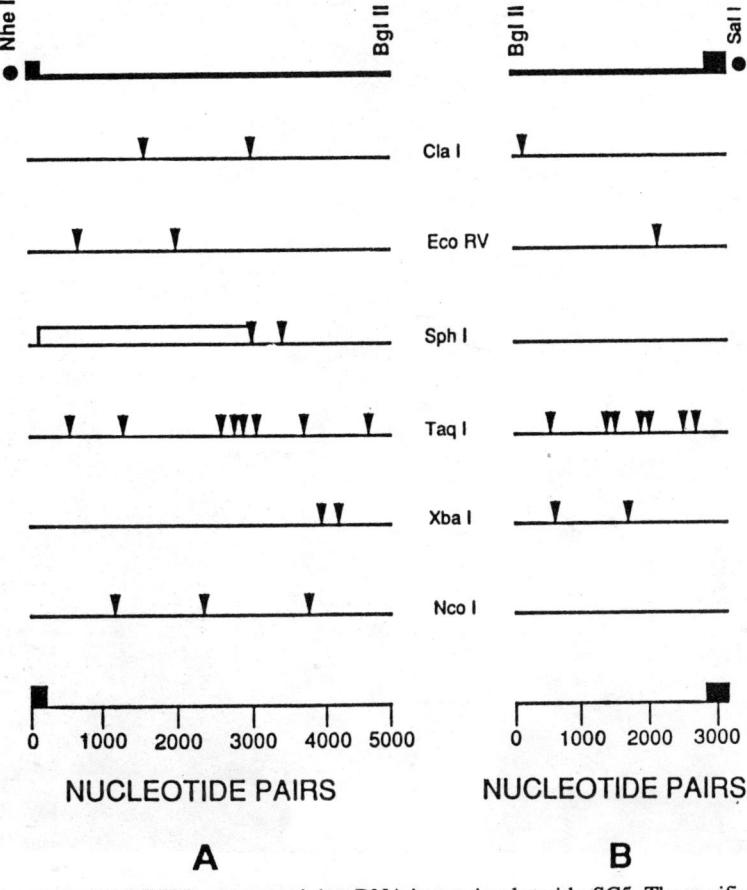

Fig. 5.4 Restriction map of the *LYS5* gene-containing DNA insert in plasmid pSC5. The purified 8.4 kb Nhe I-Sal I DNA fragment spanning the *LYS5* gene was mapped using separate ^{32}P-end labelled 5.1 kb Nhe I*-Bgl II (A)and 3.3 kb Bgl II-Sal I* (B) DNA fragments. Vector DNA is shown by elevated black boxes. Location of the insert later subcloned into pSR7 and shown to contain the full *LYS5* gene activity is shown by the box on SpH I line (fourth line from the top).

Strategy for subcloning

Five different plasmid subclones were constructed in the vector YEp24 using DNA fragments spanning segments of the 7.9 kb insert of the plasmid pSC5. The recombinant plasmids were amplified by transformation of *amp^s E. Coli* HB 101. Plasmid DNAs from the *ampr* colonies were used to transform the *LYS*5 mutants X4004-3A and AB-9, to screen for the presence of a functional *LYS*5 gene. Recombinant plasmid subclone pSR7 containing a 3.2 kb DNA segment extending from an *Sph* I site to an *Nhe* I site in the vector produced *LYS*5$^+$ transformants with the same efficiency as the parent plasmid pSC5 (Fig. 5.5). Electrophoretic analysis following restriction enzyme digestion of pSR7 plasmid with *Sph* I and *Sma* I verified the size of the *S. cerevisiae* DNA fragment (including 0.4 kb of vector DNA) containing the *LYS*5 gene to be 3.6 kb, with a deletion of a 4.7 kb *Sph* I - *Sph* I fragment from the pSC5 plasmid (Fig. 5.6).

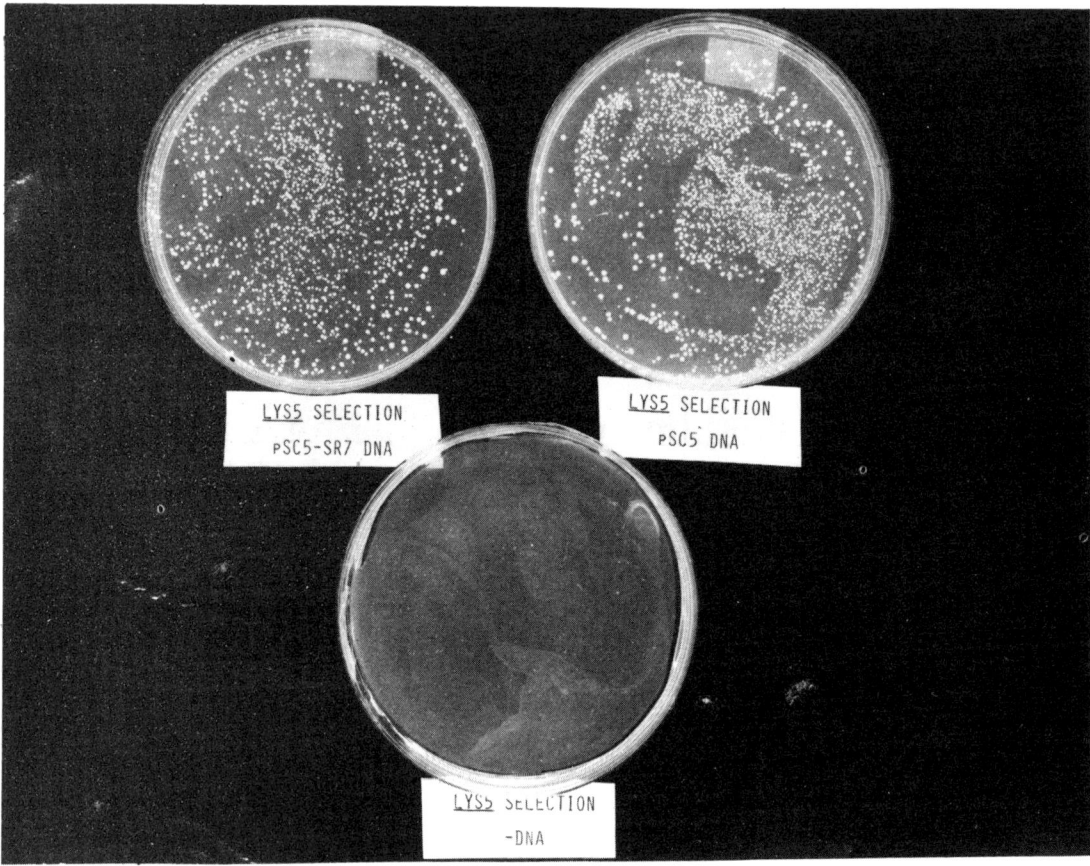

Fig. 5.5 Comparison of the transformation of *LYS*5 mutant (X4004-3A) with equivalent amounts of pSC5 and pSR7 plasmids containing the *LYS*5 gene .*LYS*5$^+$ transformants were selected on appropriately supplemented minimal medium lacking lysine: transformation with pSR7 DNA; transformation with pSC5 DNA; and negative control containing no DNA.

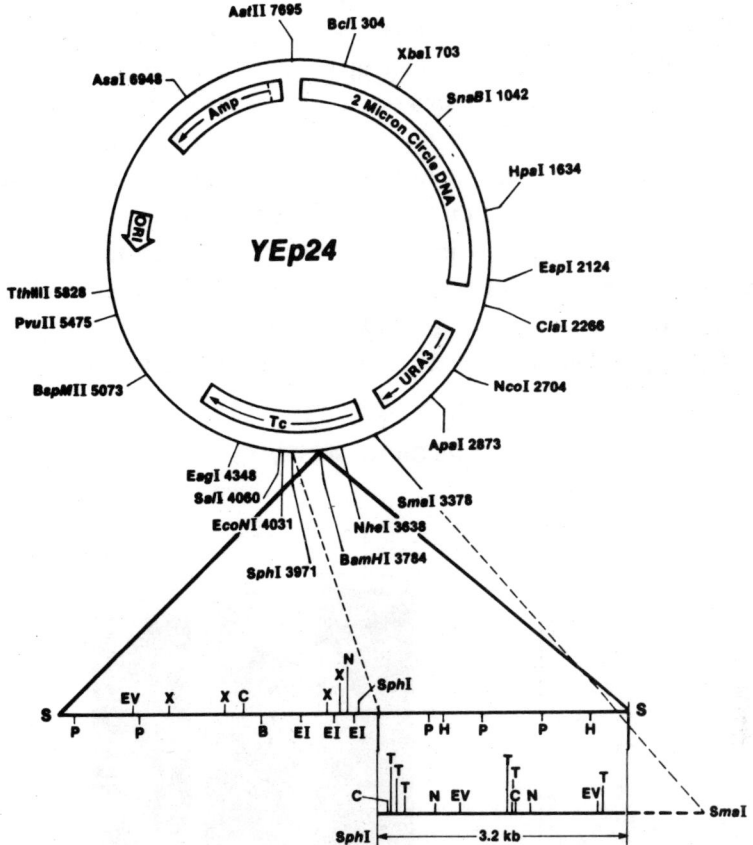

Fig. 5.6 Restriction maps of the 7.9 kb (top line) and 3.2 kb (bottom line) DNA inserts, respectively, of recombinant plasmids pSC5 and pSR7 containing the cloned *LYS5* gene of *S. cerevisiae*. Restriction enzyme sites mapped on the DNA inserts were: Sau 3A-I(S), Pvu II(P), Eco RV(EV), Xba I(X), Cla I(C), Bgl II(B), Eco RI(EI), Nco I(N), Hpa I(H), Taq I(T); and the flanking Sph I and Sma I sites in the YEp24 vector.

α-Aminoadipate reductase activity

To demonstrate and confirm that the subclone pSR7 did indeed contain the *LYS5* gene, *LYs5*⁺ transformants (containing pSC) were grown in selective minimal media, and the cell extracts assayed for α-aminoadipate reductase activity. Positive controls used were wild type *S. cerevisiae* (RC-1) and *LYS5*⁺ transformants (containing pSC5). The negative control cells grown in lysine supplemented medium included the *LYS5* mutant (X4004-3A). The results presented are an average of values obtained in five separate determinations. The wild type strain and both types of transformants (Plasmids pSC5 and pSR7) demonstrated comparable α-aminoadipate reductase activity, while minimal enzyme activity was seen in the mutant strain (Fig. 5.7). These results indicated that the DNA insert of pSR7 subclone contained the complete *LYS5* gene.

Several aspects of the α-aminoadipate pathway (enzymology, regulation and comparative molecular genetics) in the yeasts *S. cerevisiae,* and *S. pombe* have become a subject of intensive studies (Bhattacharjee, 1992). α–Aminoadipate reductase is unique among biosynthetic enzymes for its ability to catalyze the adenylation of its substrate, α-aminoadipate, prior to reduction of the same with NADPH (Sinha & Bhattacharjee, 1971). Mutations of the two distinct loci *lys2* and *lys5* lack α-aminoadipate reductase activity, suggesting the involvement of two separate genes for the biosynthesis of this enzyme (Sinha & Bhattacharjee, 1970). Elucidation of the specific contribution of each of the two genes would require purification and characterization of the enzyme in terms of molecular weight and subunit composition. It would also require the characterization of the two genes *LYS2* and *LYS5* in terms of the size of the open reading frame (ORF), regulatory elements, and the predicted molecular weight of the encoded polypeptides.

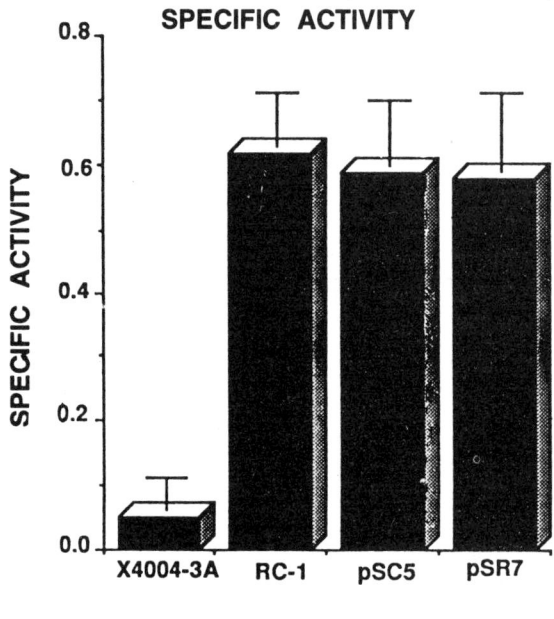

Fig. 5.7 α-Aminoadipate reductase activity (OD 460 nm/mg protein/h) of cell extracts derived from wild type (RC-1), *LYS5* mutant (X4004-3A), *LYS5*[+] (pSC5) transformant and *LYS5*[+] (pSR7) transformant.

The α-aminoadipate reductase from wild type *S. cerevisiae* has been partially purified and its molecular weight has been determined to be 180,000 daltons. Also, the *in vitro* complementation and revertant analysis of *LYS2* and *LYS5* mutants indicated both loci as structural genes for a heterodimeric enzyme (Storts & Bhattacharjee, 1989; Fjellsted & Ogur, 1970). The cloned *LYS2* gene (Eibel & Phillipsen, 1983) has been determined to encode a transcript approximately 4 kb long (Barnes & Thorner, 1986) and an ORF to encode a polypeptide 155,344 daltons (Morris & Jinks-Robertson, 1991). These results predict that *LYS5* gene must encode a much smaller transcript and an ORF to encode a small polypeptide.

The cloned *LYS5* gene of *S. cerevisiae* in the recombinant plasmid pSC5 was originally isolated from the YEp24 genomic library of wild type *S. cerevisiae* (Borell & Bhattacharjee, 1988). Detailed restriction mapping, enzyme activity of transformants, and the complementation ability (transformation of *LYS5* mutants to *LYS5+*) of the pSR7 subclone have localized the gene on a 3.2 kb *Sph* I - *Bam* HI DNA insert (Fig. 6, Rajnarayan *et al.*, 1992). Results published in a recent paper (Ohya *et al.*, 1991) indicate that a *CAL1* gene of 1.1 kb DNA is localized adjacent to the *Sal* I end of the 3.2 kb DNA insert of pSR7 subclone. A comparison of the restriction map and the ORF of the *CAL1* gene to the DNA insert of pSR7 subclone suggests that the *LYS5* gene is likely to be localized at the *Sma* I end of the insert. The complementation ability and the levels of α-aminoadipate reductase activity in *Lys5+* transformants suggest that pSR7 subclone carries the complete *LYS5* gene, which encodes a much smaller polypeptide than the *LYS2* gene. There exists approximately 1.4 kb DNA insert between the CAL1 gene and the Small end of the subclone, which must accommodate the entire *LYS5* gene. Analysis of mRNA transcripts and DNA sequence is in progress to confirm the predicted ORF of the *LYS5* gene.

Acknowledgements

We thank Mrs. Carol A. Webb for typing and Mr. Richard C. Garrad for reading the manuscript. This research was supported by Miami University, grants from Sigma Xi Society to *S. Rajnarayan*, and National Institute of Health (IR 15 GM36007) to J.K. Bhattacharjee.

References

Barnes, D.A. & Thorner, J. (1986). Genetic manipulation of *S. cerevisiae* by use of the LYS2 gene. Mol. Cell. Biol. **6**, 2828-2838.

Betterton, H., Fjellstedt, F., Matsuda, M., Ogur, M. & Tate, R. (1968). Localization of the homocitrate pathway. Biochem. Biophys. Acta. **170**, 459-461.

Bhattacharjee, J.K. & Strassman, M. (1967). Accumulation of tricarboxylic acids related to lysine biosynthesis in a yeast mutant. J. Biol. Chem. **242**, 2542-2546.

Bhattacharjee, J.K. (1983) Lysine biosynthesis in eukaryotes. In Amino Acid: Biosynthesis and Genetic Regulation. Edited by K. Herman and R.L. Sommerville, pp 229-244, Addison-Wesley, Reading, Massachusetts.

Bhattacharjee, J.K. (1985). α-Aminoadipate pathway for the biosynthesis of lysine in lower eukaryotes. CRC Crit. Rev. Microbiol. **12**, 131-151.

Bhattacharjee, J.K. (1992). Evolution of α-aminoadipate pathway for the biosynthesis of lysine among fungi. In Evolution of Metabolic Function. Edited by R.P. Mortlock, CRC Press, Boca Raton, Florida.

Birnboim, H.C. & Doly. J. (1979). A rapid alkaline extraction procedure for screening recombinant plasmid DNA. Nucl. Acids Res. **7**, 1513-1525.

Biswas, G.D. & Bhattacharjee, J.K. (1974).   Induction and complementation of lysine auxotrophs in *S. cerevisiae.* Antonie van Leeuwenhoek, J. Microbiol. Serol. **40**, 221-231.

Borell, C.W., L.A. Urrestarazu & Bhattacharjee, J.K. (1984). Two unlinked genes (*LYS*9 and LYS14) are required for the synthesis of saccharopine reductase in *S. cerevisiae.* J. Bacteriol. **159**, 429-432.

Borell, C.W. & Bhattacharjee, J.K. (1988).   Cloning and biochemical characterization of LYS5 gene of S. cerevisiae. Curr. Genet. **13**, 299-304.

Broquist, H.P. (1971). Lysine biosynthesis (yeast). Methods Enzymol. 17B, 112-129.

Botstein, D., Falco, S.C., Brennan, M., Scherer, S., Stinchcomb, D.T., Struhl, K. & Davis, R.W. (1979).   Sterile host yeast (SHY): a eukaryotic system of biological containment for recombinant DNA experiment. Gene **8,** 17-24.

Carlson, M. & Botstein, D. (1982).   Two differentially regulated mRNA with different 5'-ends encode secreted and intracellular forms of yeast invertase. Cell **28,** 145-154.

Cohen, G. (1983).   Common pathway to lysine, methionine and threonine. In Amino Acids: Biosynthesis and Genetic Regulation.   Edited by K.M. Herman and R.L. Sommerville, pp 147-172, Addison-Wesley, Reading, Massachusetts.

Eibel, H. & Phillipsen, P. (1983).   Identification of the cloned *S. cerevisiae LYS2* gene by an integrative transformation approach. Mol. Gen. Genet. **191**, 66-73.

Fjellstedt, T. & Ogur, M. (1970).   Effect of super-suppressor genes on enzymes controlling lysine biosynthesis in *S. cerevisiae.* J. Bacteriol. **101**, 108-117.

Fujioka, M. & Tanaka, Y. (1981). Role of arginine residue in saccharopine dehydrogenase (L-lysine forming) from baker's yeast. Biochem. **20**, 468-472.

Gilvarg, C. (1960). Biosynthesis of diaminopimelic acid. Fed. Proc. **19**, 948-956.

Hwang, Y.L., Lindegren, G. & Lindegren, C.C. (1966). Genetic study of lysine biosynthesis in yeast. Can. J. Genet. Cytol. **8**, 471-480.

Ito, H., Fukuda, Y., Murata, K. & Kimura, A. (1983).   Transformation of intact yeast cells treated with alkali cations. J. Bacteriol. **153**, 163-168.

Jaklitsch, W.M. & Kubicek, C.P. (1990). Homocitrate synthase from *P. chrysogenum.* Biochem. J. **269**, 247-253.

Kinzel, J.J., Winston, M.K. & Bhattacharjee, J.K. (1983).   Role of lysine--α ketoglutarate aminotransferase in catabolism of lysine as a nitrogen source for *R. glutinis.* J. Bacteriol. **155**, 417-419.

Kunze, G., Bode, R., Schmidt, H., Samsonova, I.A. & Birnbaum, D. (1987).   Identification of a *lys*2 mutant of *C. maltosa* by means of transformation. Curr. Genet. **11**, 385-391.

Lorincz, A. (1985). Quick preparation of plasmid DNA from yeast. BRL Focus. **6**, 4.

Mandel, M. & Higa, A. (1970). Calcium dependent bacteriophage DNA infection. J. Mol. Biol. **53**, 154-158.

Maniatis, T., Fritsch, E.F. & Sambrook, J. (1982). Molecular Cloning, A laboratory Manual. Cold Spring Harbor Laboratory, Cold Spring Harbor, New York.

Morris, M.E. & Jinks-Robertson, S. (1991). Nucleotide sequence of the *LYS2* gene of *S. cerevisiae*: homology to *B. brevis* tyrocidine synthase 1. Gene **98**, 141-145.

Mortimer, R.K. & Schild, D. (1985). Genetic map of *S. cerevisiae*. Microbiol. Rev. **49**, 181-212.

Norgard, M.V., Emigholz, K. & Monahan, J.J. (1979). Increased amplification of pBR322 plasmid deoxyribonucleic acid in *E. coli* K-12 strain RR1 and Chi 1776 grown in the presence of high concentrations of nucleoside. J. Bacteriol. **138**, 270-272.

Ohya, Y., Goebl, M., Goodman, L.E., Peterson-Bjorn, S., Freisen, J.D., Tamanoi, F. & Anraku, Y. (1991). Yeast CAL1 is a structural and functional homologue to the *DPR1* (RAM) gene involed in *ras* processing. J. Biol. Chem **266**, 12356-12360.

Rajnarayan, S., Vaughn, J.C. & Bhattacharjee, J.K. (1992). Physical and biochemical characterization of the cloned *LYS5* gene for α-aminoadipate reductase activity in the lysine biosynthetic pathway of S. cerevisiae. Curr. Genet. 21 (in press).

Rose, W.C., Wixon, R.L., Lockhart, H.B. & Lambert, G.F. (1955). Amino acids requirements of man XV. The valine requirement; summary and final observations. J. Biol. Chem **217**, 987-995.

Sagisaka, S. & Shimura, K. (1960). Mechanism of activation and reduction of α-aminoadipic acid. Nature **188**, 1189-1190.

Sinha, A.K. & Bhattacharjee, J.K. (1970). Control of a lysine biosynthetic step by two unlinked genes of *S. cerevisiae*. Biochem. Biophys. Res. Commun. **39**, 1205-1210.

Sinha, A.K. & Bhattacharjee, J.K. (1971). Lysine biosynthesis in S. cerevisiae. Conversion of α-aminoadipic--semialdehyde. Biochem. J. **125**, 743-749.

Storts, D. & Bhattacharjee, J.K. (1989). Properties of revertants of *lys2* and *lys5* mutants as well as α-aminoadipate-semialdehyde dehydrogenase from *S. cerevisiae*. Biochem. Biophys. Res. Commun. **161**, 182-186.

Strassman, M. & Weinhouse, S. (1953). Biosynthetic pathways III. Biosynthesis of lysine by *T. utilis*. J. Am. Chem. Soc. **75**, 1680-1684.

Strassman, M. & Ceci, L.N. (1966). Enzymatic formation of cis-homoaconitic acid, an intermediate in lysine biosynthesis in yeast. J. Biol. Chem. **241**, 5401-5406.

Strassman, M. & Ceci, L.N. (1967). A study of acetyl-CoA condensation with α-keto acids. Arch. Biochem. Biophys. **119**, 420-428.

Umbarger, H.E. (1978). Amino acid biosynthesis and its regulation. Ann. Rev. Biochem. **47**, 533-606.

Vogel, H.J. (1960). Two modes of lysine biosynthesis among lower fungi. Biochem. Biophys. Acta **41**, 172-174.

Wang, L., Okamoto, S. & Bhattacharjee, J.K. (1989). Cloning and physical characterization of linked lysine genes (*LYS*4 and *LYS*15) of *S. cerevisiae*. Curr. Genet. **16**, 7-12.

Winston, M.K. & Bhattacharjee, J.K. (1987). Biosynthetic and regulatory role of *lys*9 mutants of *S. cerevisiae*. Curr. Genet. **11**, 393-398.

Xuan, J.W., Fournier, P., Declerck, N., Chasles, M. & Gaillardin, C. (1990). Overlapping reading frames at the *LYS*5 locus in the yeast *Y. lipolytica*. Mol. Cell. Biol. **10**, 4795-4801.

Ye, Z.H. & Bhattacharjee, J.K. (1988). Lysine biosynthesis pathway and biochemical blocks of lysine auxotrophs of *S. pombe*. J. Bacteriol. **170**, 5968-5970.

6

Aspergillus — from classical to molecular genetics

A. John Clutterbuck
Genetics Department, University of Glasgow
Glasgow G11 5JS, Scotland, U.K.

ABSTRACT

Aspergillus is a model system amenable to detailed genetic and biochemical analyses. The availability of large number of morphological, and biochemical markers and the readily formed heterokaryons and unstable heterozygous diploids has allowed studies on complementation, intragenic recombination, mitotic nondisjunction and mitotic recombination. *Aspergillus* strains have been widely applied for mutagen testing and in the investigation of genetics of penicillin production with the latter showing interesting similarities with genes in *Streptomyces* species.

In molecular genetic studies, *Aspergillus* has turned out to be useful as it has a well-established transformation system, which not only facilitates gene cloning but also allows reintroduction of manipulated sequences. *Aspergillus* species have also been successfully employed to study gene regulation, development and differentiation, and in the analysis of the cell cycle.

The Early Work

Pontecorvo's original interest in launching a new genetic organism was much the same as that of Beadle and Tatum when they started work on *Neurospora*. Whereas Beadle and Tatum were fascinated by the relationship between metabolic pathways and genetics, Pontecorvo hoped to go straight to the nature of the gene. At the heart of the genetic puzzle was the dual definition of the gene as a unit of complementation and also the atom of inheritance, unsplittable by recombination. Complementation tests for auxotrophic mutants were very easy in a fungus that readily formed heterokaryons, and these heterokaryons, when marked with autonomous white, yellow or green spore colour markers had a fascination of their own (see Pontecorvo, 1947 and Fig. 6.1).

Splitting the genetic atom by recombination was also much easier to demonstrate in a microorganism than in Drosophila. Roper (1950), Pritchard (1955) and Siddiqi (1962) established beyond doubt that genes were far from indivisible. These studies initiated research into the mechanism of intragenic recombination.

In the meantime, Benzer had overtaken *Aspergillus* with bacteriophage T4, in which recombination frequencies were approximately 20 times higher (Benzer, 1955).

Pontecorvo (1958) in fact came to the surprising conclusion that differences in recombination frequencies are such that overall map lengths are similar for a wide variety of organisms, despite large differences in genome size.

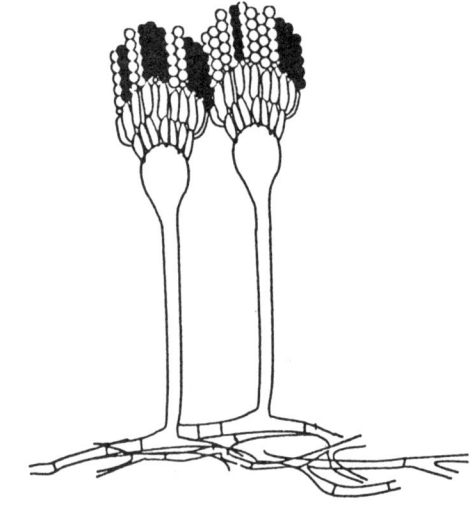

Fig. 6.1 Conidial heads from an *Aspergillus nidulans* heterokaryon.

Later estimates have increased the total map of bacteriophage T4 from 800 map units to 2500 (Stahl *et al.*, 1964) and the *Aspergillus* map has similarly grown from 660 map units to 4000 (Clutterbuck, 1990b) with a further 50% increase predicted when all chromosomes are fully mapped (Clutterbuck, 1992). Nevertheless, these map lengths are much more similar than would be predicted from a DNA difference of approximately 140-fold. If we include the mammals in the calculation, the figures are even more striking: total map length for the mouse is 1600 map units (Davisson *et al.*, 1990), only a little less than T4 or *Aspergillus*, while DNA contents per genome between the two extremes (mouse and T4) differs by a factor of 17,000.

The Parasexual cycle

Roper's discovery (1952) of vegetative diploids that could be isolated from *Aspergillus* heterokaryons provided the first step in the series of rare, may be accidental, events which, considered as a sequence, make up the parasexual cycle (Pontecorvo, 1956). *Aspergillus* diploids were inherently unstable, and analysis of their breakdown products revealed the operation of two processes: mitotic recombination (already recognized in *Drosophila*) and mitotic nondisjunction which led in some cases to homozygosity for whole chromosomes, or more importantly, to haploidization.

The elucidation and application of the parasexual cycle thus remains the most distinctive contribution of *Aspergillus* to genetics. Following the discovery of spontaneous chromosome loss (equivalent to haploidization) from interspecies cell hybrids (Migeon & Miller, 1968), Pontecorvo's hopes for parasexual gene mapping in man did become a very productive reality, while parasexual analysis of imperfect fungi, especially plant pathogens, is the only way a classical geneticist can get his hands on this important group of organisms (Tinline & McNeill, 1969).

APPLIED GENETICS

Mutagen testing

The steps of the parasexual cycle are stimulated by disturbance to the nucleus, and since two distinct processes, haploidization and mitotic recombination can be measured separately, both can be used to test for environmental agents which are likely to be mutagenic (in the broadest sense) to human cells. A whole scientific industry is based on such testing (Käfer *et al.*, 1982; Scott *et al.*, 1982). Reagents such as p-fluorophenylalanine and also chloral hydrate which was first employed by Umakant Sinha's group (Singh & Sinha, 1976, 1979; Käfer, 1986), are prime examples of agents causing chromosome malsegregation without causing damage to DNA.

Antibiotics

Another applied field in which *Aspergillus* has gained importance is the investigation of the genetics of penicillin production. Although *A. nidulans* only produces small amounts of the antibiotic when compared with wild strains of *Penicillium chrysogenum*, let alone with highly developed industrial strains, the wealth of background knowledge, genetic and molecular facilities available in this organism, make it a valuable model species in which to investigate the antibiotic pathway. So far it has been shown that three genes are clustered at the end of linkage group VI, at what was originally designated the *npe*A locus (MacCabe *et al.*, 1990). Other genes, designated *npe*B-D and *pen*A-C are now believed to affect activities which have only a secondary influence on antibiotic biosynthesis. A surprising finding is the close homology between one of the pathway genes which codes for isopenicillin N synthetase (IPNS) and similar genes from *Streptomyces* species (Weigel *et al.*, 1988; Landon *et al.*, 1990; Peñalva *et al.*, 1990). Since most bacterial and fungal species do not make penicillin, it seems highly improbable that the gene has been inherited from a common ancestor, while becoming deleted from all other bacteria and fungi. It is more likely that this is a rare example of horizontal gene transfer between species, and the preferred hypothesis is that the direction of transfer was from *Streptomycetes* which are very versatile in antibiotic production, to fungi, which produce only a few.

It is tempting to extend this hypothesis to the transfer of all three genes of the *Streptomyces* antibiotic pathway to form the *npe*A three-gene cluster in *Aspergillus* and *Penicillium*. The problem with this is that it is very difficult to see how a bacterial operon could be of any use to a eukaryote which would be expected to require separate promoters for each of the three genes. Moreover, the individual genes would presumably be of no selective value until the whole pathway was operational. One possibility is that the transfer from *Streptomyces* took place when transcription and translation were much more similar in bacteria and fungi than they are today. Another is that an intermediate stage in transfer was an endosymbiotic Streptomycete in which the antibiotic pathway was initially complete. This would allow the take-over of individual steps by copies of the genes transferred to the fungal nucleus, as each became modified to eukaryotic transcription and translation. In *P. chrysogenum* the acyltransferase protein (the third enzyme in the pathway) is confined to a vesicular microbody (Müller *et al.*, 1991): could this be the remains of the endosymbiont?

MOLECULAR GENETICS

A crucial step required to open up an organism to molecular approaches is an efficient procedure for transformation. Transformation of *Aspergillus* was first achieved by Tilburn *et al.* (1983) followed by Yelton *et al.* (1984) and Johnstone *et al.* (1985). Methods were adapted from yeast which had beaten the filamentous fungi in this field by five years (Hinnen *et al.*, 1978). At first it was assumed that efficient transformation would require a vector which could replicate in the fungus; however, transformation rates, quite good enough for most purposes, were obtained using vectors which integrate into the chromosome. The mode of integration varies with the selected marker on the vector; using the *amd*S marker, selection favours multiple integration of plasmids at nonhomologous sites, while for the *arg*B selective marker, many transformants have single copies integrated at the resident *arg*B locus on chromosome III. It appears that mitotic recombination is much more frequent in *Aspergillus* than in yeast and this allows both homologous and nonhomologous recombination of uncut plasmids with the chromosome.

Recently a vector which does replicate in *Aspergillus* has been described, and this indeed gives much more efficient transformation (Gems *et al.*, 1991). We have also found that the replicating plasmid recombines very readily with other plasmids or linear DNAs transformed into the cell, giving rise to cotransformants in which the second plasmid is maintained as a co-integrant with the replicating plasmid (D.H. Gems & A.J. Clutterbuck, unpublished).

A good transformation system makes gene cloning by complementation of mutants a routine matter. Reisolation of integrated complementing cosmids or smaller vectors is frequently successful, and depends mainly on efficient transformation of *E. coli* to recover rare plasmids released from the *Aspergillus* chromosome. Release may occur by spontaneous mitotic recombination, or may be forced by treatment with restriction enzymes. Recovery of replicating plasmids is, as a rule, easier, but we find that these plasmids are often unstable and may lose either the sequences required for maintenance in *E. coli*, or the insert to be cloned, if that is not under strict selection.

Transformation not only facilitates gene cloning, but also allows reintroduction of manipulated sequences. These may be deleted versions of a gene, which under suitable selection, can replace the resident chromosomal copy to give a null mutant. Alternatively they may be fusion constructs in which a promoter is attached to a reporter gene such as *lac*Z, or in the opposite configuration, a structural gene attached to an inducible promoter which can be switched on and off at will. The value of such constructs is seen in the impetus they have given to research, examples of which are described in the remainder of this review.

Gene regulation

The biochemistry of *Aspergillus* was for some time a neglected area. It is, therefore, ironic that this fungus has become one of the best eukaryote microorganisms for studies of metabolic regulation. The reasons for this lie in the metabolic versatility, the availability of diploids for testing dominance and gene interaction, and now the ease with which it is possible to obtain both homologous and nonhomologous integration of transforming DNA. Prominent examples of regulation studies in *Aspergillus* are the detailed analyses of acetamidase regulation (Davis & Hynes, 1989), nitrogen regulation (Arst, 1984; Scazzocchio & Arst, 1989) and the cluster of genes governing the quinate utilization pathway (Lamb *et al.*, 1990).

Development

Another example of the effective transition from classical to molecular analysis in *Aspergillus* is the study of conidial development. This began with the isolation of aconidial mutants (Clutterbuck, 1969). When the wild-type genes corresponding to these mutants were cloned, sequenced and incorporated into fusion constructs as described above, the results suggested a scheme (see Fig. 6.2) in which a cascade of regulator genes coordinates the switching on of the batteries of morphogenetic slaves (Timberlake, 1991).

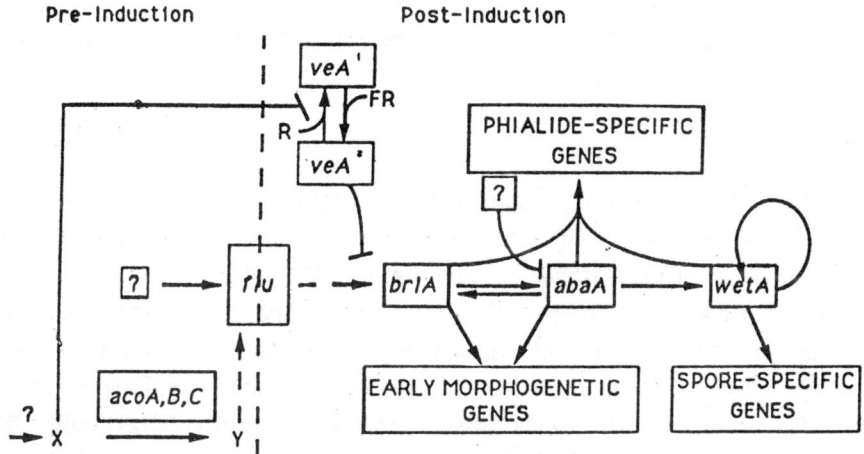

Fig. 6.2 Cascade regulation of conidiation in *Aspergillus nidulans*. Solid arrows and bars indicate positive and negative interactions, respectively. Dashed arrows are hypothetical. Red (R) and far-red light (FR) interconvert two different states of the *veA* gene product.

As in *Drosophila* morphogenesis, it turns out that the most striking *Aspergillus* aconidial mutants are in genes whose DNA sequence and function imply that they are probably regulators for the transcription of other genes. For example, mutants of the "bristle" (*brlA*) gene form only undifferentiated bristle-like conidiophores, but controlled expression of this gene when attached to an inducible promoter (Adams *et al.*, 1989) shows that it switches on both, the further regulatory genes (*abaA* and *wetA*) and other genes assumed to be the slaves that are actually responsible for morphological and physiological development. So far, only a few of these responding genes have been identified, and it is a puzzle that the earlier mutation screens failed to detect them.

Two probable examples of responder genes, which were identified by classical mutagenesis, are the conidiophore pigmentation genes *ivo*A and *ivo*B (Clutterbuck, 1990a; Birse & Clutterbuck, 1991). Another example is the *rod*A gene which, in contrast to the *ivo* genes, was discovered by reverse genetics, that is by identifying the cDNA clone of a messenger which appears only when the *brl*A gene is active. Disruption of the corresponding chromosomal gene gave a distinctive phenotype characterized by the absence of the outer rodlet layer of conidia and conidiophores (Stringer *et al.*, 1991). Although mutants with wettable conidiophores, which might have resulted from this deficiency, had been seen in surveys of conidiation mutants (Martinelli & Clutterbuck, 1971), they were not sufficiently distinctive to be given further attention. It seems likely, by comparison with other fungi (e.g. *Coprinus:* Schuren & Wessels, 1990) that one explanation for the missing mutants in responder genes is that they belong to polygenic sets, whose members have overlapping functions. In this case, mutation of only a single gene will yield a phenotype which is not very different from the wild type.

The *Schizophyllum* example also suggests a connection between the hydrophobic surface proteins of fruiting structures and the initiation of aerial growth which marks the first stage of their development (Wessels *et al.*, 1991). Saxena & Sinha (1973) described conidiation in submerged conditions under which conidiation occurred even in the absence of the normal stimulus of exposure to an aerial environment.

Cell cycle

A third example of a field in which classical genetics has provided the springboard for molecular studies is analysis of the cell cycle. Here the starting point illustrate two different ways of isolating appropriate mutants: by cunning or spadework.

The first set of mutants, which are defective in components of the spindle apparatus, all stem from *ben*A mutants resistant to the tubulin-interacting fungicide benlate (Hastie & Georgopoulos, 1971). The *ben*A gene turns out to code for beta-tubulin, and the isolation of suppressors of *ben*A led not only to two alpha-tubulin genes and a second beta-tubulin (Morris, 1986) but to a new member of the family: gamma-tubulin, which appears to attach the spindle to the spindle pole body (Oakley *et al.* 1990).

In contrast to the tubulin genes, mutants in other cell cycle genes were isolated by sheer hard work: approximately 1000 temperature-sensitive mutants yielded 45 which had cytologically detectable blocks in the cell cycle (Morris, 1976). Cloning and sequence analysis of these has uncovered a series of activities including protein phosphorylation and dephosphorylation, which interlock with those discovered in yeast, and are believed to hold the key to the operation of the cell cycle (Morris *et al.*, 1989).

References

Adams, T.H., Boylan, M.T. & Timberlake, W.E. (1989). *brl*A is necessary and sufficient to direct conidiophore development in *Aspergillus nidulans*. Cell **54**, 353-362.

Arst, H.N. Jr. (1984). Regulation of gene expression in *Aspergillus nidulans*. Microbiological Sciences **1**, 137-141.

Benzer, S. (1955). Fine structure of a genetic region in bacteriophage. Proc. Natl. Acad. Sci. USA **42**, 344-354.

Birse, C.E. & Clutterbuck, A.J. (1991). Isolation and developmentally regulated expression of an Aspergillus nidulans phenol oxidase-encoding gene, *ivo*B. Gene **98**. 69-76.

Clutterbuck, A.J. (1969). A mutational analysis of conidial development in *Aspergillus nidulans*. Genetics **63**, 317-327.

Clutterbuck, A.J. (1990a). The genetics of conidiophore pigmentation in *Aspergillus nidulans*. J. Gen. Microbiol. **136**, 1731-1738.

Clutterbuck, A.J. (1990b). *Aspergillus nidulans*. In Locus Maps of Complex Genomes, Edited by S.J. O'BrienVol 3, Lower Eukaryotes. 5th edition. Cold Spring Harbor Laboratory, pp 3.97-3.108.

Clutterbuck, A.J. (1992). Sexual and parasexual genetics of *Aspergillus species*. In *Aspergillus:* Biology and Industrial applications, Edited by J.W. Bennett & M.A. Klich, Butterworth-Heinemann, Stoneham, U.S.A. pp 3-18.

Davis, M.A. & Hynes, M.J. (1989). Regulatory genes in *Aspergillus nidulans*. Trends in Genetics **5**, 14-19.

Davisson, M.T., Roderick, T.H., Doolittle, D.P., Hillyard, A.L. & Guidi, J.N. (1990). Locus map of the mouse. In Genetic maps, Edited by S.J. O'Brien, 5th edition, Cold Spring Harbor Laboratory, pp 4.3-4.35.

Gems, D., Johnstone, I.L. & Clutterbuck, A.J. (1991). An autonomously replicating plasmid transforms *Aspergillus nidulans* at high frequency. Gene **98**, 61-67.

Hastie, A.C. & Georgopoulos, S.G. (1971). Mutational resistance to fungitoxic benzimidazole derivatives in *Aspergillus nidulans*. J. Gen. Microbiol. **67**, 371-373.

Hinnen, A., Hicks, J.B. & Fink, G.R. (1978). Transformation of yeast. Proc. Natl. Acad. Sci. USA **75**, 1929-1933.

Johnstone, I.L., Hughes, S.G. & Clutterbuck, A.J. (1985). Cloning an *Aspergillus nidulans* developmental gene by transformation. EMBO J. **4**, 1307-1311.

Käfer, E. (1986). Tests which distinguish induced crossing-over and aneuploidy from secondary segregation in *Aspergillus* treated with chloral hydrate or gamma-rays. Mut. Res. **164**, 145-166.

Käfer, E., Scott, B.R., Dorn, G.L. & Stafford, R. (1982). *Aspergillus nidulans*: systems and results of tests for chemical induction of mitotic segregation and mutation. I Diploid and duplication assay systems. Mut. Res. **98**, 1-48.

Lamb, H.K., Hawkins, A.R., Smith, M., Harvey, I.J., Brown, J., Turner, G. & Roberts, C.F. (1990). Spatial and biological characterization of the complete quinic acid utilization gene cluster in *Aspergillus nidulans*. Mol. Gen. Genet. **223**, 17-23.

Landon, G., Cohen, G., Aharonowitz, Y., Shuali, Y., Graur, D. & Shiffman, D. (1990). Evolution of isopenicillin N synthatase genes may have involved horizontal gene transfer. Mol. Biol. Evol. **7**, 399-406.

MacCabe, A.P., Riach, M.B.R., Unkles, S.E. & Kinghorn, J.R. (1990). The *Aspergillus nidulans* npeA locus consists of three contiguous genes required for penicillin biosynthesis. EMBO J. **9**, 279-287.

Martinelli, S.D. & Clutterbuck, A.J. (1971). A quantitative survey of conidiation mutants in *Aspergillus nidulans*. J. Gen. Microbiol. **69**, 261-268.

Migeon, B.R. & Miller, C.S. (1968). Human-mouse somatic cell hybrids with a single human chromosome (group E): link with thymidine kinase activity. Science **162**, 1005-1006.

Morris, N.R. (1976). Mitotic mutants of *Aspergillus nidulans*. Genet. Res. (Cambridge) **26**, 237-254.

Morris, N.R. (1986). The molecular genetics of microtubule proteins in fungi. Exp. Mycol. **10**, 77-82.

Morris, N.R., Osmani, S.A., Engle, D.B. & Doonan, J.H. (1989). The genetic analysis of mitosis in *Aspergillus nidulans*. Bioassays **10**, 196-201.

Müller, W.H., van der Krift, T.P., Krouwer, A.J.J., Wösten, H.A.B., van der Voort, L.H.M., Smaal, E.B. & Verkleij, A.J. (1991). Localization of the pathway of penicillin biosynthesis in *Penicillium chrysogenum*. EMBO J. **10**, 489-495.

Oakley, B.R., Oakley, C.E., Yoon, Y. & Jung, M.K. (1990). Gamma-tubulin is a component of the spindle pole body that is essential for microtubule function in *Aspergillus nidulans*. Cell **61**, 1289-1301.

Peñalva, M.A., Moya, A., Dopazo, J. & Ramón, D. (1990). Sequences of isopenicillin N synthetase genes suggest horizontal gene transfer from prokaryotes to eukaryotes. Proc. Royal Soc. (London) Ser. B **241**, 164-169.

Pontecorvo, G. (1947). Genetic systems based on heterokaryosis. Cold Spring Harbor Symposium on Quantitative Biology **11**, 193-201.

Pontecorvo, G. (1956). The parasexual cycle in fungi. Ann. Rev. Microbiol. **10**, 393-400.

Pontecorvo, G. (1958). Trends in genetic analysis. Columbia University Press, New York.

Pritchard, R.H. (1955). The linear arrangement of a series of alleles of *Aspergillus nidulans*. Heredity **9**, 343-371.

Roper, J.A. (1950). Search for linkage between genes determining a vitamin requirement. Nature **166**, 956.

Roper, J.A. (1952). Production of heterozygous diploids in filamentous fungi. Experientia (Basel) **8**, 14-15.

Saxena, R.K. & Sinha, U. (1973). Conidiation of *Aspergillus nidulans* in submerged culture. J. Gen. Appl. Microbiol. **19**, 141-146.

Scazzocchio, C. & Arst, H.N. Jr. (1989). Regulation of nitrate assimilation in *Aspergillus nidulans*. In Molecular Genetic Aspects of Nitrate Assimilation, Edited by Wra, J.L. & Kinghorn, J.R. Oxford University Press, Oxford, pp 299-313.

Schuren, F.H.J. & Wessels, J.G.H. (1990). Three genes specifically expressed in fruiting dikaryons of *Schizophyllum commune:* homologies with a gene not regulated by mating-type genes. Gene **90**, 199-205.

Scott, B.R., Dorn, G.L., Käfer, E. & Stafford, R. (1982). *Aspergillus nidulans*: systems and results of tests for induction of mitotic segregation and mutation. II Haploid assay systems and overall response of all systems. Mut. Res. **98**, 49-94.

Siddiqi, O.H. (1962). The fine genetic structure of the paba1 region of *Aspergillus nidulans*. Gen. Res. **4**, 12-20.

Singh, M. & Sinha, U. (1976). Chloral hydrate induced haploidization in *Aspergillus nidulans*. Experientia **32**, 1144-1145.

Singh, M. & Sinha, U. (1979). Mitotic haploidization and growth of *Aspergillus nidulans* on media containing chloral hydrate. J. Cytol. Gen. (India) **14**, 1-4.

Sinha, U. (1969). Genetic control of the uptake of amino acids in *Aspergillus nidulans*. Genetics **62**, 495-505.

Sinha, U. (1970). Competition between leucine and phenylalanine and its relation to p-fluorophenylalanine resistant mutations in *Aspergillus nidulans*. Arch. Microbiol. **72**, 308-317.

Stahl, F.W., Edgark R.S. & Steinberg, J. (1964). The linkage map of bacteriophage T4. Genetics **50**, 539-552.

Stringer, M.A., Dean, R.A., Sewall, T.C. & Timberlake, W.E. (1991). Rodletless, a new *Aspergillus* developmental mutant induced by directed inactivation. Genes and Development **5**, 1161-1171.

Tilburn, J., Scazzocchio, C., Taylor, G.G., Zabicky-Zissman, J.H., Lockington, R.A. & Davies, R.W. (1983). Transformation by integration of *Aspergillus nidulans*. Gene **26**, 205-221.

Timberlake, W.E. (1991). Molecular genetics of *Aspergillus* development. Ann. Rev. Gen. **24,** 5-36.

Tinline, R.D. & McNeill, B.H. (1969). Parasexuality in phytopathogenic fungi. Ann. Rev. Phytopathol. **7**, 147-170.

Tiwary, B.N. & Sinha, U. (1989). Isolation and preliminary characterization of a new class of amino acid uptake mutant in *Aspergillus nidulans.* Microbios Lett. **41**, 57-62.

Tiwary, B.N., Bisen, P.S. & Sinha, U. (1987a). Demonstration of an altered phenylalanyl-tRNA synthetase in an analogue-resistant mutant of *Aspergillus nidulans.* Mol. Gen. Genet. **209**, 164-169.

Tiwary, B.N., Bisen, P.S. & Sinha, U. (1987b). Genetic control of amino acid transport in *Aspergillus nidulans*: evidence for polymeric amino acid permease. Current Microbiol. **15**, 305-311.

Weigel, B.J., Burgett, S.G., Chen, V.J., Skatrud, P.L., Frolik, C.A., Queener, S.W. & Inogolia, T.D. (1988). Cloning and expression in *Escherichia coli* of isopenicillin N synthetase genes from *Streptomyces lipmanii* and *Aspergillus nidulans.* J. Bacteriol. **170**, 3817-3826.

Wessels, J.G.H., de Vries, O.M.H., Asgirsdóttir, S.A. & Springer, J. (1991). The *thn* mutation of *Schizophyllum commune*, which suppresses formation of aerial hyphae, affects expression of the *Sc*3 hydrophobin gene. J. Gen. Microbiol. **137**, 2439-2445.

Yelton, M.M., Hamer, J.E. & Timberlake, W.E. (1984). Transformation of *Aspergillus nidulans* using a *trp*C plasmid. Proc. Natl. Acad. Sci. USA **81**, 1470-1474.

7

Genetic analysis of an obligate thermophile — *Thermoactinomyces vulgaris*

**Bhupendra N. Tiwary[1], Ujjal K. Ghosh[1], Ajai K. Sharan[2],
Alik N. Singh[2], Shafique Alam[3] and Santosh Kumar[4]**
[1]*Microbial & Molecular Genetics Lab, Department of Botany*
Patna University, Patna 800 005
[2]*P.G. Centre of Botany, Maharaja College, Arrah*
V.K.S. University, Arrah
[3]*Department of Botany, Forbesganj College, Forbesganj, Bihar*
[4]*Department of Botany, L.S. College, University of Bihar*
Muzaffarpur 842 001

ABSTRACT

Thermoactinomyces vulgaris is an obligate thermophilic actinomycete. Concerted efforts have been made to develop this organism as a suitable system for gene transfer and to carry out extensive genetic and biochemical studies. A large number of auxotrophic and antibiotic resistant mutants have been reported. Treatment with different curing agents led to the co-elimination of antibiotic resistance markers at varying frequency. A plasmid band of about 24 kb has been reported in *T. vulgaris*. However restriction – digestion analysis indicates it to be a basic amplifiable unit of DNA (AUD) derived from a 4.8 kb *Sal* I unit.

Transformation is the basic mechanism of genetic recombination and various aspects of this phenomenon both in interspecific and intergeneric systems have been worked out. Subsequently the protoplast generation and re-generation have also been optimized. Few biochemical studies have been directed to understand the basis of hermophily in this organism.

INTRODUCTION

The actinomycetes constitute a remarkably distinct group of prokaryotic microorganisms. They are of economic interest because they are the producers of a large number of antibiotics (Thompson *et al.*, 1980) and other secondary metabolites of diverse nature (Hopwood & Merrick, 1977). They occupy the borderline between prokaryotes (bacteria) and eukaryotes (fungi), and hence are popularly known as "Ray fungi". Most of the members of this group give a musty odour to the soil which they inhabit and can easily be isolated from soil, manure, compost, fodders, and baggase. Actinomycetes play a pivotal role in the soil formation and in recycling of organic matters (McCarthy, 1985). Quite a large number are also concerned with the degradation of cellulose, wood, lignin, hydrocarbon, phenol, and carotenes and are thus instrumental in recycling of minerals as well. In addition, some members are reported to be efficient symbiotic nitrogen fixers (Lechevalier & Lechevalier, 1982). It is because of these potentials that various members of this group are being subjected to intensive molecular genetic analysis (Bibb *et al.*, 1983; Hopwood & Chater, 1984). Concerted efforts have been made to improve the strains and to understand the fundamentals of metabolic regulation and differentiation in these microorganisms (Hopwood *et al.*, 1985).

Thermoactinomyces vulgaris, an obligate thermophilic actinomycete grows rapidly above 50°C (Tsiklinsky, 1989). It is one of the two organisms considered to be the causative agents for a disease called "Farmer's lung" in which hypersensitivity to antigens of the organism occurs as a result of inhaling vast quantities of their spores (Cross *et al.*, 1968). Hopwood and his coworkers carried out an extensive study on this thermophile to exploit it genetically in tropical conditions. Their achievement was, however, hampered due to the fact that the organism was quite refractile to mutagenesis. Sinha and his group subsequently made concerted efforts to develop this organism as a suitable system for gene transfer and carried out extensive genetic and biochemical studies (Sinha *et al.*, 1985).

Morphological features and topography of endospores

T. vulgaris produces substrate as well as aerial mycelium. The colour of the substrate mycelium is light brown whereas that of the aerial mycelium is white. The profusely branched substrate mycelium is aseptate ranging between 0.3 to 0.4 in diameter. Heat-resistant endospores are endogenously formed both on aerial as well as substrate mycelium. These spores basically resemble the spores of bacilli and clostridia but differ from those of mesophilic *Streptomyces*. Electron microscopic studies of spores of *T. vulgaris* (McVittie *et al.*, 1972) reveals these to be polyhedral in shape with twelve pentagonal and hexagonal faces, having fibrous outer coat and inner spore core. The diameter of the spore is about 1.1. The fibrous outer coats are characteristic of Bacillus species, and hence for quite some time *T. vulgaris* remained a debatable organism so far as its taxonomic position is concerned (Goodfellow *et al.*, 1987; Priest *et al.*, 1988).

Mutational Studies in *T. vulgaris*

A perusal of literature on mutational studies in actinomycetes (Hopwood & Chater, 1984) indicates that it is only the *Streptomyces* that has received due attention, but to a limited extent. Chemical and physical mutagens, such as NTG and UV, have, however, been in frequent use (Delic *et al.*, 1970; Stonesifer & Baltz, 1982; Clarke & Hopwood, 1976; Randazzo *et al.*, 1976).

In *T. vulgaris*, mutational studies were initiated by Hopwood & Wright (1972) who reported the isolation of a few auxotrophic mutants by the treatment of the wild-type strain with UV, NTG, and heat (121°C). These workers could briefly conclude that heat is the most effective mutagen for this thermophile. The mutation frequencies observed by NTG, UV, and high temperature were 0.2 and 14 percent, respectively. The low frequency of mutation led Hopwood and Wright (1972) to postulate that the organism is possibly refractile to mutagenesis or it has a strong repair mechanism. However, Chattoo (1975) while extending the work of mutagenic action of NTG reported that mutation can be induced in *T. vulgaris* by changing the parameters of treatment of NTG for varying periods of time. Treatment of lysozyme treated spores with NTG for 120 minutes and 180 minutes could induce mutation to the extent of 0.03 percent. However the mutation frequencies could not be enhanced beyond 0.084 percent even by changing the parameters.

Sinha & Prasad (1984) extended the work with several chemical mutagens, both singly as well as in combination with various physical mutagens. Besides, these workers also tried to improve the methodology suggested by Hopwood & Wright (1972) for treatment of spores dry in *in vacuo*. Treatment of spores with 3 percent ethylmethane sulfonate (EMS) shows about 60 percent lethality. The frequency of mutation among the survivors of EMS treated spores has been reported to be 0.05 percent. However, extended length of incubation has no effect on the mutation frequency. On the other hand, dimethyl sulfonate (DMSO), an enhancer of the mutagenic effects of EMS, in a combination treatment (Sharma, 1970) could increase the frequency of mutation by 60 percent. Only a limited array of mutants could however, be isolated by such treatment. Such studies, therefore, were further extended (Sinha *et al.*, 1985).

EMS (3%) treated spores of the strain irradiated with gamma ray (50 KR) when plated on selective media could enhance the mutation quite appreciably to the extent of 18 percent. A mutation frequency of 2.6 percent could be achieved when the treatment conditions were reversed. Detailed studies carried out by Sinha *et al.* (1985) and Alam (1990) suggest that treatment of the spores of wild-type strain 1227, either with 20 percent glycerol or 1 percent DMSO, before irradiation to UV (3 minutes) induces mutation to the extent of 2.4 percent and 2.9 percent, respectively.

Prasad (1981) reported the isolation of 25 stable mutants. Biochemical characterization of these mutants indicate requirement for either amino acids or vitamins (Table 7.1); majority of them have requirements for amino acids. Among the amino acid auxotrophs, isoleucine-requirers are predominant, possibly because of the involvement of a number of structural genes in its biosynthetic pathway or due to high mutability of this particular gene (Sinha *et al.*, 1985). The physical and chemical mutagens, except heating the spores *in vacuo*, failed to yield a considerable variety of mutation in *T. vulgaris*. Further attempts have, therefore, been made to isolate antibiotic-resistance in this strain.

Table 7.1: Growth behaviour of auxotrophs isolated after treatment of the strain 1227 of *T. vulgaris* with various amino acids

Auxotroph number	Amino Acid	Vitamin	Growth on different concentration of requirement (mM)		
			0.5	1.0	1.5
1,35,36,150	Isoleucine	-	+	+	+
3,20,27,76,82, 190,201,206, 212	Isoleucine	-	-	+	+
7	Lysine	-	+	+	+
138	Glycine	-	+	+	+
162,170	Lysine	-	-	+	+
184	Glycine	-	-	+	+
15	Isoleucine	Pantothenic Acid	-	+	+
116	Isoleucine	Thiamine-HCl	+	+	+
1227 (control)	Isoleucine		+	+	+

Antibiotic resistant mutants

The first report on the isolation of an antibiotic (streptomycin) resistant mutant was by Hopwood & Wright (1972). Based on the possibility of selecting such mutants, a number of other drugs have been tested auxanographically and minimum inhibitory concentration of these been determined. Sinha *et al.* (1985) have subsequently reported different classes of mutants resistant to ampicillin, tetracycline, and chloramphenicol from the wild type 1227 strain. Interestingly, only ampicillin-resistance has been stable exhibiting cross resistance for higher concentration of tetracycline and chloramphenicol (Table 7.2).

Isolation of rifampicin and spectinomycin resistant mutants has been reported by Kumar & Sinha (1989). These mutants although genetically stable do not exhibit cross resistance to other antibiotics. Nutritional studies with these mutants reveal both to be isoleucine and lysine requirer. Kumar & Sinha (1989) have suggested that rifampicin-resistant mutants might have arisen due to mutation in the ß-subunit of RNA polymerase like those reported in *E. coli* (Doi, 1977).

Table 7.2: Cross resistance of ampicillin resistant mutants to other antibiotics.

Mutants	MMC + antibiootics (in µg/ml)			
	MMC	Ampicillin 1.0	Chloramphenicol 6.0	Tetracycline 1.0
amp 1	+	+	+	+
amp 2	+	+	±	+
amp 3	+	+	+	+
amp 4	+	+	±	+
1227 (wild type)	+	−	−	−

+ = Good growth,   ± = 50% of normal growth,   − = no grrowth

Recently, Ghosh and Tiwary (unpublished) have isolated stable nalidixic acid-resistant mutants of T. vulgaris from the wild type strain. This mutant exhibits cross resistance to ampicillin (1.2 µg/ml), chloramphenicol (8 µg/ml) and rifampicin (0.02 µg/ml). However, instead of isoleucine and lysine requirement (Kumar & Sinha, 1989), nalidixic acid-resistant mutants possess auxotrophy for asparagine and lysine. The degree of drug resistance enhances in the presence of Mg^{2+} ions (4mM). Preliminary characterization of nalidixic acid-resistant mutants show that these have an increased resistance to sucrose (osmotic shock) but a decreased resistance to detergent like sodium dodecyl sulphate (SDS).

Multiple drug resistance and spontaneous loss of antibiotic resistant characters in subcultures usually observed during isolation of mutants in *T. vulgaris* have been attributed to the involvement of extrachromosomal DNA elements (Sinha *et al.*, 1985). Natural resistance to several antibiotics also point to the possibility that the wild type 1227 strain of this thermophile harbours plasmid-borne genes for antibiotic resistance.

In a variety of systems, preliminary evidence to support the presence of an extrachromosomal DNA element has been provided by curing studies (Gregory & Huang, 1964). Characters borne on extrachromosomal DNA element may either be lost spontaneously or can be induced by treatment with curing agents (Rubin & Rosenblum, 1971).

In *T. vulgaris*, spontaneous loss of resistance to chloramphenicol and tetracycline is a matter of common occurrence, whereas some of the subcultures of ampicillin resistant mutants lose their resistance just on being maintained on MMC slants. The percent survival after 48 hours at 52°C (optimal temperature) for *T. vulgaris* decreases in subsequent subcultures on medium containing the antibiotics. Although decrease in percent survival is directly related to the loss of resistance in each subculture, the ampicillin resistance has been more stable than resistance to chloramphenicol and tetracycline (Sinha *et al.*, 1985).

Curing Experiment

A number of curing agents having diverse mechanism of action have been used to study the loss of antibiotic resistant character. These include acriflavine (Hirota & Iijima, 1957), acridine orange (Riva *et al.*, 1973), sodium dodecyl sulphate (Tomoeda *et al.*, 1968), rifampicin (Zimmerman *et al.*, 1971) as well as treatment of the strain at elevated temperature (Carlton & Brown, 1981). As a prelude to the curing experiments, Sinha *et al.* (1985) have studied the effect of acriflavine on the growth of strain 1227 by point inoculating it on plates containing different concentrations of this dye. These workers suggest that the sub-inhibitory concentration of acriflavine for a wild type strain of *T. vulgaris* is 0.001 percent. Subsequent curing studies with acriflavine on ampicillin resistant mutant (*amp*1 and *amp*4) could reveal only a marginal induction in the loss of antibiotic resistant characters (30 percent for ampicillin, 80 percent for tetracycline and 55 percent for chloramphenicol) over the spontaneous loss (0.0 for ampicillin, 0.56 for tetracycline and 0.0 percent for chloramphenicol). An enhanced loss of markers on subjecting the strains to continuous subculture on acriflavine (1μg/ml) has also been observed during this experiment.

Co-elimination of antibiotic resistant characters

The spontaneous and induced loss of characters of ampicillin resistant mutant has led Sinha *et al.* (1985) to investigate the loss of each character with respect to the other and to determine the frequency of co-elimination.

The frequency of loss of ampicillin marker from *amp*1 and *amp*4 with respect to chloramphenicol and tetracycline has been studied by analysing randomly picked-up colonies from MMC plates containing ampicillin (1 μg/ml). Ampicillin resistance was found to be lost simultaneously with other two characters in 12 percent of the colonies. Ampicillin and tetracycline (AT) resistance were lost in 8 percent of the total cases where as ampicillin and chloramphenicol

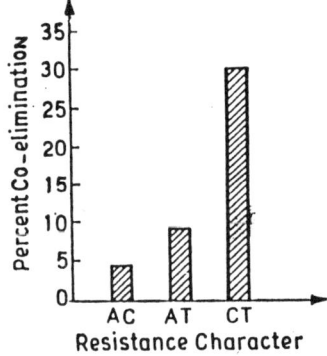

Fig. 7.1: Relative frequencies of co-elimination of pairs of characters from ampicillin-resistant mutant (amp 1)

(AC) resistance were co-eliminated in 4 percent. Chloramphenicol and tetracycline (CT) resistance were lost together in 30 percent of the cases (Figure 7.1). Surprisingly acriflavine did not have any significant effect on the co-elimination of character when used at its sub-inhibitory concentrations (Sinha *et al.*, 1985).

Since a test of natural resistance with 1227 strain revealed that this strain was itself resistant to a variety of drugs (Table 7.3), subsequent workers (Sharan & Sinha, 1989; Kumar & Sinha, 1989) extended these experiments with other curing agents. The wild type strains were continuously subcultured in the presence of subinhibitory concentrations of ethidium bromide (0.5 μg/ml), acridine orange (60 μg/ml), sodium dodecyl sulphate (8 μg/ml), rifampicin (0.01 μg/ml) and at permissive elevated temperature.

Table 3: Zone of inhibition and constitute resistance of strain 1227 of *T. vulgaris* to different antibiotiiics

Antibiotics used	Disc potency (μ g/ml)	Zone of inhibition (mm)	Constitutive resistance concentration (CRC) (μ g/ml)
Ampicillin	10	10	0.1
Tetracycline	10	13	0.05
Chloramphenicol	10	16	1.0
Kanamycin	25	18	0.5
Gentamycin	10	19	0.001
Erythromycin	10	12	0.5
Neomycin	25	10	0.2
Spectinomycin	25	10	2.0
Rifampicin	10	25	0.001
Trimethoprim	10	18	10.0
Nalidixic acid	25	12	100.0

The frequency of the loss of kanamycin resistance was to the extent of 87 percent with acridine orange, 100 percent with ethidium bromide, 0.7 percent with sodium dodecyl sulphate, and 100 percent by treatment at elevated temperature (Sharan & Sinha, 1989). Elevated temperature (60°C) had a positive response on the loss of marker when it was combined with different curing agents. Sharan & Sinha (1989) also observed induction of the loss for antibiotic resistance with the passage of subcultures in the presence of curing agents (Table 7.4).

Detailed curing studies on neomycin and rifampicin-resistant mutants by agents like ethidium bromide, rifampicin and elevated temperatures have been carried out in *T. vulgaris* by Kumar & Sinha (1987) and Kumar (1988). The results suggest that these curing agents eliminate only

Table 7.4: Extent of loss of different antibiotic resistaant phenotype exhibited by different strains of *T. vulgaris* on treatment with curing agents

Curing agent used	Strain	Extent of loss (%)		Reference
		Kanamycin	Neomycin	
Acridine orange	T 1227	87	--	Sharan & Sinha (1987)
Ethidium bromide	T 1227	100	--	Sharan & Sinha (1987)
Sodium dodecyl sulphate	T 1227	--	--	Sharan & Sinha (1987)
Treatment at elevated temperature	T 1227	100	--	Sharan & Sinha (1987)
Rifampicin	T 1227	--	--	Kumar (1988)
	UK 2001	--	--	Kumar (1988)
Ethidium bromide	T 1227	--	28	Kumar (1988)
	UK 2001	--	33	Kumar (1988)
Treatment at elevated temperature	T 1227	--	03	Kumar (1988)
	UK 2001	--	19	Kumar (1988)

neomycin-resistant character out of 17 antibiotics and 3 heavy metals tested in such a way that the resultant cured strains become sensitive to even 0.2 µg/ml of neomycin. However, neomycin sensitivity and isoleucine and lysine requirement have been found to be unstable in *T. vulgaris.*

A comparative study on the mutagenic and curing effects of rifampicin in *T. vulgaris* reveals that the drug induces loss of neomycin resistance not only in the wild type (1227) but also in rifampicin-resistant strains (Kumar & Sinha, 1987). Moreover, the levels of rifampicin and neomycin susceptibility of mutants and their cured strains are sufficient to suggest that rifampicin resistance is due to the factors on main genome of *T. vulgaris.* A similar differential effect of this drug on bacterial growth and plasmid maintenance has been speculated by Obaseiki-ebor (1984), but does not seem to affect the curing properties of rifampicin.

The elevated temperature has invariably been reported to cause genetic instability of neomycin resistance. This could be due to selective inhibition of the DNA segment responsible for resistance phenotype. However, at the low or optimal temperature about 15-20 percent of neomycin sensitive cells revert to resistant phenotype. Such instances of reversion may be interpreted by considering that at high temperature, some genetic rearrangements may take place that can be reversed at optimum temperature.

The curing studies with the wild type strain and drug resistant mutant strains of *T. vulgaris* thus provides a preliminary clue to the possible involvement of extrachromosomal DNA element and also points towards the possibility of neomycin / kanamycin resistant marker being plasmid-borne.

Possible involvement of Extrachromosomal DNA Elements

The physical presence of extrachromosomal DNA elements has been successfully demonstrated by agarose gel electrophoresis and cesium chloride-ethidium bromide density gradient centrifugation in various species of microorganisms. A similar attempt made in *T. vulgaris* has resulted into an intense band of about 24 kb that has been interpreted to be a plasmid (Sinha *et al.*, 1985; Sharan, 1987). However, even after careful isolation as well as by changing the gel concentrations, attempts to isolate the three forms i.e. OC, linear and CCC of the plasmid in subsequent preparations of *T. vulgaris* remained unsuccessful (Kumar, 1988). Even gentle lysis of the mycelia of *T. vulgaris* showed chromosomal trailing without any distinct band in the gel electrophoretic profile of the wild type 1227. On the other hand, the same treatment with *E. coli* V517 strain, which harbours 8 plasmids, showed distinct bands embedded in the background of diffused DNA of the main genome. Moreover, the possibility of the existence of a linear plasmid in *T. vulgaris* as reported in *Streptomyces remosus* (Chardon-Loriaux, 1986), is also ruled out because such plasmids are prone to alkaline denaturation whereas, in *T. vulgaris*, the band on gel becomes more compact after alkaline treatment (Kumar, 1988; Kumar & Sinha, 1989).

Reiterated DNA sequence in *T. vulgaris*

Reiteration of a specific DNA fragment, although unusual and rare in prokaryotes (Anderson & Roth, 1977), is a phenomenon of common occurrence in actinomycetes. It has been reported in several species of *Streptomyces* (Ono *et al.*, 1982). In *T. vulgaris* also, the appearance of bands on the gels after plasmid isolation can be explained if the organism is considered to possess a repeated DNA sequence in its genome. In general, amplification is the main cause of reiteration of specific DNA sequences. High temperature treatment (62-65°C) of the alkaline DNA preparations denatures the genomic DNA. As a result, the repeat units, if present, may lose their stability and remain as a single helix in the solution. However, with the return of renaturation factors (low pH and low temperature), these repeat units may form duplexes because of complimentarity in the strands and, precipitate after ethanol treatment. In *T. vulgaris*, the wild-type and the revertants (neomycin-sensitives) exhibit a sharp band of about 24 kb after selective isolation of multicopy DNA. This band has been found to be missing in the cured strains (Kumar & Sinha, 1987), which are phenotypically neomycin-sensitive and double auxotrophs for isoleucine and lysine. This phenomenon could be attributed either to deletion of a part of the genome or to deamplification of the repeat units in the presence of curing agents.

Although unequivocal evidence for amplifiable unit of DNA (AUD) in the genome of *T. vulgaris* is not available in the absence of DNA-DNA hybridization studies. The work of Kumar &

Sinha (1987) and Kumar (1988) strongly suggests the existence of such an amplifiable unit, which has been proposed to have a saltatory type of replication as suggested by Hasegawa *et al.* (1985). The saltatory replication allows jumping of replication units from one place to another as against the progressive form of replication. It is most likely that curing agents do not allow such replications of AUD above a particular concentration. Evidence to support this contention comes from the finding that withdrawl of curing agents in *T. vulgaris* leads to reversion of antibiotic resistance and amino acid auxotrophy (Kumar, 1988).

Restriction Enzyme Analysis

Preliminary restriction enzyme analysis of putative plasmid DNA of *T. vulgaris* 1261 strain was carried out with *Bam*HI and *Eco*RI (Sinha *et al.*, 1985; Sharan, 1987). *Bam*HI digested DNA produced two bands whereas *Eco*RI could produce only one band.

Evidence for the presence of AUD in *T. vulgaris* genome has subsequently been presented by Kumar (1988) who observed tandemly reiterated stretches after restriction enzyme analysis. The DNA preparations of wild type 1227 and mutant strains revealed ladder-liked discrete bands on agarose gels whereas, cured strains did not exhibit any banding pattern. The revertants, on the other hand, on restriction digestion showed ladder like bands on agarose gels. The deletion of amplifiable unit, which subsequently showed deamplification has been suggested to be one of the possible reasons for the absence of reiterations in the cured strains of *T. vulgaris* (Kumar, 1988).

Partial digestion of the genomic DNA has identified different AUDs in different organisms (Ono *et al.*, 1982; Altenbuchner & Cullum, 1984). In *T. vulgaris* also, one of the *SalI* digestions has identified the AUD of 4.8 kb, which upon amplification gives rise to the multimers of 9.6, 14.0, 19.0 and 24.0 kb. Kumar (1988) has further suggested that even the 24 kb DNA segment might be considered as the basic amplifiable unit of DNA, similar to those reported in the *range* of 2.9 to 35 kb in the main genome of *S. glaucescens* (Hasegawa *et al.*, 1985).

Methylation pattern of *T. vulgaris* genomic DNA

Restriction enzymes generally fail to cut at their recognition sites if specific adenine or cytosine residues are methylated. Although such methylated residues complicated the restriction enzyme mapping, they may prove to be advantageous during analysis of gene regulation. Nevertheless, there are evidences to suggest that specific methylation of cytosine residues may be involved in rendering genes transcriptionally inactive. The use of isoschizomer enzymes like *Mbo*I, *Sau*3A, *Dpn*I, *Taq*I etc. (Table 7.5) in molecular biology has yielded valuable information regarding changes in methylation pattern (Lacks & Greenberg, 1977). For example, *Mbo*I, irrespective of cytosine methylation, cuts GATC but is sensitive to adenine methylation (Dreiseikelmann et al., 1979). On the other hand, *Sau*3A cut GATC, irrespective of adenine methylation but is simultaneously sensitive to cytosine methylation. *Dpn*I, an isoschizomer enzyme, however, cleaves GATC sequence only when both the strands have methylated adenine

Table 7.5 : Effect of methylation on restriction enzyme site recognition in *T. vulgaris*

Restriction enzyme	Recognition sequence	Methylated sequence cleaved	Methylated sequence not cleaved
Mbo 1	GATC	GAT*C	G*ATC
Sau 3A	GATC	G*ATC	GAT*C
Dpn 1	G*ATC	G*ATC	GAT*C
Taq 1	TCGA	T*CGA	TCG*A

(McClelland, 1983). Using three isoschizomer enzymes (*Mbo*I, *Sau*3A, *Dpn*I) an extensive investigation on the modification of GATC site in the genome of wild type strain 1227 of *T. vulgaris* have been carried out (Kumar, 1988) which indicates the distribution of only modified cytosine bases and not of any adenine residues. Furthermore, since *T. vulgaris* genome is prone to digestion both by A+T rich (*Eco*RI, *Hind*III) and G+C rich (*Sal*I, *Bam*HI) restriction enzymes, Kumar (1988) has inferred that this thermophile possess a relatively low G+C content (49.5 percent) as compared to *Streptomyces*, the genome of which is frequently digestible by G+C recognising enzymes (Fig. 7.2).

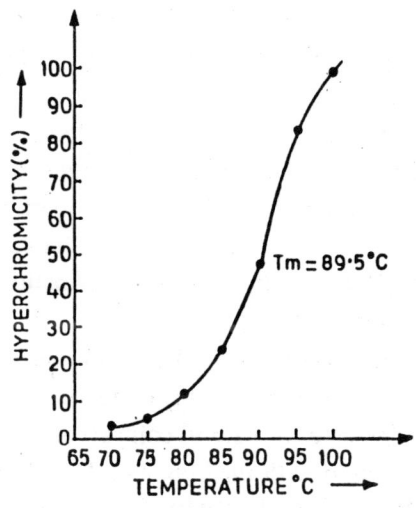

Fig. 7.2. Melting curve of native unsonicated DNA of the strain 1227 of *Thermoactinomyces vulgaris*

Transformation in *T. vulgaris*

Genetic recombination through transformation in *T. vulgaris* has been extensively studied by Hopwood & Ferguson (1970) and Hopwood & Wright (1972). Although only some of the actinomycetes have been reported to be naturally competent for transformation, competence has also been induced by manipulating the culture conditions during growth by using several divalent cations (Hopwood & Wright, 1972; Prasad, 1981; Sinha & Singh, 1989). Studies on the

Table 7.6 : Effect of various metal ions on transformation frequencies in *T. vulgaris*

Cation added (20 mM)	Total number of viable spores tested x 10^4	Frequencies of transformants/10^4 survivors	
		thi^+	nic^+
None	8.80	13.86	13.00
Ca^{2+}	9.60	133.75	128.83
Mg^{2+}	8.80	106.70	98.70
Ba^{2+}	5.60	64.23	60.71
Fe^{3+}	5.20	54.60	49.20
Cu^{2+}	13.60	47.05	31.76
Cd^{2+}	7.40	44.70	33.00
Zn^{2+}	9.20	20.00	26.52
Sr^{2+}	18.80	29.70	26.32

effects of divalent cations (Table 7.6) on the transformation frequency of nic^+, thi^+ markers in *T. vulgaris* 1261 strain suggest that Ca^{2+} has a pronounced effect on enhancing the frequency of transformation followed by $Mg^{2+}Ba^{2+}Fe^{2+}Cu^{2+}Cd^{2+}Zn^{2+}$ Sr^{2+} (Sinha & Prasad, 1984).

Different aspects of transformation under liquid shake conditions have been optimised by Sinha & Prasad (1980, 1984) and Prasad (1981) in *T. vulgaris*. It has been reported that cells become competent for transformation after about 130 minutes of growth and optimum transformation is achieved after about 150 minutes. The optimum temperature for the uptake of transforming DNA has been found to be 40°C (Table 7.7). Sinha & Prasad (1980) suggested that the saturating concentration of DNA required for transformation in *T. vulgaris* is 1 µg/ml and that the uptake is completed within 15 minutes.

Optimisation of competence by manipulating the environmental conditions and physiological parameters have yielded valuable information regarding transformation in *T. vulgaris* (Prasad, 1981; Singh & Sinha, 1991). The level of natural competence in this thermophile gets enhanced due to the presence of Ca^{2+} ions in growth medium and the maximum transformation frequency achieved is 16 percent (Sinha & Prasad, 1980). Subsequent work of Singh (1988) on the development of competence in *T. vulgaris* suggests that addition of $CaCl_2$ (0.6 mM) in the growth medium not only reduces the time to achieve optimal competence from 150 minutes to 75 minutes but also enhances the transformation frequency from 16 to 80 percent. Also, the pH for the development of optimum competence has been found to be slightly acidic (6.4). A drastic reduction in the levels of competence to the extent of 50 percent transformation frequency by addition of antibiotics like chloramphenicol, tetracycline, and rifampicin either just before or at the onset of competence suggests that continuous synthesis of proteins and mRNA during the acquisition of competence is essential (Singh, 1988; Singh & Sinha, 1991). In addition to antibiotics, EDTA, sodium periodate and 2,4-dinitrophenol (DNP) also reduce the transformation frequency in *T. vulgaris*. However, the inhibitory effects of these complexing agents can be reversed by higher concentrations of Ca^{2+} ions (Singh, 1988). This finding supports the earlier

Table 7.7 : Effect of different cooncenttration of calcium chlooride and temperature on the frequencies of transformation in *T. vulgaris*

Parameter		Viable spores tested x 10^4	Frequencies of transformants/10^4 survivors	
			thi[+]	nic [+]
Ca^{2+} (mM)				
	0.0	7.50	66.00	60.00
	10	7.50	400.00	365.00
	20	7.50	540.00	400.00
	40	7.50	500.00	460.00
	80	7.50	400.00	500.00
	150	7.50	345.00	240.00
Temperature (°C)				
	0	4.56	303.00	280.00
	20	4.92	570.00	500.00
	30	4.44	655.00	600.00
	40	3.48	910.00	802.00
	45	2.72	870.00	720.00
	50	2.32	700.00	700.00
	55	2.72	580.00	540.00

contention that Ca^{2+} ions might be participating in the protection of competence-specific protein and/or transforming DNA (Mandel & Higa, 1970; Singh & Sinha, 1982a).

In addition to specific proteins, the production of another protein having the ability to induce competence has been suggested on the basis of observation of stimulation of competence by supernatants from the competent cultures of *T. vulgaris* (Singh, 1988). The restorative activity of the supernatants, however, has been reported to be lost by heat treatment. Such stimulation of competence has generally been attributed to an extracellular protein, the competence factor (CF), which is excreted in the medium by competent culture (Morrison, 1981). The actual mechanism of action of such activator substances has although not been clearly understood, it has been suggested that the latter might either help the binding of the transforming DNA to receptor sites or inducing permeability changes in plasma membrane thus facilitating the entry of DNA (Morrison, 1981; Singh & Sinha, 1991).

Heterospecific transformation

The ability of plasmids containing cloned DNA segments to transform naturally competent cells has become a useful tool for recombinant DNA research. There have been a number of reports pertaining to interspecific transformation of genetic material among gram-positive bacteria (Ehrlich, 1977; Gryczan et al., 1978; Espinosa et al., 1982). It has generally been observed that the major barrier in such interspecific gene transfer is the lack of sequence homology rather than enzymatic restriction of the foreign DNA. Stewart & Carlson (1986) have opined that apart from the size of plasmid and readily selectable genetic marker (preferably antibiotic resistance), a broad host range is also an essential requirement for interspecific transformation. Development of such techniques have thus opened up a broad and promising field for rationally combining desirable properties from incompatible lines. Although competence mediated heterospecific transformation has not been reported in actinomycetes, T. vulgaris has been shown to be an ideal organism to study the details of such gene transfer in thermophiles, in particular, and filamentous bacteria in general (Sinha & Singh, 1989).

Intergeneric transformation of highly competent cells of T. vulgaris with plasmids pIJ61 and pIJ486 from S. lividans has been found to be inefficient. The work of Singh (1988) shows that pIJ486 (6.2 kb) DNA fails to enter the cells of T. vulgaris as evidenced by the failure of cells to grow on neomycin. In addition, there is a loss of neomycin resistance after two subcultures. The low efficiency and instability of heterologous plasmids in T. vulgaris has been explained due to sequence non-homology and non-recognition of Streptomyces promoters by T. vulgaris DNA, like other prokaryotic systems (King & Chater, 1986). Differences in optimal temperatures for the growth of Streptomyces (30°C) and T. vulgaris (52°C) might also account for the inefficient heterospecific gene transfer since there are reports of the loss of plasmid-coded characters on shifting the temperature beyond normal (Birch & Cullum, 1985).

Protoplast formation and regeneration

Attempts have also been made to obtain maximum protoplast yield and their subsequent regeneration in T. vulgaris by optimising the culture conditions like suitable buffer and appropriate regeneration time (Singh, 1988). Addition of glycine (0.4 percent) in the growth medium gives maximum protoplast yield. Glycine being a partial growth inhibitor facilitates the action of lysozyme by interfering with cross-linking of cell walls and by replacing D-alanine in peptidoglycan (Keller et al., 1983). In addition to glycine, cations like Mg^{2+} and Ca^{2+} also exhibit stimulatory roles in protoplast formation in this thermophile. Presence of 0.6 mM of $CaCl_2$ and 0.75 mM of $MgCl_2$ makes the conditions favourable for the action of lysozyme.

Whereas, Ca^{2+} provides stability to membrane bound enzymes, Mg^{2+} stimulates the acid phosphatases in *T. vulgaris* (Singh & Sinha, 1984).

Regeneration of protoplasts on agar media depends mainly on the composition, drying of the regeneration medium, and temperature. Singh (1988) developed a regeneration medium (RMMC) by modifying the normal growth medium (MMC) used for *T. vulgaris* and has achieved regeneration upto 1.3 percent. The sucrose concentration optimum for regeneration (0.15 M to 0.2 M), appears to be low due to a low requirement of sucrose (0.08 M) by this thermophile for its normal growth as compared to *Streptomyces* and *Bacillus*. The most significant observation in the regeneration of protoplast in *T. vulgaris* is the optimal temperature i.e. 52°C (Singh 1988), which happens to be its optimum growth temperature too. In general, actinomycetes require lower temperatures for regeneration (Schupp & Divers, 1986). Need for an optimum temperature has been suggested to be related to normal metabolic activities which are nevertheless important for wall synthesis, and further indicates the intrinsic property of thermostability in *T. vulgaris* as evidenced by viable and regenerating protoplasts at 52°C (Singh, 1988).

Biochemical Studies in *T. vulgaris*

A number of biochemical investigations have been made to explain the mechanism of thermophily and the rapid turnover of macromolecules in thermophilic actinomycetes in particular and in thermophiles in general (Singh, 1988). Inherent macromolecular stability at elevated temperatures, although not universal, is a widely accepted phenomenon for thermophilic existence (Barker, 1978; Singh & Sinha, 1982b). Thermostability of macromolecules, especially of proteins is, however, not restricted to those derived from thermophiles only. Even in mesophiles, a few enzymatic proteins exhibit remarkable thermostability having transition temperatures of 60°C - 70°C (Cass & Stellwagen, 1975). The properties of certain extracellular proteins (enzymes) of *T. vulgaris* like -amylases (Khandeparker & Dhala, 1973) and proteases (Ruttloff, 1978) have though been worked out, do not provide much insight into the thermophilic mechanisms. Amongst the intracellular key enzymes, the most extensive investigation is that on phosphatases which perform diverse metabolic functions (Mukhopadhyay & Sinha, 1979; Sinha & Singh, 1980; Singh & Sinha, 1982 a,b, 1984).

The growth pattern and phosphatase formation in *T. vulgaris* in response to various physiological concentration of phosphate sources suggest that both acid and alkaline phosphatases are synthesized constitutively and their amounts in vivo are only partially sensitive to phosphate in the culture medium (Sinha & Singh, 1980). In addition, Mg^{2+} (10mM) enhances the activities of both the phosphatase isozymes (Table 7.8). Singh & Sinha (1982a) have further observed that *T. vulgaris* possesses thermolabile acid and alkaline phosphatases which are constantly replenished as a result of rapid turnover (resynthesis) and which, in turn, counteracts the destructive effects of heat permitting the organism to grow at high temperature. However, rapid turnover of macromolecules rather than thermostability appears to be only restricted to spore forming actinomycetes (Singh & Sinha, 1984).

In another report on Ca^{2+} - dependent growth and metabolic status of *T. vulgaris*, Singh & Sinha (1982a) have reported that this divalent cation is associated with increased hydrolysis of ATP. Whereas, Mg^{2+} stabilized the nonspecific acid and alkaline phosphatases, Ca^{2+} protects the membrane bound ATPases in obligate thermophiles. Subsequent observation on delayed senescence (arrival of stationary phase) in the presence of Ca^{2+} (Kumar & Sinha, 1989) further suggest the protective nature of this cation.

Antibiotic resistant mutants having isoleucine and lysine auxotrophy have been found to

Table 7.8 : Effect of various compounds on specific activities of acid and alkaline phosphatases in *T. vulgaris*

Compounds added	Concentration (mM)	Specific activity of (n mol of p-nitrophenol released/min./ mg of protein)			
		Acid phosphatases		Alkaline phosphatase	
		$-Mg^{2+}$	$+Mg^{2+}$	$-Mg^{2+}$	$+Mg^{2+}$
None	-	43.2	150.7	26.1	65.2
Sn-glycero-2 phosphate	1	76.2	256.7	43.4	107.2
	5	47.7	171.1	26.4	57.4
KH_2PO_4	1	76.0	267.9	50.4	124.4
	5	53.9	192.5	25.6	81.5
Glycerol	1	25.7	104.5	11.0	34.8
	5	30.9	115.7	15.4	40.1
Glycerol + KH_2PO_4	1 (each)	62.3	223.1	40.3	103.9
	5 (each)	44.3	160.4	27.5	72.6
Glycerol + Sn-glycero-2-phosphate	1 (each)	65.7	236.8	42.8	114.5
	5 (each)	57.5	188.1	36.5	86.9

leach out brown coloured secondary metabolite in the medium (Kumar & Sinha, 1989). PAGE and isoelectric focussing of this secondary metabolite suggests it to be proteinaceous in nature having a molecular weight of about 70 kd (Alam, 1990). The nature of this proteinaceous secondary metabolite as an antimicrobial agent is yet to be established. Earlier reports mention that this thermophile produces an antibiotic thermorubin effective against *Escherichia coli* (Pirali *et al.*, 1974; Hopwood & Merrick, 1977).

Autolysis is of common occurrence in this group, particularly in thermophiles (Krassilinikov, 1981). The lysis has been observed in the old cultures of the majority of strains of *T. vulgaris* and this can be further induced by the elevated temperature, amino acid starvation, nalidixic acid, as well as other stress conditions. *T. vulgaris* mutant resistant to this drug show a clear zone in the centre of the colony (Fig. 7.3). There is no evidence so far to suggest whether autolysis is due to involvement of an actinophage as reported in *T. vulgaris* (Kurup & Heinzen, 1978) or any other agent (Ghosh & Tiwary, unpublished).

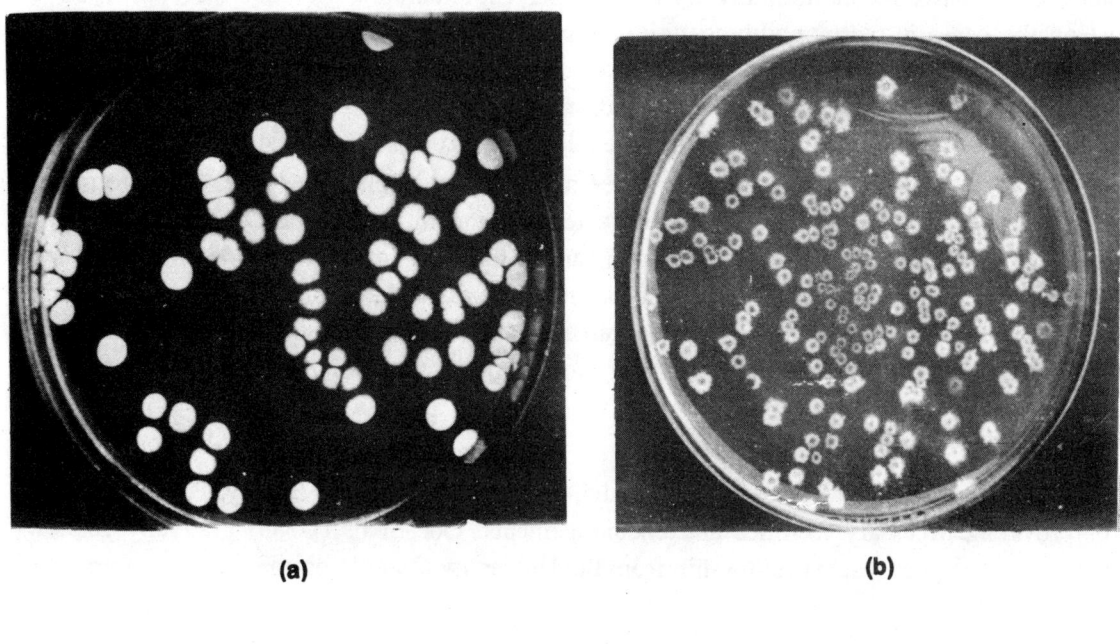

(a)   (b)

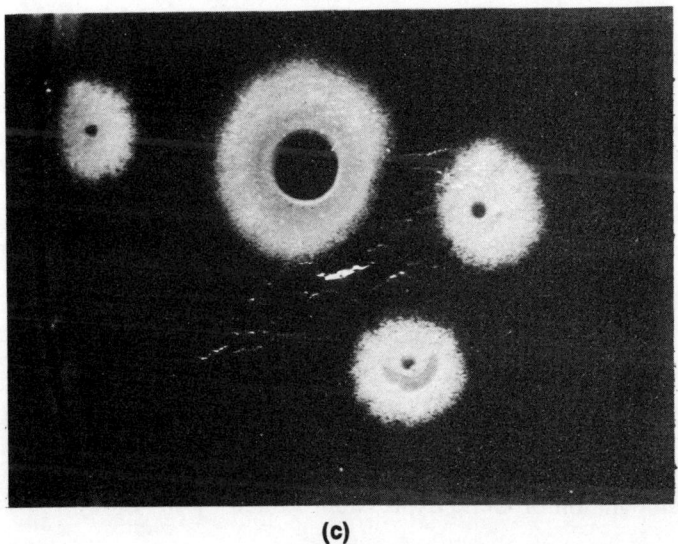

(c)

Fig. 7.3. (a) Growth of the wild type strrain 1227 on MMC, (b) Spontaneous autolysis of the wild type on MMC and (c) nalidixic acid-resistant mutant showing clear zone in the centre (magnified) of the colony in the presence of minimal inhibitory concentration of the drug

Epilogue

T. vulgaris, a spore-forming actinomycete, seems to be an ideal organism for investigating the biochemical basis of thermophily apart from its exploitation as a host for expression of mesophilic genes in thermophilic condition. Such studies have their industrial applications because thermophily forms a model system for elucidating the high temperature chemical catalysis of industrialized enzyme processes. Many industrial companies around the world are engaged in isolating and characterising thermophiles to exploit the enzymes of specific interest.

In the light of the foregoing review it seems worth developing the genetic transformation system further wih the help of partially characterized tandemly reiterated DNA sequences of *T. vulgaris.* The latter might be used as a cloning and/or expression vector provided it is fused with suitable multiple cloning sites and strong promoter to drive the expression in a thermophilic environment. Conditions for heterologous transformation need to be standardized further so as to make this thermophile a model system for biotechnological exploitation.

Acknowledgements

We are thankful to Professor B.M.B. Sinha, Head, Department of Botany, Patna University for providing necessary facilities and encouragements. One of us (UKG) gratefully acknowledges the receipt of teacher fellowship from the University Grants Commission, Government of India.

References

Akagawa, H., Okanishi, M. & Umezawa, H. (1975). A plasmid involved in Chloramphenicol production in *Streptomyces venezuelae:* evidence from genetic mapping. J. Gen. Microbiol. **90,** 336-356.

Alam, S. (1990). Studies on transformation in *Thermoactinomyces vulgaris.* Ph.D. Thesis, Patna University, Patna, India.

Altenbuchner, J. & Cullum, J. (1984). DNA amplification and unstable arginine gene in *Streptomyces lividans.* Mol. Gen. Genet. **201,** 192-197.

Anderson, R.P. & Roth, J.R. (1977). Tandem genetic duplication in phage and bacteria. Ann. Rev. Microbiol. **31,** 473-505.

Barker, S.A. (1978). Industrialized enzymes processes. Sci. Prog. Oxf. **63,** 477-505.

Bibb, M.J., Chater, K.F. & Hopwood, D.A. (1983). Developments in *Streptomyces* cloning. In experimental manipulation of Gene expression, Edited by M. Inouye. Acad. Press, N.Y. pp. 53-82.

Birch, A.W. & Cullum, J. (1985). Temperature sensitive mutants of the *Streptomyces* plasmid pIJ 702. J. Gen. Microbiol. **131,** 1299-1303.

Carlton, B.C. & Brown B.J. (1981). Gene mutation: In Manual of Methods for General Bacteriology, Edited by P. Gerhardt, R.G.E. Murray, R.N. Costilow, E.W. Nester, W.A. Wood, N.R. Kreiz & G.B. Phillips. Am. Soc. Microbiol. Washington, D.C. pp. 222-242.

Cass, K.H. & Stellwagen, E. (1975). A thermostable phospho-fructokinase from the extreme thermophile. *Thermus* X-1. Arch. Biochem. Biophys. **171,** 689-694.

Chattoo, B.B. (1975). Observations on the lethal and mutagenic effects of NTG on *Aspergillus* and *Thermoactinomyces*, Ph.D. Thesis, University of Delhi, Delhi, India.

Chaturvedi, S.N. & Singh, V.P. (1978). Increased mutagenic effects of EMS in Moong Bean, (*Phaseolus aureus* Roxb.) by DMSO. J. Cytol. Genet. **13,** 116-119.

Clarke, C.M. & Hopwood, D.A. (1976). Ultraviolet mutagenesis in *Streptomyces coelicolor* induction of reversion in polyauxotrophic strain. Mut. Res. **41,** 201.

Cross, T., Walker, P.D. & Gould, G.W. (1968). *Thermophilic actinomycetes* producing resistant endospores. Nature (London) **220,** 352-354.

Chardon-Loriaux, J., Charpentier, M. & Perchoren, F. (1986). Isolation and characterization of a linear plasmid from *Streptomyces rimosus*. FEMS Microbiol. Lett. **35,** 151-155.

Delic, V., Hopwood, D.A. & Friend, E.J. (1970). Mutagenesis of N-methyl-N'-nitro-N-nitrosoguanidine (NTG) in *Streptomyces coelicolor*. Mut. Res. **9,** 167.

Dreiseikelmann, B., Eichenlaub, R. & Wackernagel, W. (1979). The effect of differential methylation by *Escherichia coli* of plasmid DNA and phage T7 and *Mbo* I from *Moraxella bovis*. Biochem. Biophys. Acta. **562,** 418-594.

Doi, R.H. (1977). Role of ribonucleic acid polymerase. Bacteriol. Rev. **41:** 568-594.

Ehrlich, S.D. (1977). Replication and expression of plasmids from *Staphylococcus aureus* in *Bacillus subtilis*. Proc. Natl. Acd. Sci. U.S.A. **74,** 1680-1682.

Espinosa, M., Lopez, P., Perez Urena, M.T. & Lacks, S.A. (1982). Interspecific plasmid transfer between *Streptococcus pneumoniae* and *Bacillus subtilis*. Mol. Gen.Genet. **188,** 195-201.

Goodfellow, M., Lacey, J. & Todd, C. (1987). Numerical classification of Thermophilic Streptomycetes. J. Gen. Microbiol. **133,** 3135-3149.

Gregory, K.F. & Huang, J.C.C. (1964). Tyrosinase inheritance in *Streptomyces scabies* II. Induction of tyrosinase deficiency by acridine dyes. J. Bacteriol. **87,** 1287-1294.

Gryczan, T.J., Contente, S. & Dubnau, D. (1978). Characterization of *Staphylococcus aureus* plasmids introduced by transformation into *Bacillus subtilis*. J. Bacteriol. **134,** 318-329.

Hasegawa, M., Hinterman, G., Simonet, J.M., Crameri, H., Piret, J. & Hutter, R. (1985). Certain chromosomal regions in *Streptomyces glaucescens* tend to carry amplification and deletion. Mol. Gen. Genet. **200,** 375-384.

Hirota, Y. & Iijima, T. (1957). Acriflavine as an effective agent for eliminating F factor in *Es-

cherichia coli K-12. Nature (London) **180**, 655-656.

Hopwood, D.A. & Ferguson, H.M. (1970). Genetic recombination in a thermophilic actinomycete *Thermoactinomyces vulgaris.* J. Gen. Microbiol. **63**, 133-136.

Hopwood, D.A. & Wright, H.M. (1972). Transformation in *Thermoactinomyces vulgaris.* J. Gen. Microbiol. **71**, 383-398.

Hopwood, D.A. & Merrick, M.J. (1977). Genetics of antibiotic production. Bacteriol. Rev. **41**, 595-635.

Hopwood, D.A. & Chater, K.F. (1984). In Genetics and Breeding of Industrial Microorganisms, Edited by B. Christopher. C.R.C. Press. Inc. Boca Raton. Florida.

Hopwood, D.A., Bibb, M.J., Chater, K.F., Kieser, T., Bruton, C.J., Kieser, H.M., Lydiate, D.J., Smith, C.P., Ward, J.M. & Schrempf, H. (1985). Genetic manipulation of *Streptomyces:* A Laboratory Manual. Pub. The John. Innes Foundation, Norwich, England.

Keller, U., Poschman, S., Krengil, U., Kleinkauf, H. & Kraepelin, G. (1983). Studies of protoplast fusion in *Streptomyces chrysomallus.* J. Gen. Microbiol. **129**, 1725-1731.

Khandeparkar, V.G. & Dhala, S.A. (1973). Amylase from thermophilic actinomycetes I. Selection of strain. Ind. J. Microbiol. **13**, 73-80.

King, A. & Chater, K.F. (1986). The expression of *Escherichia coli lac Z* gene in *Streptomyces.* J. Gen. Microbiol. **132**, 1739-1752.

Krassilinikov, N.A. (1981). Ray fungi (Higher forms). Amerind Publishing Co. Pvt. Ltd. New Delhi, Vol. I, pp. 137-139.

Kumar, S. (1988). Identification and characterization of a cloning vehicle in *Thermoactinomyces vulgaris.* Ph.D. Thesis, Patna University, Patna, India.

Kumar, S. & Sinha, U. (1987). Plasmid mediated neomycin resistance in *Thermoactinomyces vulgaris.* DAE. Symposium on Biology and Molecular Biology of Streptomycetes. NCL. Pune, pp. 48-51.

Kumar, S. & Sinha, U. (1989). Delayed senescence in a thermophilic actinomycete by calcium ions in shake culture. Microbios. Lett. **42**, 55-59.

Kurup, V.S. & Heinzen, R. (1978). Isolation and characterization of actinophages of *Thermoactinomyces* and *Micropolyspora.* Can.J. Microbiol. **24**, 794-797.

Lacks, S.A. & Greenberg, B. (1977). Complementary specificity of restriction endonulceases of *Diplococcus pneumoniae* with respect to DNA methylation. J. Mol. Biol. **114**, 153-168.

Lechevalier, M.P. & Lechevalier, H.A. (1982). Taxonomy of *Frankia.* Fifth International Symp. on Actinomycetes. Mexico, 3.

Mandel, M. & Higa, A. (1970). A calcium dependent bacteriophage DNA infection. J. Mol. Biol. **53**, 159-162.

McCarthy, A.J. (1985). Development in the taxonomy of isolates of thermophilic actinomycetes. In Frontiers in Applied Microbiology, Edited by K.G. Mukherjee, N.C. Pathak & V.P. Singh. Print House Lucknow (India) Vol-I, pp. 1-14.

McClelland, M. (1983). The effect of site specific methylation on restriction endonuclease cleavage (Update). Nucl. Acid. Res. **11,** 169.

McVittie, A. Wildermuth & Hopwood, D.A. (1972) Fine structure and surface topography of endospores of *Thermoactinomyces vulgaris.* J. Gen. Microbiol. **71,** 367-381.

Morrison, D.A. (1981). Competence specific protein synthesis in *Streptococcus pneumoniae.* In Transformation, Edited by M. Polsinelli & G. Mazza. Cotzwold Press, Oxford. pp. 39-54.

Mukhopadhyay, A. & Sinha, U. (1979). Phosphatases of *Thermoactinomyces vulgaris.* Acta Bot. Indica **7,** 151-154.

Obaseiki-ebor, E.E. (1984). Rifampicin curing of plasmid in *Escherichia coli* K12 rifampicin resistant host. J. Pharm. Pharmacol. **36,** 467-470.

Ono, H., Hinterman, G., Crameri, R., Wallis, G. & Hutter, R. (1982). Reiterated DNA sequence in a mutant strain of *Streptomyces glaucescens* and cloning the sequence in *Escherichia coli.* Mol. Gen. Genet. **186,** 106-110.

Pirali, G., Somma, S., Lancini, G.C. & Sala, F. (1974). Inhibition of peptide chain initiation in *Escherichia coli* by thermorubin. Biochem. Biophys. Acta. **366,** 310-318.

Prasad, U. (1981). Studies on mutation and transformation in *Thermoactinomyces vulgaris.* Ph.D. Thesis, University of Delhi, Delhi, India.

Priest, F.G., Goodfellow, M. & Todd, C. (1988). A numerical classification of the genus *Bacillus.* J. Gen. Microbiol. **134,** 1847-1882.

Randazzo, R., Sciandrello, G., Carere, A., Bignami, M., Vellich, A. & Sermonti, G. (1976). Localized mutagenesis in *Streptomyces coelicolor* A3(2). Mut. Res. **36,** 291.

Riva, S., Fietta, A., Berti, M., Silvesti, L.G. & Romero, E. (1973). Relationship between curing of the F episome by rifampicin and by acridine orange in *Escherichia coli.* Antimicrob. Agents. Chemotherap. **3,** 456-462.

Rubin, S.J. & Rosenblum, E.D. (1971). Effects of ethidium bromide on growth and loss of the penicillinase plasmid of *Staphylococcus aureus.* J. Bacteriol. **108,** 1200-1204.

Ruttloff, H. (1978). Proteases aus thermophilen microorganismen. Wiss. Z. Univ. Halle. **27,** 123-133.

Schupp, T. & Divers, M. (1986). Protoplast preparation and regeneration in *Nocardia mediter-ranei.* FEMS Microbiol. Lett. **36,** 159-162.

Sharan, A.K. (1987). Genetics and expression of plasmids in *Thermoactinomyces vulgaris.* Ph.D. Thesis, Patna University, Patna, India.

Sharan, A.K. & Sinha, U. (1989). Extrachromosomal determinants for kanamycin resistance in *Thermoactinomyces vulgaris.* In Perspective in Cytology and Genetics, Edited by G.K. Manna & U. Sinha, Vol **6**, 85-93.

Sharma, R.P. (1970). Combined effect of physical and chemical mutagens on mutation frequency in *Aspergillus nidulans.* Indian J. Genet. Pl. Breed. **30**, 404.

Singh, A.N. (1988). Mutation and transformation studies in *Thermoactinomyces vulgaris.* Ph.D. Thesis, Patna University, Patna, India.

Singh, A.N. & Sinha, U. (1991). Factors affecting competence in *Thermoactinomyces vulgaris* strain 1261, a thermophilic actinomycete. Ind. J. Exp. Biol. **29**, 249-251.

Singh, V.P. & Sinha, U. (1982a). Ca^{2+} dependence and metabolic status of an obligate thermophile, *Thermoactinomyces vulgaris* under shake culture conditions. Experientia **38**, 670-671.

Singh, V.P. & Sinha, U. (1982b). Thermostability and turnover of phosphatases in the obligate thermophile *Thermoactinomyces vulgaris.* Ind. J. Exp. Biol. **20**, 26-30.

Singh, V.P. & Sinha, U. (1984). Calcium dependent growth and metabolism of *Thermoactinomyces vulgaris.* In Recent Trends in Botanical Researches, Edited by R.P. Sinha. Spectrum Publishing House, Patna & Delhi. pp. 229.

Sinha, U. & Singh, V.P. (1980). Phosphate utilization and constitutive synthesis of phosphatases in *Thermoactinomyces vulgaris.* Tsiklinsky. J. Biochem. **190**, 457-460.

Sinha, U. & Prasad, U. (nee Gulati) (1980). Calcium stimulate transformation in *Thermoactinomyces vulgaris.* J. Cyto. Genet. **15**, 113-115.

Sinha, U. & Prasad, U. (nee Gulati) (1981). *Thermoactinomyces vulgaris* - A system refractile to physical and chemical mutagens. In Perspective in Cytology and Genetics, Edited by G.K. Manna & U. Sinha, Vol 3, pp. 184-187.

Sinha, U. & Prasad, U. (nee Gulati) (1984). Studies on transformation in *Thermoactinomyces vulgaris.* In Perspectives in Cytology and Genetics, Edited by G.K. Manna & U. Sinha, Vol **4**, pp. 185-190.

Sinha, U., Prasad, U., Sinha, S. & Sharan, A.K. (1985). Possible plasmid involvement in antibiotic resistance in *Thermoactinomyces vulgaris.* In Trends in Molecular Genetics, Edited by U. Sinha & W. Klingmúuller. Spectrúm Publishing House, Patna & Delhi. pp. 103-120.

Sinha, U. & Singh, A.N. (1989). Genetic manipulation of *Thermoactinomyces vulgaris.* In Plant Sciences Research in India, Edited by M.L. Trivedi, B.S. Gill & S.S. Saini. Today and Tomorrow Printers & Publishers, New Delhi, India, pp. 383-389.

Stewart, G.J. & Carlson, A.C. (1986). The biology of natural transformation. Ann. Rev. Microbiol. **40**, 211-235.

Stonesifer, J. & Baltz., R.H. (1982). Genetic control of spontaneous and induced mutagenesis in *Streptomyces fradiae*. In 13th. Int. Congr. Microbiology (Abstract), Boston, U.S.A., The American Society for Microbiology, Washington D.C. pp. 148.

Thompson, C.J., Ward, J.M. & Hopwood, D.A. (1980). DNA cloning in Streptomyces, resistant gene from antibiotic producing species. Nature (London) **286**, 525-527.

Tomoeda, M., Inuzuka, M., Kubo, N. & Nakamura, S. (1968). Effective elimination of drug resistance and sex factors in *Escherichia coli* by sodium dodecyl sulphate. J. Bacteriol. **95**, 1078-1089.

Tsiklinsky, P. (1899). Sur les mucedinces thermophiles. Ann. Inst. Pasteur (Paris) **13**, 1078-1089.

Zimmerman, W., Rosselet, A. & Knusel, F. (1971). Effect of rifampicin and selected rifampicin derivatives on staphylococcal ß-lactamase plasmids. Annals. NYA Sci. **182**, 329-341.

8

Para-fluorophenylalanine resistance in *Aspergillus nidulans*: mode of action of antimetabolite

Bhupendra N. Tiwary[1] and Sheela Srivastava[2]
[1]*Microbial & Molecular Genetics Lab, Department of Botany*
Patna University, Patna 800 005 (India)
[2]*Department of Genetics, University of Delhi*
South Campus, Benito Juarez Road, New Delhi 110 021 (India)

ABSTRACT

In *Aspergillus nidulans* tyrosine is synthesized by two pathways, i.e., by the transamination of p-hydroxy-phenylpyruvic acid (known in microorganisms and plants) and by the hydroxylation of phenylalanine (known in animals). p-Fluorophenylalanine (FPA) an analogue of amino acid phenylalanine, is a potent growth inhibitor in *A. nidulans*, and this action is based on its incorporation into proteins. A large number of FPA-resistant mutants isolated and studied by us and other workers have led to the identification of 24 genetically unlinked loci in this fungus. Many of the amino acid uptake mutants show interesting interaction amongst them. A model has been proposed to explain this interaction on the basis that in the recombinants, the synthesis of an active permease through complementary polypeptides renders the double mutants sensitive to FPA. Further studies have established that the transport of amino acids is regulated by a more intricate general control circuit. An amino-acid binding protein has also been identified. Phe-tRNA synthetase has been purified and partially characterized. A large number of other FPA-resistant mutants are available where the mechanism is not very well understood.

INTRODUCTION

Ever since the discovery of parasexuality (Pontecorvo et al., 1953), the ascomycetous fungus *Aspergillus nidulans* has been the favourite choice of geneticists and biochemists. One of the main reasons for this fungus to attract is the fact that it represents a class of eukaryotic microorganisms which are more evolved and complex in their molecular organisation than the lowly bacteria (Clutterbuck, 1974; Smith & Pateman, 1977). In addition, *A. nidulans* excels other microbial systems because of its inherent attributes like (a) availability of the unicellular haploid as well as diploid conidia which facilitate the study of the allelic relationship as well as the genotoxic effects of various chemicals, (b) availability of detailed genetic map (Clutterbuck, 1981), coupled with stocks of various strains with different markers thus providing a wide scope for obtaining stable heterozygous diploids of various strains of various combinations, (c) existence of parasexuality in addition to normal sexual cycle thus opening-up vistas for novel recombination, particularly mitotic recombination, and (d) well defined methods for inducing haploids from artificially synthesized heterozygous diploids (McCully & Forbes, 1965; Singh & Sinha, 1976). These features coupled with various interesting findings have made this fungus a model genetic test system for analysing the growth pattern, differentiation, gene organisation and regulation, and biochemical events occurring in this habitually haploid fungus (Sinha & Jha, 1985).

Although rapid advances made during last two decades in the field of molecular biology of *A. nidulans* have led to a much better understanding of the intricate metabolic pathways operating in eukaryotes, the biochemical genetics of this fungus was initiated only in early fifties (Pontecorvo *et al.*, 1953; Pontecorvo & Käfer, 1958, also reviewed by Fincham *et al.*, 1979). In this review, we mainly intend to summarise the detailed growth studies, genetics and biochemistry of aromatic amino acid biosynthesis and resistance to para-fluorophenylalanine, a toxic analogue of the aromatic amino acid, phenylalanine.

RESULTS

[A] Aromatic amino acid biosynthesis in *A. nidulans*

Prior to the reports made by Sinha (1967a,b), there was no report on the analysis of the pathways leading to the synthesis of aromatic amino acids in *A. nidulans*. In contrast to microorganisms and plants which are able to synthesize aromatic amino acids from simpler compounds, animals require dietary phenylalanine and tryptophan but can synthesize tyrosine by hydroxylation of phenylalanine (Kaufman, 1963). In most of the organisms, in general, there is only one pathway for tyrosine synthesis, i.e., either by the transamination of p-hydroxyphenylpyruvic acid (known in microorganisms and plants) or by the hydroxylation of phenylalanine (known in animals). The observations made in *A. nidulans* suggest that both transamination and hydroxylation pathways for tyrosine synthesis occurs in this fungus (Fig. 8.1) and that a metabolic block in the transamination pathway results in a partial tyrosine-requirement which is invariably associated with resistance to FPA due to overproduction of phenylalanine (Sinha, 1967a,b). This pioneer work subsequently led to detailed studies on the growth and inhibition by several antimetabolites in *A. nidulans* (Sinha 1970a,b, 1972, Verma, 1973; Srivastava & Sinha, 1975; Chattoo & Sinha, 1974; Sinha & Chattoo, 1977; Singh *et al.*, 1977; Saxena & Sinha, 1977; 1978a,b; Singh & Sinha, 1979a,b; Tiwary & Sinha, 1985a,b, 1989; Tiwary *et al.*, 1987a,b; Singh *et al.*, 1990; Singh & Tiwary, 1992a,b).

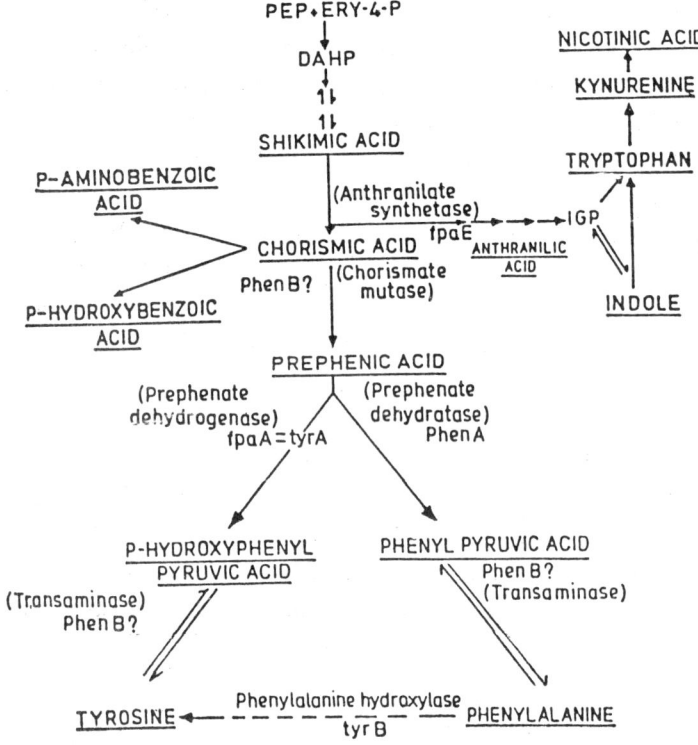

Fig. 8.1 Pathways of aromatic amino acid biosynthesis in *Aspergillus nidulans*

[B] Growth inhibition by amino acid analogues

A number of structural analogues of essential metabolites have been shown to be of immense importance in unravelling of the structure-function relationship of the genetic machinery operative at the molecular level (van Tuyl, 1977; Wheatley, 1978). The history of the use of antimetabolites in microbial systems dates back to 1940 when D.D. Woods made his first observation that sulphonilamides exert their inhibitory effects on the growth of bacteria because of their structural analogy to para-aminobenzoic acid, a common vitamin. Since this report, a variety of analogues of essential metabolites such as amino acids, nitrogenous bases, and vitamins have been found to be inhibitory to the growth of a wide range of microorganisms (Richmond, 1962; Umbarger, 1971; Verma & Sinha, 1973; van Tuyl, 1977).

A perusal of the literature indicates that not all amino acid analogues are inhibitory to the growth of microorganisms. Amongst several structural analogues of amino acids used only the analogues of two amino acids, namely phenylalanine (aromatic) and methionine (sulphur-containing) have contributed to our understanding of metabolite synthesis and regulation to a greater extent (Wheatley, 1978).

Phenylalanine contributes on an average about 3-4 percent of the total **residues in proteins** and its clustering and interaction with other aromatic groups produces hydrophobic areas in the polypeptide chain, generating ultimately alpha-helical chains, beta-pleated sheets and association between various subunits of proteins. A substitution of the structural analogue of phenylalanine may lead to alteration in protein configuration without necessarily destroying its essential function. FPA, although is the most extensively exploited analogue of aromatic amino acid, has long been in dispute as to whether it is an analogue of phenylalanine or tyrosine (Umbarger, 1971). It resembles both of these amino acids structurally except the para position of the aromatic ring (Fig. 8.2). Whereas phenylalanine has one H-atom at para-position, tyrosine has a OH-group at this position. Other biophysical characteristics like C-H and C-F bonds and electronic properties of H and F atoms clearly suggest that FPA could only be the structural analogue of phenylalanine and not of any other amino acid (Richmond, 1966; Sinha, 1967a). The finding of Arnstein and Richmond (1964) that the utilization of FPA for protein synthesis is only at the expense of phenylalanine also supports this contention.

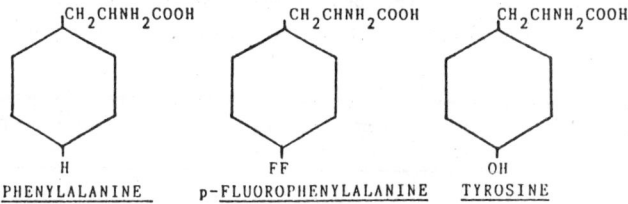

Fig. 8.2 Structural analogy between phenylalanine, p-fluoro-phenylalanine (FPA) and tyrosine

Methionine, on the other hand, is of significance because it is used as the first amino acid in most of the developing polypeptide chains. Ethionine, a structural analogue of this amino acid, has also received considerable attention because it not only replaces methionine in the polypeptide chain but also causes ethylation of other macromolecules especially DNA, thus leading to mutation, liver injury, and even carcinogenesis (Wheatley, 1978).

Auxanographic test with a number of amino acid analogues - (2-thienylalanine, 5-methyltryptophan, norvaline, homoserine, canavanine, ethionine and FPA) have revealed that a wild type strain of *A. nidulans* is quite resistant to most of the analogues except to FPA and ethionine. FPA has been shown to be the most potent inhibitor of growth in this fungus, while the effect of ethionine is much less pronounced (Verma and Sinha, 1973). Apart from *A. nidulans*, FPA is also toxic to the growth of *Escherichia coli* (Cohen and Munier, 1959), *Salmonella typhimurium* (Kerridge, 1960), *Neurospora crassa* (Beeman and Smith, 1969), *Schizophyllum commune* (Hannan, 1972), *Saccharomyces cerevisiae* (Stromnaes, 1968), *Dictyostelium discoideum* (Sinha and Ashworth, 1978), and *Anacystis nidulans* (Fiske and Kane, 1984; Kane and Fiske, 1985). The tryptophan analogues (4-methyltryptophan and 5-methyltryptophan) have however, been found to inhibit the growth of *S. typhimurium* (Ames, 1964) and *Bacillus subtilis* (Jensen, 1969). Among the tyrosine analogues, D-tyrosine has been shown to be potent growth inhibitor of *B. subtilis* (Champney and Jensen, 1969) and *N. crassa* (Horowitz *et al.*, 1970a,b).

Detailed growth kinetic studies carried out in *A. nidulans* in the presence of FPA indicates that there is no change in the growth pattern of this fungus even on addition of the analogue (Verma and Sinha, 1973; Sinha and Tiwary, 1986; Singh and Tiwary, 1992b). Although the analogue has been shown to be incorporated into total proteins (Sinha and Verma, 1972) and more specifically in phosphatases (Singh and Sinha, 1979b), it fails to transform the exponential growth to a linear one suggesting thereby that it is inhibitory not only because it gets incorporated into proteins but also because it affects some vital biosynthetic steps. Under such circumstances where growth inhibitory effects of the analogue are overcome by the natural amino acid, it is most likely that the antimetabolite affects some essential steps in the biosynthetic pathway of natural amino acid, the scarcity of which leads to growth inhibition (Tiwary and Sinha, 1985b). A severe growth inhibitory effect of FPA on a phenylalanine-auxotrophic (phenA3) strain of *A. nidulans* supports this contention (Verma, 1973; Tiwary *et al.*, 1987b; Singh and Tiwary, 1992b) (see Table 8.1).

Table 8.1 Growth of a wild type (*biA1*; *phenA3*) strain of *A. nidulans* in the presence of minimum supporting concentrations (MSC) of phenylalanine and varying concentrations of FPA

MSC of Phenylalanine (mM)	Concentrations of FPA (mM)	Total protein yield (mg)[*]
0.136	0.000	7.40
	0.038	5.15
	0.077	3.55
	0.115	1.70
	0.154	0.18
	0.192	0.00
	0.230	0.00

[*]Total protein yield/1×10^8 conidia grown in 50ml of MM

Although growth inhibition by amino acid analogues has been investigated in lower as well as higher eukaryotic systems, the interpretation and correlation of results are still more difficult due to inherent complexity of the systems. FPA has also been tested for its effects on fungi (Käfer and Upshall, 1973; Morris and Oakley, 1979), higher plants (Balakrishnan and Sinha, 1976; Sinha and Bhojwani, 1976; Noda, 1982; Banks *et al.*, 1982) and animal cells (Sisken & Wilkes, 1967).

Most of the amino acid analogues exert their inhibitory effect on the growth by interfering with the protein synthetic machinery or by affecting the proteins synthesized in their presence. In other words, they either stop or reduce the normal protein synthesis or alter the properties of proteins synthesized in their presence. Cumulative experience of various workers with a variety of test systems reveals that these effects are caused as a consequence of competition between the antimetabolite and the natural amino acid at various steps of metabolism (Sinha, 1970a).

[C] Mechanisms of action of amino acid analogues

The biochemical effects exerted by an amino acid analogue on the cellular system has been a subject of investigation in various microorganisms. The manner in which these toxic analogues become involved in protein synthesis and eventually inhibit the growth has been explained by studying the uptake of analogues and their consequent incorporation.

(a) Uptake of analogues in cellular systems

One of the vital sites at which analogues might interfere with normal amino acids is at the cell membrane during uptake from the external medium. The receptor sites of specific membrane permeases of amino acids can be blocked by varying external concentrations of analogues. Several hypotheses have been proposed to explain the mechanisms of uptake and the subsequent pathways of an amino acid in different organisms. In *E. coli*, Britten & McClure (1962) have suggested that an amino acid must pass through the intracellular pool before being incorporated into proteins. On the other hand, van Venrooij *et al.* (1974) is of the opinion that HeLa cells probably incorporate amino acids more directly from the external medium and do not preferentially incorporate them from an intracellular pool. In general, bacteria can concentrate amino acids from the external medium, especially when the medium contains very low levels of amino acids or their analogues, for example, less than 10^{-4}M of FPA.

Studies on the assimilation of ^{14}C-FPA into the amino acid pool in *E. coli* and the effects of phenylalanine on this assimilation suggest that in the presence of phenylalanine, uptake of antimetabolite into the pool is extremely low until the natural metabolite is exhausted (Cowie, 1959). A quantitative measurement of the uptake of FPA and phenylalanine reveals that p-fluorophenylalanine rapidly enters the cell but only when its concentration is at least four times the concentration of phenylalanine. This, in turn, suggests that the "concentrating system" has more affinity for the natural amino acid than for its analogue.

Investigations on the uptake of natural amino acids and their analogues in eukaryotic systems suggest that most of the analogues are transported into the cell by the same uptake system(s) as their natural counterparts (Guroff *et al.*, 1969; Benko *et al.*, 1969; Sinha, 1969, 1970a; Kinghorn & Pateman, 1975; Srivastava & Sinha, 1981; Tiwary, 1989a; Tiwary & Sinha, 1985a, 1989 & Tiwary *et al.*, 1987a). The competition between normal amino acid and the analogue is usually so much drastic that in some cases the uptake of normal amino acid is completely inhibited. An approximately 90 percent inhibition of phenylalanine uptake by FPA in *A. nidulans* which allows entry of enough analogue into the cell supports this contention (Sinha, 1969; Singh & Tiwary, 1992a,b).

A detailed study on competition between leucine and phenylalanine in *A. nidulans* reveals that a phenylalanine-auxotroph (*phen*A3) is competitively inhibited by leucine but vise-versa is not true (Sinha, 1970a), indicating there by that leucine competes with phenylalanine. On the basis of interaction between FPA-resistance and amino acid-requirements, Sinha (1970a) has postulated that there are at least two sites in *A. nidulans* for which leucine might compete with phenylalanine, one of these being amino acid permease. The other site for leucine-phenylalanine -FPA interaction is yet to be identified. Both of these sites are, perhaps, common for leucine-phenylalanine and phenylalanine-FPA interaction. Boyaval *et al.* (1984), on the other hand have studied the specificity of phenylalanine and tyrosine carriers in actively metabolising cells of *Brevibacterium linens* and have reported that the fluorinated analogues of these amino acids are potent competitive inhibitors for these carriers.

(b) Incorporation of analogues into proteins

The endogenous synthesis of amino acids is under two main types of control: feed-back inhibition and repression. Both of these controls are brought about only when the amino acid is present in excess. Amino acid analogues, because of their structural similarities with the natural amino acids mimic the latter and cause either false feed-back inhibition or repression of the biosynthetic enzymes. It is, nevertheless, established that all amino acids as well as their analogues have to undergo an activation steps prior to incorporation into proteins, although activation does not necessarily ensure their incorporation. The steps involved are, first, activation to the S-adenosyl derivatives, followed by charging of the transfer RNA by specific aminoacyl-tRNA synthetases (Fig. 8.3). A great deal of information is available suggesting that various analogues are activated by purified or even crude form of aminoacyl-tRNA synthetases (Dunn & Leach, 1967; Tiwary *et al.*, 1987b, Singh *et al.*, 1990; Singh & Tiwary, 1992b).

Enzyme + ATP ⇌ E.ATP

E.ATP + Phe ⇌ E.ATP.Phe

E.ATP.Phe ⇌ E.AMP ~ Phe + ppi

E.AMP + Phe + tRNA ⇌ E + AMP + Phe.tRNA

Fig. 8.3 Steps involved in the activation of phenylalanine and aminoacylation of the Phe-tRNA by phenylalanyl-tRNA synthetase

With regards to analogues, the question arises whether the same mechanism(s) of activation do operate. Though the majority of amino acid analogues can be activated by the enzyme active on natural amino acids (Richmond, 1962), it remains to be conclusively established whether this type of competition between amino acids and their analogues for activating enzyme can itself lead to growth inhibition.

The pioneer work of Arnstein and Richmond (1964) on phenylalanine/FPA activation ratio of 1:3 by enzymes extracted from rabbit reticulocytes suggests little selectivity against the activation of the analogue. This ratio has been reported to be equivalent to the ratio of phenylalanine and FPA incorporated into proteins. A significant variation in results has, however, been observed when the natural amino acid and the analogue were simultaneously used in the same reaction for competing with the same enzyme. The activation of FPA was observed to be significantly reduced in the presence of as low as 0.2M phenylalanine. On the other hand, phenylalanine-activation was reduced to 50 percent only when FPA was present in 10-fold excess. In *E. coli* also, it has been demonstrated that there is natural preference for phenylalanine over FPA right from the time of transport (Cowie, 1959).

Recently, Singh and Tiwary (1992b) have undertaken a detailed competition studies with the help of a *phen*A3 auxotroph of *A. nidulans* and report that in the presence of equimolar concentrations of phenylalanine and its analogue, the natural amino acid is taken-up rapidly. The complete inhibitory effect of FPA and its natural counterpart by activating enzyme isolated from the *phen*A3 auxotroph has also been observed (Singh *et al.*, 1990; Singh & Tiwary, 1992b).

(c) Effect of analogues on structural proteins

Since the first report of incorporation of an amino acid analogue (ethionine) into total proteins by Levine and Tarver (1951), extensive investigations have been carried out to study the effects of amino acid analogues on various types of structural proteins (Richmond, 1962; Sinha 1970a; Sinha & Verma, 1972; Balakrishnan & Sinha 1976; Morris & Oakley 1979; Boyaval *et al.*, 1984). These reports mention that a number of analogues, e.g. norleucine, 2-thienylalanine, FPA, 7-azatryptazan, and selenomethionine can get incorporated into structural proteins in place of their natural counterparts. FPA has been found to get incorporated into total proteins of *N. crassa* (Brooks *et al.*, 1972), *A. nidulans* (Verma & Sinha, 1971, 1972; Morris & Oakley, 1979) and other eukaryotic microbes. The incorporation of FPA in place of phenylalanine is generally random, there being no report of any preferential effect for a particular site in a protein. Furthermore, there is no change in the composition of proteins except amino acid-amino acid analogue transposition. The latter nevertheless leads to the formation of defective proteins, ultimately leading to growth inhibition (Sinha & Verma, 1972; Sinha & Tiwary, 1986).

(d) Effect of analogues on enzymatic proteins

Amino acid analogue-incorporation into enzymatic proteins is likely to result either into (a) synthesis of a normal enzyme molecule, (b) production of an enzyme molecule with altered or reduced activity, or (c) the derepression or repression of an enzyme or a group of enzymes. The third possibility, however, does not result due to incorporation of analogue into enzyme molecule per se but due to its involvement in some of the regulatory molecules, like repressors or effectors (Singh & Sinha, 1979b).

Instance of incorporation of FPA in α-amylase of *B. subtilis* and exopenicillinase of *E. coli* leading to reduction in their specific activities to the extent of 70-80 percent have been demonstrated by Yoshida (1960) and Richmond (1960), respectively. Non-recovery of the altered enzyme molecule has, however, made the result difficult to be interpreted. Had it been possible to separate the impaired enzyme molecules in the two experiments, it would have been a direct demonstration of the effect of FPA on these enzymes.

Reports of inhibition of activities of enzymes due to incorporation of analogues in eukaryotes are few and far in between. Singh and Sinha (1979b) have studied the action of FPA on synthesis and activity of phosphatase isoenzymes of *A. nidulans*. Characterisation of phosphatase enzyme reveals five isoenzymes (Table 8.2). These workers have also demonstrated that FPA-incorporation causes inactivation of alkaline phosphatase isoenzyme III of *A. nidulans*. Since the sequence of amino acids at active site is usually highly conserved and remains the same for different isoenzymes of a single enzyme, the inactivation of isoenzyme III has been suggested to be due to incorporation of FPA at a position other than the active site. On the basis of observations that FPA does not derepress any phosphatase isoenzyme and that constitutive synthesis of enzyme remains unaffected even in the presence of the analogue, Singh and Sinha (1979b) further suggested that FPA need not necessarily affects all the proteins synthesized in its presence. On the other hand, FPA containing tyrosinases of *N. crassa* has been found to be normal in its specific activity and thermostability (Horowitz *et al.*, 1970a). These workers have also reported derepression of tyrosinase activity by some protein synthesis inhibitors (FPA, D-tyrosine, ethionine) by stifling the synthesis of an active repressor (Horowitz *et al.*, 1970b). A similar derepression of tyrosinase activity by FPA has also been found in *A. nidulans* (Sinha & Srivastava, 1974-75).

Table 8.2 Properties of five isoenzymes of phosphatase enzyme of A. nidulans observed on polyacrylamide discgel electrophoresis

Isoenzyme number	Nature	Molecular weight[*]
I	Acidic, derepressible	[#]
II	Acidic, constitutive	[#]
III	Alkaline, derepressible	150000
IV	Alkaline, constitutive	150000
V	Acidic, derepressible	100000

[*] (After Dorn, 1965, 1967, 1968)
[#] Not estimated

(d) Resistance to amino acid analogues

The development of resistance to inhibitory action of antimetabolites occurs spontaneously and frequently in a sensitive microbial population. Generalizations derived from biochemical studies in a variety of biological systems (Georgopoulos, 1977; van Tuyl, 1977; Sinha & Tiwary, 1986; Tiwary & Singh, 1992) suggest that the sensitivity of living cells to exogenously supplied toxicants depends mainly upon:

i.   The permeability of cell membrane to the toxic chemical, *ii.* The availability of compounds which may activate, inactivate or antagonise the effect of toxic compound,

iii. The accessibility and nature of target sites, and

iv. The existence of alternate pathways bypassing the *inhibited* reactions.

Although a large number of mutants resistant to analogues of aromatic amino acids have been reported for a variety of microorganisms, the mechanism of growth inhibition and eventual development of resistance is clearly understood only for a few. Reports, mainly confined to bacterial and only a few for eukaryotic systems, reveal that the action of analogues is largely concerned either with the protein synthetic machinery or the regulation of biosynthetic pathways. It seems most likely that there exists a direct correlation between mechanisms of resistance of the mutants and the site of action of amino acid analogues. It seems that bacterial systems, in general, show predominance of resistance mechanisms involving derepression or feedback-inhibition. So far as eukaryotic systems are concerned, perhaps there is no report of derepressed or feedback-insensitive mutants. A thorough understanding of the mechanisms of resistance in eukaryotes comes from the isolation and genetic as well as biochemical characterization of FPA-resistant mutants of *A. nidulans*.

(e) Genetics of FPA-resistant mutants of *A. nidulans*

One of the most extensively studied aspects in *A. nidulans* is the isolation and characterization of mutants resistant to FPA. Concerted efforts made during last three decades by various workers have led to the identification of 24 genetically unlinked loci in this fungus (Table 8.3), mutations at which lead to resistance to FPA (Morpurgo, 1962; Warr & Roper, 1965; Sinha 1967a,b, 1969, 1970a; Verma, 1973; Srivastava & Sinha, 1975; Kinghorn & Pateman, 1975; Pitrowska *et al.*, 1976; Singh & Sinha, 1979a; Tiwary & Sinha, 1985a,b, 1986, 1987, 1989; Tiwary *et al.*, 1987a,b; Tiwary, 1989a,b; Singh *et al.*, 1990; Singh & Tiwary, 1992a). Srivastava & Sinha (1975) have calculated the probable number of mutable loci for FPA-resistance in this fungus to be 28. It is, therefore, apparent that more than 80 percent of the expected FPA-resistant loci have been identified. These loci are distributed over six out of eight linkage groups of *A. nidulans*. So far no *fpa* locus has been identified on linkage groups IV and VII. Nevertheless, it is the only report amongst eukaryotes where such a large number of mutants for one particular analogue of an amino acid have been recovered.

It is significant to note that different selection techniques (media composition) have been employed by various workers for the selection of FPA-resistant mutant in *A. nidulans*. Interestingly, each technique preferentially selects for one class of mutation and provides ample idea about the nature of mutants. For example, use of only one analogue (FPA) in the presence of glucose as the carbon source (MM) leads to the isolation of mutants having overproduction of the natural metabolite associated with partial requirement for other amino acids (Sinha,

Table 8.3 FPA-Resistant loci in *Aspergillus nidulans*

S.No	Locus symbol	Selected on	Dominant/ Recissive	Linkage group	Reference
1.	*fpaA*	CM+FPA	Recessive	I	Morpurgo, 1962; Sinha, 1967 a,b Sinha, 1967a, 1970a
2.	*fpaB*	CM+FPA	Recessive	I	Sinha, 1967a, 1969
3.	*fpaD*	CM+FPA	Dominant	VIII	Sinha, 1967a, b
4.	*fpaE*	CM+FPA	Recessive	II	Srivastava & Sinha, 1975
5.	*fpaF*	CM+FPA+	Recessive	VI	Srivastava & Sinha, 1975
6.	*fpaG*	ethionine	Recessive	V	Srivastava & Sinha, 1975
7.	*fpaH*	CM+FPA+ ethionine	Semi-dominant	VI	Srivastava & Sinha, 1975
8.	*fpaI*	CM+FPA+ ethionine	Dominant	I	Srivastava & Sinha, 1975
9.	*fpaJ*	CM+FPA+ ethionine	Dominant	VI	Srivastava & Sinha, 1975
10	*fpaK*	CM+FPA+ ethionine	Dominant	VIII	Kinghorn & Pateman, 1975
11	*aauC*	CM+FPA+ ethionine	Dominant	II	Kinghorn & Pateman, 1975
12	*aauD*	Glutamate as carbon & N source	Dominant	VIII	Pitrowska et al., 1976
13	*nap3*	CM+FPA + seleno-methionine	Dominant	*	Singh & Sinha, 1979a
14	*fpaL*	Acetate+ FPA+ ethionine	Recessive	VI	Singh & Sinha, 1979a
15	*fpaM*	Acetate+ FPA+ ethionine	Recessive	I	Singh & Sinha, 1979a
16	*fpaN*	Acetate+ FPA+ ethionine	Recessive	VIII	Tiwary & Sinha, 1985a, 1989
17	*fpaO*	Acetate+ FPA+ ethionine	Recessive	I	Tiwary et al., 1987a
18	*fpaP*	Acetate+ FPA+ ethionine	Recessive	II	Tiwary et al., 1987a
19	fpaQ	Acetate+ FPA+ ethionine	Recessive	II	
20	*fpaR*	Acetate+ FPA+ ethionine	Dominant	I	Tiwary et al., 1987a
21	*fpaS*	Acetate+ FPA+ ethionine	Dominant	II	Tiwary et al., 1987a
22	*fpaT*	Arginine+ FPA+ ethionine	Dominant	I	Tiwary et al., 1987a
23	*fpaU*	MM+FPA	Recessive	V	Tiwary et al., 1987a
24	*fpaV*	Aspartate+F PA	Recessive	III	Singh et al., 1990 Singh & Tiwary, 1992b

1967a,b). The presence of two amino acid analogues (FPA and ethionine) either in normal (glucose) or reduced carbon (acetate) or nitrogen (arginine) flow enables one to isolate preferentially mutants defective in specific and/or general amino acid permeases (Srivastava & Sinha, 1975; Tiwary & Sinha, 1985a, 1989; Tiwary *et al.*, 1987a). The rationale behind the use of acetate or arginine in the selection medium is based upon the finding that on a poor source of carbon and/or nitrogen the aromatic amino acid permease gets derepressed (Singh *et al.*, 1977), making the cells sensitive due to excessive entry of the analogue. Under such circumstances only those cells having a defect/deficiency in the permease(s) are likely to survive and eventually selected as resistant mutants. A third procedure involving a *phen*A3 auxotroph as the wild type has been found preferentially selecting FPA-resistant mutants defective in phenylalanyl-tRNA synthetase enzyme (Tiwary *et al.*, 1987b; Tiwary, 1989b; Singh *et al.*, 1990; Singh & Tiwary, 1992a).

The analogue-resistant mutants exhibit different degrees of resistance to FPA. Also, some of the mutants isolated appear to be dominant while others are recessive or semi-dominant (Verma & Sinha, 1971; Tiwary & Singh, 1992). In general, the extent to which a mutant shows resistance to FPA is assumed to be related to the severity of the metabolic block. The dominant mutants were initially considered to be regulatory in nature (Sinha, 1969), however, subsequent reports have discounted this hypothesis (Kinghorn & Pateman, 1975; Tiwary *et al.*, 1987a).

The ease with which FPA-resistant mutants of *A. nidulans* have been genetically characterized is mainly due to the availability of different colour markers, well marked auxotrophs and master strains having each of the eight linkage groups represented by at least one marker (McCully & Forbes, 1965), and techniques to develop heterozygous diploids and hybrid cleistothecia. In addition, techniques to induce haploids (Morpurgo, 1961; Singh & Sinha, 1976) have been of advantage in assigning genes to respective linkage groups through mitotic cycles.

(f) Biochemical analyses of FPA-resistant mutants

Although more than 80 percent of the total expected number of *fpa* loci have been identified and located on respective linkage groups by mitotic and meiotic analyses and characterized extensively by genetic interaction studies in *A. nidulans*, the biochemical characteristics have, however, been clearly understood only for a few. Unlike prokaryotes where resistance to analogue (FPA) is acquired mainly by producing inhibition-resistant or repression-resistant enzymes of the phenylalanine or tyrosine biosynthetic pathways (Gollub & Sprinson 1969; Im & Pittard 1971), development of resistance to FPA in *A. nidulans* has been found to be due to either of three mechanisms, namely over-production of phenylalanine, uptake deficiency/defect, and alteration in phenylalanyl-tRNA synthetase. It is pertinent to note that in one particular organism only one of these mechanisms has been established for amino acid analogue-resistance whereas in *A. nidulans* all the three mechanisms have been found to be operative (Tiwary & Singh, 1992).

(i) Overproduction of phenylalanine

The biochemical basis of FPA-resistance in *fpa*A and *fpa*E mutants of *A. nidulans* has been explained to be due to an over-production of phenylalanine (Sinha, 1967a,b; Verma & Sinha, 1971). Over-production by these two mutants is not because of insensitivity to feedback-inhibition or repression rather it is due to mutations leading to reduced or altered prephenate dehydrogenase (*fpa*A) and anthranilate synthetase (*fpa*E) in tyrosine and tryptophan biosynthetic pathways, respectively. The pioneering work of Sinha (1967a,b) on aromatic amino acid biosynthesis suggests that both transamination and hydroxylation pathways for tyrosine biosynthesis are operative in this fungus and that a metabolic block in the transamination pathway leads to a partial tyrosine-requirement and FPA-resistance (*fpa*A=*tyr*A) due to an overproduction of prephenic acid and ultimately of phenylalanine. Similarly, a block in the activity of anthranilate synthetase channelises the chorismic acid back to prephenic acid and ultimately to phenylalanine. The mutants of *fpa*E class do always reveal a complete requirement for tryptophan. On the basis of their biochemical analogy with those reported in mammals, Sinha and Verma (1971) have suggested that these mutants can be better termed as "phenylketonuric".

Isolation and characterization of antimetabolite-resistant mutants having an enhanced production of the natural metabolite like those reported in *A. nidulans* is of industrial importance. It is possible to manipulate the growth conditions especially the carbon and/or nitrogen sources to obtain significantly high levels of excretion of amino acids or other metabolites by these mutants.

(ii) Defect/deficiency in the amino acid uptake system(s)

Selection of mutants resistant to structural analogues of essential metabolites has been widely used in a variety of microorganisms in order to investigate the nature of defect and/or deficiencies in specific as well as general permeases. Such mutants have provided ample information regarding the genetic control and specificity of transport mechanisms (Anraku, 1980; Wolfinbarger, 1980). The ultimate aim of such studies is to make accurate location of the structural genes of permeases and to unravel their regulation (Sinha, 1969; Pall, 1969; Tiwary et al., 1987a; Tiwary, 1989a).

The first report on the isolation of an amino acid uptake mutant in *A. nidulans* was that of Sinha (1969) who identified a locus *fpa*D. The mutant showed reduced levels of uptake of ^{14}C - phenylalanine as well as ^{14}C-FPA in addition to other amino acids. Direct measurements of the uptake of ^{3}H-amino acids by the wild type and mutants show that a mutation at the locus *fpa*D distinctly retards the uptake of tyrosine, tryptophan, methionine, leucine, and aspartic acid and slows down the uptake of histidine, glycine, glutamic acid, and lysine. Cross resistance of *fpa*D mutants to analogues of other amino acids supports this contention. The results further indicate that phenylalanine enters the cell by only one system, the Km and Vmax of which is about 2×10^4 and 5.2×10^4 µmoles/mg dry weight/minute, respectively. On the basis of detailed genetic analysis in heterozygous diploids and heterokaryons, Sinha (1970b) further points out towards the possible regulatory role of *fpa*D locus.

Since the identification of *fpa*D locus, so far six more loci (Table 8.4), mutation at any one of which leads to FPA-resistance and defect or deficiency in the uptake of amino acids have been reported in *A. nidulans* (Srivastava & Sinha, 1975; Kinghorn & Pateman, 1975; Tiwary & Sinha, 1985a, 1989; Tiwary *et al.*, 1987a). Characterization of these mutant loci have yielded valuable information regarding the nature of amino acid permeases in this fungus. In general, it has been observed that the wild type strain produces an active permease which concentrates FPA

Table 8.4 Characteristics of amino acid uptake mutants of *A. nidulans*

Loci/ mutants	Defective/deficient for the uptake of amino acids	Dominant/ Recessive	Linkage group	References
*fpa*D	Aromatic	Dominant	VIII	Sinha, 1969
*fpa*K	Aromatic	Dominant	VIII	Srivastava & Sinha, 1975
*aau*C	Aromatic, acidic and neutral	Dominant	II	Kinghorn & Pateman, 1975
*aau*D	Aromatic, acidic and neutral	Dominant	VIII	Kinghorn & Pateman, 1975
*fpa*O	Acidic, neutral and basic	Recessive	I	Tiwary & Sinha, 1985a, 1989
*fpa*P	?	Recessive	II	Tiwary et al., 1987a
*fpa*Q (Q79)	Aromatic, acidic, neutral and basic	Semi-dominant	II	Tiwary et al., 1987a
*fpa*Q	Aromatic, acidic and neutral	Dominant	II	Tiwary et al., 1987a
nap3	Phenylalanine, methionine, serine and leucine	Recessive	**	Pitrowska et al., 1976

?   The *fpa*P locus has been considered to be involved in the regulation of permease since it utilizes all amino acids as the nitrogen source but interacts positively with uptake mutant *fpa*K yielding FPA-sensitive recombinants.

**   Not determined

in the mycelium. FPA-resistant mutants having defect in the transport of phenylalanine and some other amino acids lack this concentrating system for natural as well as their analogues (Sinha & Tiwary, 1986). The observation made by Sinha (1969, 1972) that recombinants fpaD11; phenA3 do not grow at all even in the presence of phenylalanine suggests only one uptake system for this amino acid which is not handled efficiently by any non-specific or general transport system, even if it exists in A. nidulans. However, subsequent reports of Srivastava and Sinha (1975), Kinghorn and Pateman (1975) and Tiwary *et al.* (1987a) on the identification of four more loci (*fpa*K69, *aau*C1, *aau*D1, *fpa*Q79/80) showing defect in the uptake of phenylalanine in addition to some other classes of amino acids rules out this hypothesis. Non-recovery of recombinants between analogue resistance and amino acid auxotrophy could well be due to other reasons, the possibility of which has been explained by subsequent workers (Tiwary *et al.*, 1987b; Singh *et al.*, 1990; Singh & Tiwary, 1992a).

Dominance of transport mutants

A perusal of FPA-resistant amino acid transport mutants in *A. nidulans* reveals that six out of seven uptake defective loci are either dominant or semi-dominant in heterozygous diploid. This is in contrast to those reported in bacterial and other fungal systems where such mutants have been characterized as recessive to their wild type alleles. Tiwary and Sinha (1985b) have suggested one plausible reason to explain by considering that the heterozygous diploid containing the wild type and mutant alleles synthesizes suboptimal levels (50 percent or even more) of active permeases which is able to transport only sub-inhibitory concentrations of FPA into the mycelium. As a result the diploid is not inhibited by the concentration of FPA normally used and the mutation appears dominant. In a situation where a mutation affects the permease partially, the level of enzyme synthesized in heterozygous diploid could be more than 50 percent, as a consequence of which the mutant would appear semi-dominant. In a homozygous sensitive diploid, the permease will be fully transcribed leading to an active transport of FPA and sensitivity to it. This also explains the differential behaviour of different mutants at the locus *fpa*Q. The isolate *fpa*Q80 appears to be more severely affected than the isolate *fpa*Q79 as a result of which the former behaves as dominant and the latter as a semi-dominant (Tiwary *et al.*, 1987a).

Interaction between *fpa* loci: concept of polymeric permease

Genetic interaction studies between mutants deficient in the regulation of particular steps in the metabolism provide an insight into the nature of such mutants. Such interactions between genes involved in the amino acid uptake have also been reported in N. crassa (Verma, 1973; Tiwary, 1989a). Theoretically, a cross involving two non-allelic, freely-recombining and non-interacting FPA-resistant loci should yield only 25 percent FPA sensitive progeny. The number of resistant progeny is expected to decrease if the loci concerned are linked, and increase (uptake 50 percent) if there is interaction leading to sensitivity of the double mutant class. Results obtained by crossing four fpa loci of *A. nidulans* (Table 8.5) indicate that amino acid uptake mutants interact among themselves leading to sensitivity of double mutant class (*fpa*D11, *fpa*K69; *fpa*D11, *fpa*Q79; *fpa*K69, *fpa*P77) to FPA. A similar pattern of utilization of amino acids as the sole source of nitrogen by sensitive wild-type and sensitive recombinants (double mutants) has also been observed (Tiwary, 1989a).

Earlier works provided evidence to suggest that different FPA-resistant loci (concerned with amino acid uptake) code for different structural proteins, the latter being shared by different amino acid permeases (Srivastava & Sinha, 1975; Kinghorn & Pateman, 1975). Subsequent

Table 8.5 Genetic Interaction between FPA-resistant locl (*fpaD*, *fpaQ*, *fpaK* and *fpaP*)
concerned with amino acid uptake in *A. nidulans*

Pairs of loci considered	Number of progeny	Segregation of markers			
		FPA-Sensitive	FPA-resistant	Yellow	Green
fpaD11-fpaK69	200	98	102	99	101
fpaD11-fpaQ79	355	169	186	184	171
fpaD11-fpaQ80	198	101	97	98	100
fpaK69-fpaP77	271	141	130	139	132

isolation of some more fpa loci involved in the uptake of aromatic amino acid in *A. nidulans* clearly demonstrates that loci *fpaD*, *fpaK*, *fpaP* and *fpaQ* in addition to *aauC* and *aauD* transcribe and translate different but intimately associated polypeptides. The product of these genes is needed for the normal transport function, a mutation in any one of which leads to a defect in the uptake of analogue as well as natural amino acid, eventually leading to analogue resistance. Complementary polypeptides in recombinants lead to the synthesis of an active permease rendering the double mutants sensitive to FPA. These observations have led to suggest that aromatic permease at least in *A. nidulans* is a polymeric protein (Tiwary *et al.*, 1987a). A model first proposed (Verma, 1973), explaining the behaviour of uptake genes in combination, has been further elaborated by Tiwary (1989a) (Fig. 8.4a, b). However, involvement of common carriers or overlapping control switches as reported in *N. crassa* (DeBusk & DeBusk, 1980) can also not be ruled out.

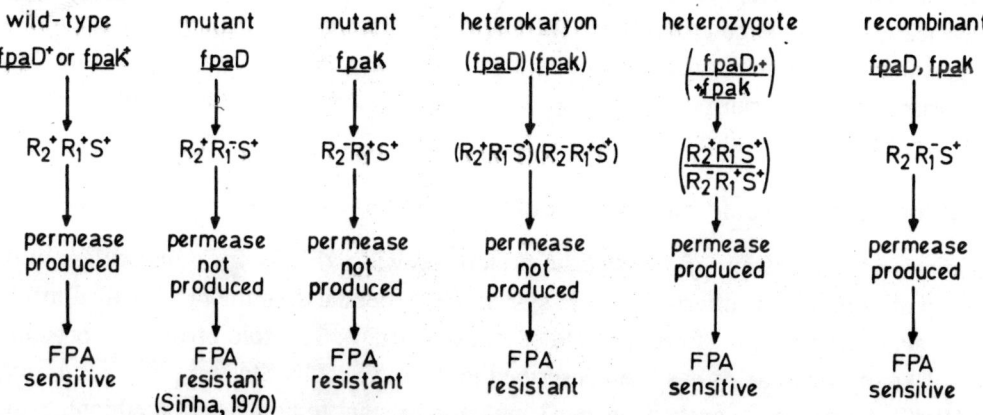

Fig. 8.4a The interaction between two permease-defective mutant loci *fpaD* and *fpaK*. While *fpaD* is a mutation in the first regulatory gene (R_1) Sinha, 1970 b) *fpaK* appears to have lesion in the second order regulatory gene(R_2). Both, therefore, do not produce the permease and are resistant to FPA. The differencial effects of these mutations in combination producedifferent results. In heterokaryon, gene products are not able to complement but in diploids they freely interact thus leading to FPA resistance and senitivity respectively. In double recombinants, the mutated R_2 and R_1 cross see each other resulting in FPA-sensitive phenotype.

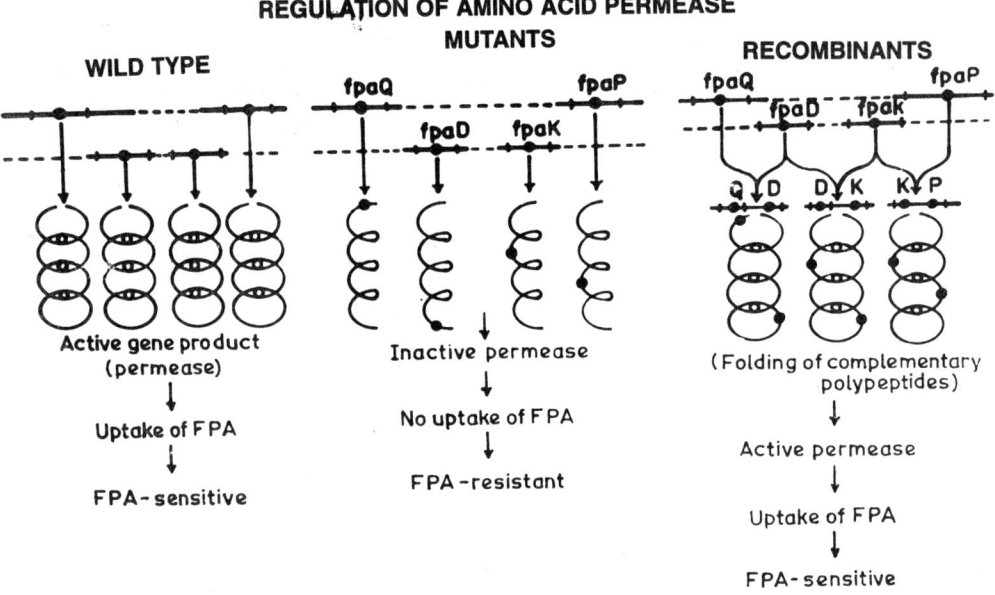

Fig. 8.4b Model explaining the regulation of amino acid permease in *Aspergillus nidulans*

The most significant aspect of the genetic interaction studies carried out with uptake mutants is the interaction between *fpa*P (showing a normal pattern of amino acid utilization) and *fpa*K (defective in aromatic amino acid uptake). Tiwary (1989a) has suggested that *fpa*P locus might be controlling the synthesis of a transcription regulator protein which has to act on a number of different promoters, or an activity regulator which interacts with a number of different uptake complexes. Nevertheless, scattering of different genes concerned with the synthesis of amino acid permease(s) on linkage group II and VIII of *A. nidulans* (Fig. 8.5) provides ample evidence to suggest that an "operon" organisation, as in case of bacteria, is not involved in the synthesis of permeases and that transport of amino acids in this fungus is regulated by more intricate genetic control circuits (Tiwary, 1989a).

Purification of amino acid binding protein

Isolation and purification of an extrinsic protein showing affinity for a particular amino acid or a group of amino acids offers a suitable system for understanding the biochemical mechanism of amino acid transport (Wolfinbarger, 1980; Anraku, 1980). The role of binding protein in the active uptake of amino acids has been described in *A. nidulans* by Stepien (1976) and Singh and Sinha (1980). Biochemical analysis of *nap*3 mutants resistant to both selenomethionine and FPA reveals that one protein (of 15.5 kd) which is invariably present in the wild-type and binds specifically to methionine is absent in the mutant (Stepien, 1976). Similarly, partial purification of phenylalanine-binding protein from the wild type and *fpa*D11 mutant using Triton X-100 extraction and affinity chromatography on L-phenylalanine CH-sepharose (Singh & Sinha,

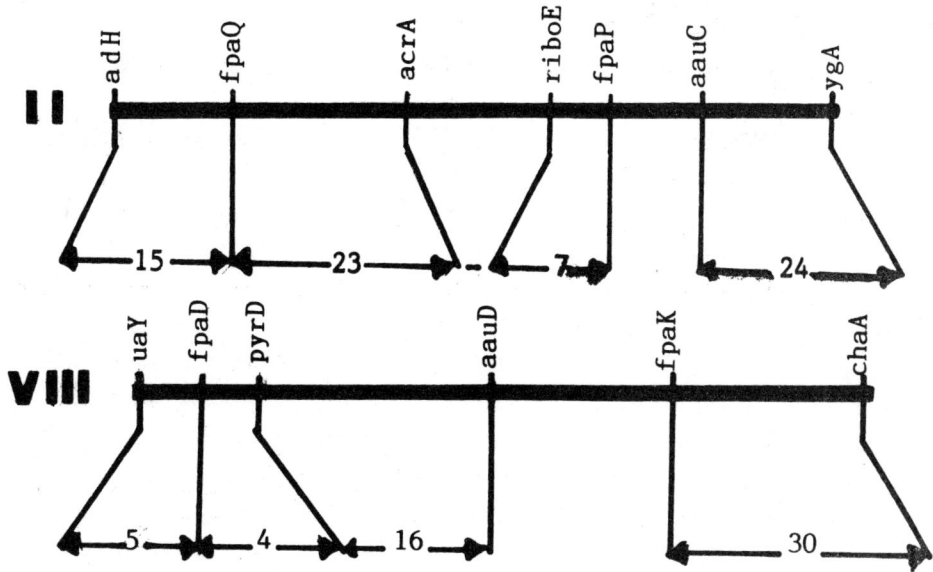

Fig. 8.5 Chromosomal location of *fpa* genes involved in the synthesis and/or regulation of amino acid permeases in *Aspergillus nidulans*

1980) shows that the mutant has a significantly reduced level of phenylalanine-binding protein as compared to the wild-type strain (Fig. 8.6). It has been suggested that *fpa*D locus might be controlling the synthesis of this binding protein, a lack of which results in the failure of phenylalanine-uptake and ultimately FPA-resistance.

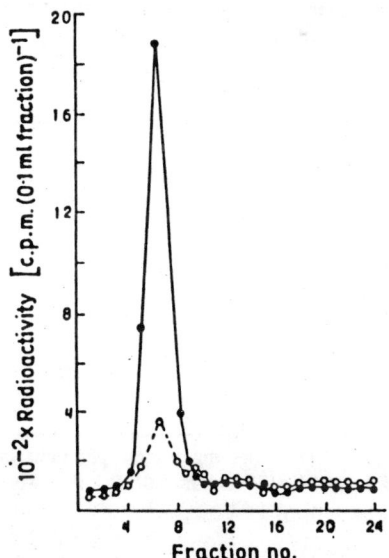

Fig. 8.6 Elution profile of proteins of *A. nidulans* bound to L-phenylalanine-CH-Sepharose gel and eluted with guanidine hydrochloride: wile type (O) and *fpaD11* mutant (O).

Fig. 8.7 SDS-polyacrylamide gel electrophoresis of phenylalanyl-tRNA synthetase from *biA1, phenA3* strain of *A. nidulans* after affinity chromatographic purification on Sepharose 4B column.

iii. Altered phenylalanyl-tRNA synthetase

Since obligatory steps in the incorporation of amino acids into polypeptides are their activation and loading onto specific tRNAs, it is essential at least for those analogues which inhibit growth due to their incorporation into proteins to be recognised by corresponding activating enzyme (aminoacyl-tRNA synthetase). If the enzyme fails to bind the amino acid analogue to corresponding tRNA, the organism will develop resistance. Such mutants have been known in bacterial systems since 1960s (see review Schimmel & Soll, 1979). As far as the eukaryotes are concerned there are only limited number of reports of isolation of analogue-resistant aminoacyl-tRNA synthetase mutants (Lewis, 1963; Staheli *et al.*, 1981). In *A. nidulans, fpa*B mutant was previously suspected to be a phenylalanyl-tRNA (Phe-tRNA) synthetase mutant (Sinha, 1972), but subsequent work has shown the incorporation of FPA into its proteins (Verma, 1973).

Earlier reports of Sinha (1967a) mention that the inhibitory effects of FPA on *A. nidulans* can be competitively reversed only by phenylalanine, and that recombinants between *phen*A auxotrophs and FPA-resistance become inviable even on a medium supplemented with phenylalanine possibly because mutations leading to FPA-resistance prevent incorporation of the analogue into proteins (Sinha, 1969). Tiwary (1985), however, developed a method to isolate mutants using a *phen*A auxotroph as the wild-type strain by optimising the concentrations of phenylalanine and FPA so that their combined effect allows both growth of the auxotroph and selection for FPA-resistance. This technique has led to the identification of two loci *fpa*U (Tiwary *et al.*, 1987b; Tiwary, 1989b) and *fpa*V (Singh *et al.*, 1990; Singh & Tiwary, 1992a) on linkage groups V and III, respectively. Characterization of these mutants, (Table 8.6) reveals that

Table 8.6 Specific acivity of phenylalanyl-tRNA synthetase from the wild type and FPA-resistant mutants with phenylalanine and FPA as substrates

Strains	Specific activity[*]	
	Phenylalanine	FPA
biA1;phenA3	7.3×10^4	4.8×10^4
biA1;phenA3;fpaU85	7.3×10^4	7.1×10^2
biA1;phenA3;fpaV86	7.4×10^4	6.6×10^4

[*]Specific activity was measured in terms of μM ppi released per mg protein per hour in a reaction mixture containing 1mM phenylalanine/FPA, 1mM ATP-sodium salt, 0.5 mM $MgCl_2$ and 10 μg of purified enzyme.

whereas wild-type strain produces Phe-tRNA synthetase having the ability to activate both phenylalanine and FPA to an approximately equal extent, the mutant enzyme retains the ability to activate phenylalanine but shows a diminished activity with the analogue (Tiwary *et al.*, 1987b; Singh & Tiwary, 1992a). Based upon the findings of Dunn and Leach (1967) in *E. coli* that FPA is utilized by only AAA-tRNA synthetase whereas phenylalanine is the substrate for both GAA- as well as AAA-tRNA synthetases, the possibility of an alteration in the synthesis and/or activity of AAA-tRNA synthetase due to mutation in the *fpa*U gene has been suggested (Tiwary *et al.*, 1987b).

The molecular weight of Phe-tRNA synthetase purified from *A. nidulans* by affinity chromatography has been estimated in native (by gel filtration chromatography) as well as denatured (by SDS-polyacrylamide gel electrophoresis) states and has been reported to have a quarternary structure of about by 250kd with an $\alpha_2 \beta_2$-subunit composition (Fig. 8.7) where α_1-and β-subunits, respectively, are of about 59 and 66kd (Singh *et al.*, 1990). On the other hand, gel electrophoretic profile of Phe-tRNA synthetase purified from *fpa*V mutants reveal the appearance of an additional polypeptide band migrating slightly faster than the normal β-subunit. Singh *et al.* (1990) have suggested that a modified β-subunit might be resulting into functional inactivity of the enzyme in the mutant strain. It is most likely that as a consequence of modification in one of its subunits, there occurs a drastic decrease in the turnover number of enzyme molecules for the analogue due to lack of anticooperativity between domains. However, whether modification in the β-subunit takes place during activation or loading of the analogue onto tRNAPhe by Phe-tRNA synthetase is yet to be established.

Uncharacterized FPA-resistant mutants of *A. nidulans*

Except those belonging to so far described three classes, none of the other FPA-resistant mutants have been well characterized. Singh *et al.* (1977) have made an attempt to characterize some of the mutants by studying their resistance pattern on a poor source of carbon (acetate) or nitrogen (arginine) and have found that *fpa*D, *fpa*F, *fpa*G and *fpa*J lose their resistance while *fpa*K and *fpa*I maintain their resistance to FPA. On the basis of contrasting pattern of resistance of *fpa*D and *fpa*K (both permease-deficient/ defective), it has been suggested that the effect in *fpa*K is more drastic than that in *fpa*D. The permease of *fpa*D gets derepressed under limited condition of carbon or nitrogen. On the other hand, the transport mutant *fpa*K is so much defective that even under derepressed condition, significant amount of FPA is not taken in and the strain maintains its resistance to the analogue. Regarding the mechanisms of resistance of mutants *fpa*L, *fpa*M, *fpa*N, *fpa*R and *fpa*S, Singh and Sinha (1979a) and Tiwary and Sinha (1987) have suggested that these mutants possess normal transport systems for phenylalanine and have high degrees of resistance to FPA on glucose as well as on acetate or arginine. Perhaps, these mutants have different mechanisms, yet to be understood, for acquiring resistance. On the other hand *fpa*T mutants which have been selected on a medium containing arginine utilize all amino acid as the nitrogen sources except arginine on which its growth is significantly poor (Tiwary *et al.*, 1987a). This is perhaps because arginine blocks its own transport system and that basic amino acids can be transported by a general permease as has been reported in other fungal systems (Wolfinbarger, 1980).

Epilogue

Aspergillus nidulans is one of the most extensively exploited fungus after *Neurospora crassa*, for which detailed basic information is available. Ever since the discovery of parasexuality by Pontecorvo *et al.* (1953) and elucidation of the fact that p-fluorophenylalanine (FPA) leads to a wholesale segregation in the treated diploids by Morpurgo (1961), this fungus has reached the zenith of genetic and molecular analysis. FPA, apart from acting as a haploidising agent, gets readily incorporated into proteins. Isolation and characterization of various mutants resistant to this analogue clearly establish three modes for acquiring resistance to toxic antimetabolites by *A. nidulans* in particular, and eukaryotes in general. Analysis of transport mutants provide the means of locating structural and regulatory genes on the linkage map. Excretory (overproducing) mutants can be profitably exploited for the production of phenylalanine and tyrosine at industrial level. Detailed analyses of mutants having altered aminoacyl-tRNA synthetases are likely to unravel the intricate mechanisms of protein biosynthesis and the mechanisms of aminoacylation, the accuracy of which is the prime need for the regulation of gene expression.

The recently developed fungal transformation and expression systems (Fincham, 1989) mainly relies upon the selectable markers. In this respect, *fpa* genes, specially dominant ones, can be used as selectable marker if suitably cloned and expressed under the control of regulated promoters. In addition, cloning of *fpa*U and *fpa*V genes affecting the two subunits of phenylalanyl-tRNA synthetase is likely to throw some more light on our understanding of the gene organisation and regulation of synthetases in eukaryotic systems.

Acknowledgements

We are grateful to late Professor Umakant Sinha for initiating us in the field of biochemical genetics of *Aspergillus nidulans*. His memory has been the guiding spirit in writing this review. We gratefully dedicate this paper to him.

References

Ames, G.P. (1964).   Uptake of amino acids by *Salmonella typhimurium*. Arch. Biochem. Biophys. **104**, 1-18.

Anraku, Y. (1980).   Transport and utilization of amino acids by bacteria. In Microorganisms and nitrogen sources, Edited by J.W. Payne, Wiley & Sons, pp. 9-30.

Arnstein, H.R.V. & Richmond, M.H. (1964). The utilization of p-fluorophenylalanine for protein synthesis by the phenylalanine incorporation system for rabbit reticulocytes. Biochem. J. **91**, 340-346.

Balakrishnan, S. & Sinha, U. (1976).   Macromolecular changes induced by p-fluorophenylalanine in *Triticale*. J. Ind. Bot. Soc. **55**, 175-181.

Banks, P., Britten, E.J. & Byth, D.E. (1982). Heritable para-fluorophenylalanine-induced aneuploidy in maize. J. Heredity. **73**, 465-466.

Beeman, E.A. & Smith, R.C. (1969). Inhibition by ethionine of the growth of *Neurospora crassa*. Can. J. Microbiol. **15**, 445-449.

Benko, P.V., Woods, T.C. & Segel, I.H. (1967). Specificity and regulation of methionine transport in filamentous fungi. Arch. Biochem. Biophys. **122**, 783-803.

Boyaval, P., Moreira, E. & Desmazeaud, M.J. (1984). Le transport de la phenylalanine et de la tyrosine chez Brevibacterium linens, specificite et incorporation dans les proteines. Can. J. Microbiol. **30**, 430-438.

Britten, R.J. & McClure, F.T. (1962). The amino acid pool in *Escherichia coli*. Bacteriol. Rev. **26**, 292-335.

Brooks, C.J., DeBusk, B.G., DeBusk, A.G. & Catcheside, D.E.A. (1972). A new class of p-fluorophenylalanine-resistant mutant in *Neurospora crassa*. Biochem. Genet. **6**, 239-254.

Champney, W.S. & Jensen, R.A. (1969). D-tyrosine as a metabolic inhibitor of *Bacillus cereus*. J. Bacteriol. **98**, 205-213.

Chattoo, B.B. & Sinha, U. (1974). Mutagenic activity of N-methyl-N-nitro-N-nitrosoguanidine (NTG) and N-methyl-N'nitrosourea (NMU) in *Aspergillus nidulans*. Mut. Res. **23**, 41-49.

Clutterbuck, A.J. (1974). *Aspergillus nidulans*. In Handbook of Genetics. Plenum Press, Vol. I, pp. 447-510.

Clutterbuck, A.J. (1981). Loci and linkage map of *Aspergillus nidulans*. Aspergillus News Lett. **15**, 58-72.

Cohen, G.N. & Munier, R. (1959). Effects des analogues structuraux la synthese d' enzymes chez *Escherichia coli*. Biochem. Biophys. Acta **31**, 347-456.

Cowie, D. (1959). In Carnegie Institute of Washington. YB 58, 294.

DeBusk, R.M. & DeBusk, A.G. (1980). Physiological and regulatory properties of the general amino acid transport system of *Neurospora crassa*. J. Bacteriol **143**, 188-197.

Dorn, G. (1965). Genetic analysis of the phosphatases in *Aspergillus nidulans*. Genet. Res. **6**, 13-26.

Dorn, G. (1967). Purification of two alkaline phosphatases from *Aspergillus nidulans*. Biochem. Biophys. Acta **132**, 190-193.

Dorn, G. (1968). Purification and characterization of phosphatase I from *Aspergillus nidulans*. J. Biol. Chem. **243**, 3500-3506.

Dunn, T.F. & Leach, F.R. (1967). Incorporation of p-fluorophenylalanine into proteins by cell-free systems. J. Biol. Chem. **242**, 2693-2699.

Fincham, J.R.S. (1989). Transformation in fungi. Microbiol Rev. **53**, 148-170.

Fincham, J.R.S., Day, P.R. & Radford, A. (1979). In Fungal genetics, Blackwell Scientific Publication, Oxford, pp. 1-636.

Fiske, M.J. & Kane, J.F. (1984). Regulation of phenylalanine biosynthesis in *Rhodotorula glutinis*. J. Bacteriol. **160**, 676-681.

Georgopoulos, S.G. (1977). Development of fungal resistance to systemic fungicides. In Antifungal compounds, Vol. 2, Interactions in Biological and Ecological Systems, Edited by M.R. Seigel & H.D. Sisler, Marcel Dekker Inc., New York & Basel, pp. 439-495.

Gollub, E. & Sprinson, D.B. (1969). A regulatory mutation in tyrosine biosynthesis. Biochem. Biophys. Res. Commun. **35**, 389-395.

Guroff, G., Bromwell, K. & Abramovitz, A. (1969). Mode of action of p-chlorophenylalanine in *Pseudomonas* species (11299a). Arch. Biochem. Biophys. **131**, 543-550.

Hannan, M.A. (1972). Mutation in *Schizophyllum commune* for resistance to p-fluorophenylalanine. Experientia **26**, 1242-1243.

Horowitz, N.H., Fling, M., Feldman, H.M., Pall, M.L. & Frohener, S.C. (1970a). Derepression of tyrosinase synthesis in *Neurospora* by amino acid analogs. Dev. Biol. **21**, 147-156.

Horowitz, N.H., Feldman, H.M. & Pall, M.L. (1970b). Derepression of tyrosinase synthesis in *Neurospora* by cycloheximide, actinomycin D and puromycin. J. Biol. Chem. **245**, 2784-2788.

Im, S.W.K. & Pittard, J. (1971). Phenylalanine biosynthesis in *Escherichia coli* K12, Mutants derepressed for chorosmate mutase-P-prephenate dehydratase. J. Bacteriol. **106**, 748-790.

Jensen, R.A. (1969). Antimetabolite action of 5-methyltryptophan in *Bacillus subtilis*. J. Bacteriol. **97**, 1500-1501.

Käfer, E. & Upshall, A. (1973). The phenotypes of eight disomics and trisomics of *Aspergillus nidulans*. J. Heredity **64**, 35-38.

Kane, J.F. & Fiske, M.J. (1985). Regulation of phenylalanine ammonia lyase in *Rhodotorula glutinis*. J. Bacteriol **97**, 1114-1117.

Kaufman, S. (1963). The structure of the phenylalanine-hydroxylation cofactor. Proc. Nat. Acad. Sci, USA **50**, 1085-1093.

Kerridge, D. (1960). The effects of inhibitors on the formation of flagella by *Salmonella typhimurium*. J. Gen. Microbiol **23**, 519-538.

Kinghorn, J.R. & Pateman, J.A. (1975). Mutations which affect amino acid transport in *Aspergillus nidulans*. J. Gen. Microbiol **86**, 174-184.

Levine, M. & Tarver, H. (1951). Incorporation of ethionine into rat proteins. J. Biol. Chem. **192**, 835-850.

Lewis, D. (1963). Structural genes for the methionine activating enzymes and its mutation as a cause of ethionine. Nature (London) **200**, 151.

McCully, K.S. (1964). Unpublished results.

McCully, K.S. & Forbes, E. (1965). The use of p-fluorophenylalanine with master strains of *Aspergillus nidulans* for assigning genes linkage groups. Genet. Res. **6**, 352-359.

Morpurgo, G. (1961). Somatic segregation induced by p-fluorophenylalanine. Aspergillus News Lett. **2**, 10.

Morpurgo, G. (1962). Resistance to p-fluorophenylalanine. Aspergillus News Lett. **3**, 11.

Morris, N.R. & Oakley, C.E. (1979). Evidence that p-fluorophenylalanine has a direct effect on tubulin in *Aspergillus nidulans*. J. Gen. Microbiol. **114**, 449-454.

Noda, K. (1982). Chromosome reduction in root meristems of triticale "Carman" by p-fluorophenylalanine. Seiken. Ziho **30**, 9-15.

Pall, M.L. (1969). Amino acid transport in *Neurospora crassa*. I. Properties of two amino acid transport systems. Biochem. Biophys. Acta. **173**, 113-127.

Pitrowska, M., Stepien, P.P., Bartnik, P. & Zarkzewaska, E. (1976). Basic and neutral amino acid transport in *Aspergillus nidulans*. J. Gen. Microbiol **92**, 89-96.

Pontecorvo, G. (1953). The genetic of *Aspergillus nidulans*. Adv. Genet. **5**, 141-238.

Pontecorvo, G. & Käfer, E. (1958). Genetic analysis based on mitotic recombination. Adv. Genet. **9**, 71-104.

Richmond, M.H. (1960). Incorporation of DL-B-(p-fluorophenylalanine) (14C) alanine into ex-openicillinase by *Bacillus cereus*. Biochem. J. **77**, 122-135.

Richmond, M.H. (1962). The effect of amino acid analogues on growth and protein synthesis in microorganisms. Bacteriol Rev. **26**, 398-420.

Richmond, M.H. (1966). Structural analogy and chemical reactivity in the action of antibacterial compounds. In Biochemical Studies of Antibacterial Compounds, Cambridge Univ. Press, pp. 301-335.

Sanchez, S. Martinez, L. & Mora, J. (1972). Interaction between amino acid transport systems in *Neurospora crassa* J. Bacteriol. **112**, 276-284.

Saxena, R.K. & Sinha, U. (1977). Thermoregulation of abnormal sexual differentiation in *Aspergillus nidulans*. Trans. Mycol. Soc. Japan **18**, 264-269.

Saxena, R.K. & Sinha, U. (1978a). Metabolic status of temperature induced swollen conidia of *Aspergillus nidulans*. Experientia **34**, 43.

Saxena, R.K. & Sinha, U. (1978b). Laccase (p-diphenol oxidase) activity in shake cultures of *Aspergillus nidulans*. Ind. J. Exptl. Biol. **16**, 458-460.

Schimmel, P.R. & Soll, D. (1979). Aminoacyl-tRNA synthetases, general features of recognition of transfer RNAs. Ann. Rev. Biochem. **48**, 601-648.

Singh, M. & Sinha, U. (1976). Chloral hydrate induced haploidization in *Aspergillus nidulans.* Experientia **32**, 1144-1145.

Singh, M. & Sinha, U. (1979a). Isolation and characterization of a new class of amino acid-analogue-resistant mutants in *Aspergillus nidulans* using reduced carbon flow. Genet. Res. **34**, 121-130.

Singh, M. & Sinha, U. (1979b). Mode of action of p-fluorophenylalanine in *Aspergillus nidulans,* Effects on the synthesis and activity of phosphatase isoenzymes. J. Gen. Microbiol. **115**, 101-110.

Singh, M. & Sinha, U. (1980). A mutant of *Aspegillus nidulans* with reduced level of phenylalanine-binding protein. J. Gen. Microbiol. **120**, 549-552.

Singh, M., Srivastava, S., & Sinha, U. (1977). Carbon and nitrogen utilization and p-fluorophenylalanine-resistance in *Aspergillus nidulans.* Trans. Mycol. Soc. Japan **18**, 257-263.

Singh, N.K. & Tiwary, B.N. (1992a). Isolation and characterization of an analogue-resistant aminoacyl-tRNA synthetase mutant in *Aspergillus nidulans.* Ind. J. Exptl. Biol. **30**, 94-98.

Singh, N.K. & Tiwary, B.N. (1992b). Modelling for competition between phenylalanine and its toxic analogue in a *phen*A auxotroph of *Aspergillus nidulans.* Acta Bot. Indica **20**, 177-181.

Singh, N.K., Tiwary, B.N. & Sinha, U. (1990). Domain-inhibition-linked low activity of Phe-tRNA synthetase for para-fluorophenylalanine in a mutant of *Aspergillus nidulans.* Fungal Genet. News Lett. **37**, 38-39.

Sinha, U. (1967a). Genetics of aromatic amino acid biosynthesis in *Aspergillus nidulans.* Ph.D. Thesis, University of Glasgow, U.K.

Sinha, U. (1967b). Aromatic amino acid biosynthesis and p-fluorophenylalanine-resistance in *Aspergillus nidulans.* Genet. Res. **10**, 261-272.

Sinha, U. (1969). Genetic control of the uptake of amino acids in *Aspergillus nidulans.* Genetics **62**, 495-505.

Sinha, U. (1970a). Competition between leucine and phenylalanine and its relation to p-fluorophenylalanine resistant mutations in *Aspergillus nidulans.* Arch. Microbiol. **72**, 308-317.

Sinha, U. (1970b). Cascade regulation in *Aspergillus nidulans.* In Symposium on "Micromolecules in storage and transfer of biological information". Department of Atomic Energy, Govt. of India, Bombay, pp. 367-371.

Sinha, U. (1972). Studies with p-fluorophenylalanine resistant mutants of *Aspergillus nidulans.* Beitr. Biol. Pflanzen **48**, 171-180.

Sinha, U. & Ashworth, J.M. (1978). Effects of p-fluorophenylalanine on the growth and differentiation of *Dictyostelium discoideum.* Phytomorphology **23**, 210-215.

Sinha, U. & Bhojwani, S.S. (1976). Cytomorphological effects of p-fluorophenylalanine on *Allium cepa*. Acta Bot. Indica **4**, 26-29.

Sinha, U. & Chattoo, B.B. (1977). Nitrosoguanidine mutagenesis in relation to genome replication in *Aspergillus nidulans*. J. Cytol. Genet. 12, 92-98.

Sinha, U. & Jha, S.N. (1985). *Aspergillus nidulans* as a genetic test system. In Frontiers in Applied Microbiology, Edited by K.G. Mukerji, N.C Pathak & V.P. Singh, Prints House (India). Lucknow, pp. 137-160.

Sinha, U. & Srivastava, S. (1974-75). Derepression of tyrosinase in *Aspergillus nidulans*. J. Cytol. Genet. 14 & 15, 58-61.

Sinha, U. & Tiwary, B.N. (1986). Genetics of antimetabolite resistance in *Aspergillus*. In Aspects of Plant Sciences, Edited by S.S. Bir, Vol 9, pp. 35-62.

Sinha, U. & Verma, S. (1971). A phenylketoneuric mutant of *Aspergillus nidulans*. J. Cytol. Genet. (Congress Supple). 193-296.

Sinha, U. & Verma, S. (1972). Incorporation of p-fluorophenylalanine into proteins of *Aspergillus nidulans*. Current Trends in Plant Sciences, Unviersity of Delhi, Golden Jubilee Symposium, pp 39-40.

Sisken, J.E. & Wilkes, E. (1967). The time of synthesis and the conservation of mitosis-related proteins in cultured human amnion cells. J. Cell. Biol. **34**, 97-110.

Smith, J.E. & Pateman, J.A. (1977). Genetics and physiology of *Aspergillus*. Academic Press, London.

Srivastava, S. & Sinha, U. (1975). Six new loci controlling resistance to p-fluorophenylalanine in *Aspergillus nidulans*. Genet. Res. **25**, 29-38.

Srivastava, S. (nee Verma). & Sinha, U. (1981). Uptake patterns in p-fluorophenylalanine-resistant mutants of *Aspergillus nidulans*. In Perspectives in Cytology and Genetics, Edited by G.K. Manna & U. Sinha, Vol. **3**, pp. 531-534.

Staheli, P., Kradolfer, P., Neiderberger, P. & Hutter, R. (1981). Inhibition of Yeast tRNA[TRP] aminoacylation by 4-methyltryptophan. Arch. Microbiol. **129**, 146-149.

Stepien, P.P. (1976). Purification of methionine-binding protein from *Aspergillus nidulans* by affinity chromatography. Biochem. Biophys. Acta **439**, 154-159.

Stromnaes, O. (1968). Genetic changes in *Sachharomyces cerevisiae* grown on media containing DL-para-fluorophenylalanine. Hereditas **59**, 197-220.

Tiwary, B.N. (1985). Genetics of aromatic amino acid transport in *Aspergillus nidulans*. Ph.D. Thesis, Patna University (India).

Tiwary, B.N. (1989a). Interaction between *fpa* genes of *Aspergillus nidulans* concerned with amino acid uptake. Biologischez. Zbl. **108**, 77-81.

Tiwary, B.N. (1989b). Isolation and characterization of a mutant with an altered tRNA synthetase in *Aspergillus nidulans*. In Perspectives in Cytology and Genetics, Edited by G.K. Manna & U. Sinha, Vol. **6**, pp. 107-114.

Tiwary, B.N., Bisen, P.S. & Sinha, U. (1987a). Genetic control of amino acid transport in *Aspergillus nidulans*, Evidence for polymeric amino acid permease. Curr. Microbiol. **15**, 305-311.

Tiwary, B.N., Bisen, P.S. & Sinha, U. (1987b). Demonstration of an altered phenylalanyl-tRNA synthetase in an analogue-resistant mutant of *Aspergillus nidulans*. Mol. Gen. Genet. **209**, 164-169.

Tiwary, B.N. & Singh, N.K. (1992). Molecular mechanisms of antimetabolite resistance in *Aspergillus*. In Microbes and Environment, Edited by A.B. Prasad & K.S. Bilgrami, Narendra Publishing House, New Delhi (in Press).

Tiwary, B.N. & Sinha, U. (1985a). Preferential selection and genetic characterization of an amino acid uptake mutant in *Aspergillus*. Curr. Sci. **54**, 244-245.

Tiwary, B.N. & Sinha, U. (1985b). Aromatic amino acid analogue-resistance in *Aspergillus nidulans*. Ind. Bot. Cont. **2**, 127-137.

Tiwary, B.N. & Sinha, U. (1986). Genetic characterization of mutants resistant to an aromatic amino acid analogue in *Aspergillus nidulans*. In Perspectives in Cytology and Genetics, Edited by G.K. Manna & U. Sinha, Vol. **5**, pp. 741-746.

Tiwary, B.N. & Sinha, U. (1987). Characteristics of six new p-fluorophenylalanine-resistant loci of *Aspergillus nidulans*. Fungal Genet. News Lett. **34**, 56-58.

Tiwary, B.N. & Sinha, U. (1989). Isolation and preliminary characterization of a new class of amino acid uptake mutant in *Aspergillus nidulans*. Microbios. Lett **41**, 57-61.

Trehan, K. & Sinha, U. (1984). Isolation and characterization of p-fluorophenylalanine-resistant mutant in the cyanobacterium *Anacystis nidulans*. In Perspectives in Cytology and Genetics, Edited by G.K. Manna & U. Sinha, Vol. **4**, pp. 199-203.

Umbarger, H.E. (1971). Metabolic analogues as genetic and biochemical probes. Adv. Genet. **16**, 119-140.

Van Tuyl, J.M. (1977). Genetics of fungal resistance to systemic fungicides. Meded Landbourwhogeschool, Wageningen, Netherland **77**, 1-126.

Van Venrooij, J., Moonen, H. & Van Loonklassen, L. (1974). Source of amino acids used for protein synthesis in HeLa cells. Eur. J. Biochem. **50**, 297-304.

Verma, S. (1973). Studies on the effects of p-fluorophenylalanine and resistance to this analogue in *Aspergillus nidulans* (Eidam) Winter. Ph.D. Thesis, University of Delhi (India).

Verma, S. & Sinha, U. (1971). Genetic analysis of p-fluorophenylalanine-resistant mutants of *Aspergillus nidulans*. J. Cytol. Genet. (Congress Suppl.) 297-300.

Verma, S. & Sinha, U. (1972). Inhibition of growth by amino acid analogues in *Aspergillus nidulans*. Beitr. Biol. Planzen. **49**, 47-58.

Warr, J.R. & Roper, J.A. (1965). Resistance to various inhibitors in *Aspergillus nidulans*. J. Gen. Microbiol. **40**, 273-281.

Wheatley, D.N. (1978). Biological and biochemical effects of phenylalanine analogues. Int. Rev. Cytol. **55**, 109-164.

Wolfinbarger, L. (1980). Transport and utilization of amino acids by fungi. In Microorganisms and Nitrogen Sources, Edited by J.W. Payne, John Wiley & Sons, pp. 63-87.

Woods, D.D. (1940). The relation of p-aminobenzoic acid to the mechanisms of action of sulphonilamide. Brit. J. Exptl. Pathol. **21**, 74-90.

Yoshida, A. (1960). Studies on the mechanisms of protein synthesis, Incorporation of FPA into -amylase of *Bacillus subtilis*. Biochem. Biophys. Acta **41**, 98-104.

<u>9</u>

Polyene antibiotics as tools in the control of systemic fungal infections

Mona Moonis and Bimal Kumar Bachhawat
Department of Biochemistry
University of Delhi, South Campus, New Delhi 110021

ABSTRACT

Polyene antibiotics are groups of macrolide lactones which are produced by different species of *Streptomyces*. The most common polyene antibiotics are amphotericin B, nystatin, candidin, hamycin, and filipin. Considering the importance of polyene antibiotics in the control of systemic fungal infections, their physical properties as well as mechanism of action is discussed in this review. Over the years, liposomes, the phospholipid vesicles, have attracted attention as drug carriers in a wide variety of fungal infections. The major advantage in the use of liposomes as drug carriers is the reduction in the toxicity of the liposomal drug that allows much larger doses of drug administration.

Results from our laboratory have shown that liposomes had a profound effect in reducing the toxicity of amphotericin B as well as in enhancing the *in vivo* antifungal potential in the control of aspergillosis virulence in Balb/c mice. Recent work indicates that although the toxicity of liposomal hamycin, an aromatic polyene, was high as compared to non aromatic polyenes, it led to a significant augmentation in its *in vivo* antifungal activity in controlling experimental aspergillosis in mice. These studies suggest that appropriate modification of the liposome surface may further reduce the toxicity of hamycin which may have potential applications for clinical studies.

INTRODUCTION

Opportunistic fungal infections are the major causes of morbidity and mortality in immunocompromised host (Cohen *et al.*, 1981; Frazer *et al.*, 1979). The most common organisms causing fungal disease in immunocompromised patients include species of *Aspergillus* and *Candida*. Polyene antibiotics are commonly used in the treatment of such systemic fungal infections. Polyene antibiotics are produced mostly by soil actinomycetes of the genus *Streptomyces*. They exert their antifungal activity by forming transmembrane pores which leads to the leakage of vital metabolites resulting in cell death (Bolard, 1986), and exhibit a broad spectrum of activity against dermatophytes, fungi, and protozoa. Nystatin was the first antibiotic discovered in this class of compounds (Omura & Tanaka, 1984). Since then over 200 polyenes have been described and over 40 have been elucidated for their structures. Polyene antibiotics are characterized by large 20-40 membered lactone ring containing 3-8 conjugated double bonds. This group of macrolide antibiotics has been classified according to the number of conjugated double bonds as trienes (aureofuscin), tetraenes (etruscomycin), pentaenes (filipin), hexaene, (dermostatin) and hepataenes (amphotericin B, hamycin). The heptaene group of polyene antibiotics is classified further into aromatic (hamycin) and non-aromatic polyenes (amphotericin B). The aromatic ring is usually p-amino acetophenone (hamycin) or p-(N-methylamino)-acetophenone (aureofungin A). Other important polyene heptaenes include nystatin, candidin and mycoheptin. X-ray crystallographic studies of amphotericin B (Amp B) have shown that all the double bonds are in the trans configuration (Mechlinski, *et al.*, 1970). The second characteristic feature of the polyene antibiotics is the presence of a large number of hydroxyl groups on the molecule. They are usually present on alternate carbon atoms on the macrolide ring.

The net charge conferred on the polyene molecule is due to the amino group of the hexose or carboxyl group. When both the charges are present on the polyene it becomes zwitterionic in character. The amino group is associated with an amino sugar that is linked glycosidically to the macrolide. In most cases the sugar molecule is mycosamine whose structure was established as 3-amino, 3,6-dideoxymannose (Medoff, *et al.*, 1983). Polyene macrolides exhibit a characteristic absorption spectrum, that ranges between 200-405 nm. Extensive work on the absorption spectrum of the polyenes has shown that the characteristic spectrum is due to the presence of conjugated double bond system (Mazerski, *et al.*, 1982). The diminution in the spectra of polyenes in aqueous solvents was attributed to micelle or aggregate formation. Polyene antibiotics exhibit poor solubility in water, alcohols, and esters but dissolve readily in dimethyl sulfoxide/dimethylformamide. The stock solution of polyenes remain stable in organic solvents provided they are shielded from light and oxygen to which they are susceptible.

Mechanism of action of polyene antibiotics

Work done by various workers on yeasts such as *Saccharomyces cerevisiae* and various species of *Candida* has shown that polyenes induce leakage of ions and small molecules such as K^+, Na^+, phosphates etc. Even at a concentration as low as 1×10^{-6}M, the polyenes cause a rapid decrease in the dry weight of mycelial material and induce leakage of sugars, amino acids and metabolites from the cytoplasm into the culture medium. Similar results were obtained with filipin on *S. cerevisiae* and *Neurospora crassa*. Bacteria as intact cell or protoplast do not take up polyenes and are unaffected by them. This suggests that the mode of action of polyene antibiotics is via membrane sterols which are present in eukaryotic membranes but absent in bacterial cell walls or membranes. The first convincing evidence for the role of sterol in the mode of action of polyene antibiotics came from the studies on model membranes of *Acholeplasma laidlawii*. This organism is unable to synthesize the sterols *de novo*, but when grown in sterol rich medium, can incorporate the sterol in the cell membrane. Filipin showed no effect on the growth of *A. laidlawii* cells in a sterol-deficient medium, however, when grown in a sterol rich medium both filipin and Amp B could inhibit the fungal growth (De Kruijff *et al.*, 1974).

A. laidlawii cells when grown in the presence of sterol and tested for 50% release of K^+ in 10 minutes by a spectrum of polyene antibiotics showed the following order of decreasing effect, Amp B > nystatin > filipin > etruscomycin > primaricin. The effect of filipin was most drastic i.e. it led to the release of cytoplasmic components such as inorganic phosphates from yeast and in addition proteins from mycoplasmas. Amp B and nystatin, in contrast, caused release of K^+ and inorganic phosphates. The effect of filipin could be due to the large pore size formed by it (Bolard, 1986).

Addition of sterol into the medium was seen to shield the polyene sensitive cells from the action of polyenes (Hammond, 1977). The reversal of polyene effect could be due to a secondary physicochemical interaction between the polyenes and the added sterol which prevents the interaction of antibiotics with sterol of the organism. Dose response studies of Amp B on *S. cerevisiae* cells showed that at low concentrations, Amp B caused enhanced reversible cell permeability and was fungistatic whereas, at high concentrations it was fungicidal (Bolard, 1986).

A number of factors have been identified which play an important role in causing antibiotic resistance. Alteration in sterol content of *S. cerevisiae* from ergosterol to zymosterol led to nystatin resistance (Gale, 1984). In some cases, there was reduction in the ergosterol content e.g., in case of yeast mutants as compared to wild-type (Gale, 1984). Methylated sterols have also been shown to render *S. cerevisiae* cells resistant to polyene action (Gale, 1984). These modifications could lead to the decreased binding of polyenes to sterol.

A large body of work has been done on liposomes (the phospholipid vesicles) as model membranes for studying the polyene-sterol interactions. De Kruijff and Demel (1974) have proposed mechanisms for induction of permeability changes for various polyenes based on polyene-sterol interactions. They proposed the Amp B-cholesterol complex as a circular arrangement of 8 Amp B molecules interdigitated with 8 cholesterol molecules. The outside portion is hydrophobic whereas the inside portion is hydrophilic due to the presence of the hydroxyl groups of Amp B. Two such complexes i.e. half pores generate a full pore which traverses the membrane. The pore size of such complexes is about 8 Å. Similar pores were visualized for nystatin-cholesterol and etruscomycin-cholesterol complexes. The filipin-cholesterol complex was visualized to be 150-250 Å in diameter, oriented towards the hydrophobic core of the membrane. The aggregate was conceived to consist of $2°$ regular arrays of filipin molecules, so stacked that the exterior is hydrophobic due to the presence of double bonds to which equal amounts of cholesterol were bound. The formation of this aggregate causes membrane fragmentation.

Based on their mode of action on membranes, the polyenes have been classified into 2 groups, "large polyenes" (mostly, heptaenes) which produce reversible permeability changes at low concentration and cell lysis at higher concentration and "small polyenes" which produce lytic effects by an "all or none phenomenon" and cause more damage to the cell membrane as compared to large polyenes (Kotler-Brajtburg *et al.*, 1979). The large polyenes, such as Amp B, have been reported to form half and full pores in the vesicles whereas small molecules, e.g. filipin do not form pores but orient in the hydrophobic area of the liposomal membrane which is perpendicular to the array of phospholipid molecules (De Kruijff & Demel, 1974). Orientation of small polyenes in liposomes without cholesterol incorporation may be unstable due to the organization of the hydrophobic side of the molecules being towards the hydrophilic surface. Probably due to this and the spanning of small polyenes only through half the bilayer, the polyene molecule may be drawn towards the cholesterol of the RBC membrane more rapidly than the large polyenes (Mehta, 1989). Due to this higher affinity of small polyenes to cholesterol of RBC's, the drug may be extruded from the liposomes, thereby disrupting them, and as a result cause RBC lysis. In contrast, the half pores formed by large polyenes have a very short half life. Due to the lateral movement of the phospholipid bilayer, two half pores come together and traverse the whole bilayer i.e. lead to the formation of full pores. This organization of large polyenes is more stable and hence reduces the chances of transfer of large polyenes to RBC's (Mehta, 1989).

Bolard (1986) have suggested the use of large unilamellar vesicles (LUV) as an ideal model membrane for studies of polyene-sterol interaction. They showed by circular dichroism studies that the interaction of Amp B with LUV is different from that of small unilamellar vesicles (SUV). In addition, Amp B showed differential spectra with cholesterol and ergosterol containing membranes. The physicochemical studies by Bolard (1991) were done with preformed liposomes and the drug was incorporated in the bilayer after the liposomal preparation. Bolard hypothesized that the entire amount of Amp B associated with the liposomes does not exist in the bound form. Some amount of drug remains free and that the same may be responsible for the killing process. Once the unbound Amp B attaches to sterol of the fungus, it shifts the drug which is in bound form in liposomes to free form and in turn increases the concentration of Amp B attached to the fungus, augmenting the killing process (Bolard, 1991).

Based on the experimental data for Amp B-cholesterol complex, Andreoli (1974) proposed a model for the pore formation (Fig. 9.1). The model envisaged the C-1-C-13 and C-20-C-33 segments of Amp B molecule as rod like array of 20-29 Å in length. The C-15 hydroxyl group, C-16 carboxyl group and C-19 mycosamine group are present on one side of the rod located at the water bilayer interphase. The rod is embedded in the interior parallel to the hydrophobic chains of phospholipid as well as to planar cyclopentano phenanthrene skeleton and C-17 acyl residues of cholesterol. According to this model, 2 kinds of hydrogen bonds are involved between polar group of Amp B and 3-OH proton and a carboxyl oxygen of C-16 carboxyl group whereas the second is between 3-OH oxygen and C-17 hydroxyl proton of Amp B.

Recent observations from various laboratories including that of Bolard suggest that aggregation of Amp B molecules may be playing an important role in the formation of pores. Cholesterol may be stabilizing the aggregation of Amp B-complex which in effect may be stabilizing the pore formation. The role of cholesterol as presented in the schematic diagram may thus have to be re-examined in the light of the above observations.

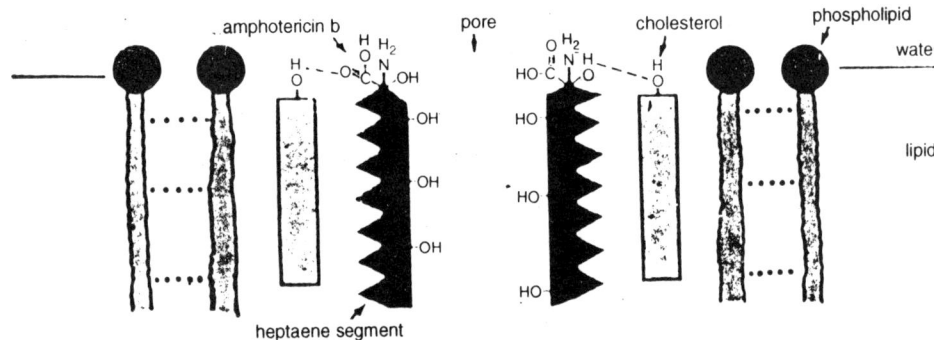

Fig. 9.1: A schematic representation of the hypothetical model of amphotericin B - cholesterol complex. The dotted lines represent London - Vander Waals forces whereas the dashed lines represent hydrogen bonds (Adapted from Andreoli, 1974). Recent studies have demonstrated that cholesterol stabilizes the aggregation of Amp B molecules which in effect may be stabilizing the pore complex. The role of cholesterol as presented in the schematic diagram may thus have to be reexamined in the light of the new observations.

Aromatic polyenes

Aromatic polyenes are the sub group of large ring macrolide antibiotics which are characterized by a biological activity on pathogenic yeast like organisms that is 2-3 orders of magnitude higher than that shown by their non aromatic counterparts, e.g. Amp B and nystatin (Gale, 1984; Hammond, 1977). Well known examples of aromatic polyenes include hamycin, candicidin D (levorin A_2), gedamycin (partricin A), mepartricin A and B, vacidin A (partricin B) and perimycin A (Fig. 9.2). The aromatic polyenes induce permeability in all sterol containing membranes by forming reversible cation selective channels which exhibit intercationic selectivity (Cybulska *et al.*, 1981). The hemolytic effect of these polyenes results in uncontrolled solute distribution leading to an osmotic imbalance. The aromatic heptaenes thus induce a colloidosmotic type of lysis.

Although the toxic effects of aromatic polyenes on yeast as well as RBC are due to alterations produced by membrane permeability but the mechanisms of these two effects is dissimilar. Inhibition of yeast growth is caused by metabolic imbalance generated by decrease in the intracellular potassium ion concentration whereas hemolysis results from an osmotic imbalance due to ion flux which is a physicochemical effect (Hammond, 1977).

The enhanced biological activity of aromatic polyenes appears to be related to the stability life time and intrinsic permeability of the polyene-sterol complex. Studies on the polyene-sterol complex of candicidin D revealed that the cistrans chromophore of candicidin D produced much more longer lasting channels as compared to non aromatic polyene antibiotics (Liras & Lampen, 1974). In most biological models, aromatic heptaenes induce the formation of most stable and specific cation channels. The aromatic polyene levorin A_2 forms ion channels which are permeable to only monovalent cations in both cholesterol as well as ergosterol containing membranes (Cybulska *et al.*, 1984). It was hypothesized earlier that anion selectivity is due to positive potential generated by OH dipoles inside the channels (Kasumov *et al.*, 1981). The positive potential produced by OH groups inside the channel increases the anion selectivity and reduces the cation selectivity. The reduction in the number of OH groups in the hydrophilic chain of the macrolide can thus alter the selectivity of the channels from an anionic to cationic character.

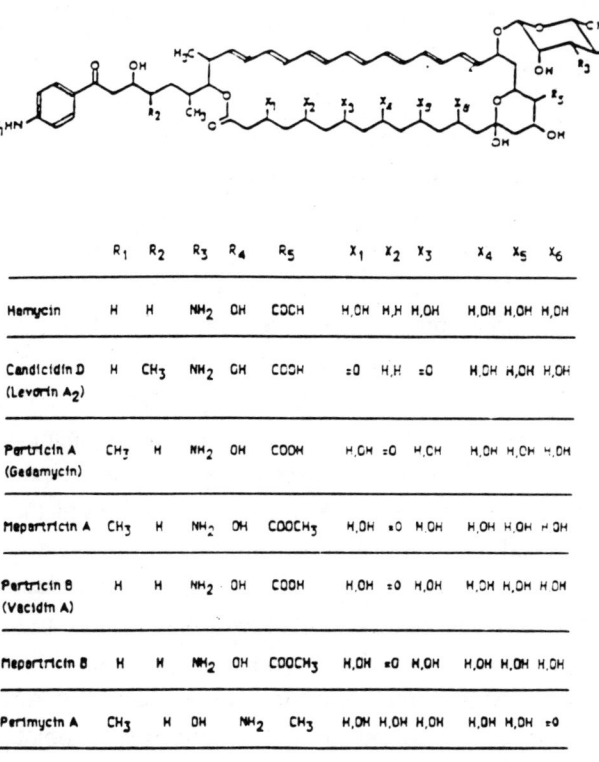

	R_1	R_2	R_3	R_4	R_5	X_1	X_2	X_3	X_4	X_5	X_6
Hamycin	H	H	NH_2	OH	COCH	H,OH	H,H	H,OH	H,OH	H,OH	H,OH
Candicidin D (Levorin A_2)	H	CH_3	NH_2	OH	COOH	=O	H,H	=O	H,OH	H,OH	H,OH
Partricin A (Gedamycin)	CH_3	H	NH_2	OH	COOH	H,OH	=O	H,OH	H,OH	H,OH	H,OH
Mepartricin A	CH_3	H	NH_2	OH	$COOCH_3$	H,OH	=O	H,OH	H,OH	H,OH	H,OH
Partricin B (Vacidin A)	H	H	NH_2	OH	COOH	H,OH	=O	H,OH	H,OH	H,OH	H,OH
Mepartricin B	H	H	NH_2	OH	$COOCH_3$	H,OH	=O	H,OH	H,OH	H,OH	H,OH
Perimycin A	CH_3	H	OH	NH_2	CH_3	H,OH	H,OH	H,OH	H,OH	H,OH	=O

Fig. 9.2: Chemical structures of aromatic heptaene antibiotics

Comparative studies of the ionophoric action of a series of natural or semisynthetic derivatives of aromatic polyene, Vac A, revealed that the hemolytic activity of this compound was governed by the presence of a free carboxyl group at C-18 position of the macrolide (Cybulska *et al.*, 1980). Vac A (presence of free carboxyl group) showed hemolytic activity on RBC whereas Vac A methyl esters (absence of free carboxyl groups) showed very little hemolytic activity. While studying the ionophoric and hemolytic activities of Vac A and perimycin A, Cybulska *et al.* (1984) observed that both the polyenes were similarly efficient in inducing K^+ permeability but showed a 100-fold difference in their hemolytic activity. The hemolytic activity of the above compounds were found to be related to their protonophoric activity, which in turn required the presence of a free carboxyl group in the polyenes as in Vac A. Moreover, the hemolytic activity was found to be governed by the efficiency of K^+/H^+ exchange induced by these polyenes. Perimycin A which lacks the free carboxyl group is dependent on Cl^- which slows down the K^+ flux Ce.

The cation permeability induced by Vac A and candicidin D on both ergosterol- and cholesterol-containing membranes (10 mol % sterol) were observed to elicit cation permeability in similar concentration range i.e. they were as active in cholesterol as in ergosterol containing membranes (Cybulska *et al.*, 1981). Sterol-free vesicles in the same study were found to be insensitive to both aromatic polyenes but sensitive to Amp B which exhibited significant proton release at even low concentrations i.e., 0.2 $\times$ 10^{-3} mol/mol lipid (Cybulska *et al.*, 1981). From the discussion above it is clear that unlike non-aromatic polyenes, aromatic polyenes are cation selective.

Zwitterionic or negatively charged aromatic polyenes which possess a free ionizable carboxyl group exhibit similar efficiency (either high or low) on both cholesterol-and ergosterol-containing membranes (Cybulska *et al.*, 1985). On the other hand, positively charged polyenes which lack the free carboxyl group show a differential behaviour on these two types of membranes, being less efficient on cholesterol-than ergosterol-membranes. In ergosterol-containing membranes the positively charged polyenes behave like channel formers, whereas in cholesterol containing membrane they cause slow and progressive permeability at high concentrations.

Studies on aromatic polyenes have revealed that the absence of carboxyl group or its esterification consequently leads to very low hemolytic and unchanged enhanced activity on yeasts. The differential toxicity of aromatic polyenes presumably requires a specific conformation of the sugar moiety (Cybulska *et al.*, 1985), and the freedom of the amino sugar moiety to rotate seems to play a pivotal role (Fig. 9.3). In the case of zwitterionic or negatively charged polyenes, the movement of the amino sugar moiety is hindered by electrostatic interaction with the carboxyl group, whereas in the positively charged polyenes due to lack of COO$^-$ group, the amino sugar moiety is free to rotate. The interaction of ergosterol with polyenes does not seem to be influenced by hindered or free rotation of the amino sugar. In contrast, cholesterol shows a clear preference for fixed conformation of the amino sugar moiety. From these studies, Cybulska *et al.* (1985) concluded that selective toxicity of polyenes in ergosterol - and cholesterol-containing membranes is governed by the presence or absence of carboxyl group at C-18 rather than the net charge of the polyene.

Liposomes as carrier for polyene antibiotics

Specific drug delivery at therapeutic sites circumventing the toxic sites has been a coveted goal in clinical therapeutics. Liposomes offer an advantage as drug carriers because they protect the drug from exerting its action before reaching the desired site and enhance the accessibility, recognition, and selective interaction with target tissues (Poznasky & Juliano, 1984). In certain disease conditions, the innate defence barrier of macrophages gets impaired so much so that macrophages themselves become the primary host cells harbouring intracellular pathogens. The preponderant uptake of liposomes by the reticuloendothelial system can be exploited to passively target them to macrophages for the treatment of various macrophage-associated disease (Alving, 1983).

Some of the important macrophage-associated diseases include leishmaniasis, tuberculosis, salmonellosis, hepatitis, and encephalitis. The conceivable goals for liposomal drug delivery to macrophage include intracellular destruction of organisms such as *Leishmania donovani* and *Trypanosoma cruzei* which successfully parasitise the macrophages. Alving *et al.* (1978) and New *et al.* (1981) have exploited the passive uptake of liposomes by macrophage rich organs for the treatment of leishmaniasis in animals. Experiments from our laboratory with mannose ligated on the surface of liposomes showed a significant decrease in the toxicity and an increase in the therapeutic efficacy of Amp B in the control of experimental aspergillosis in Balb/c mice (Ahmad *et al.*, 1991). Recently liposomal Amp B has been successfully used for the treatment of hepatosplenic candidiasis in a human patient (Bjorkholm *et al.*, 1991).

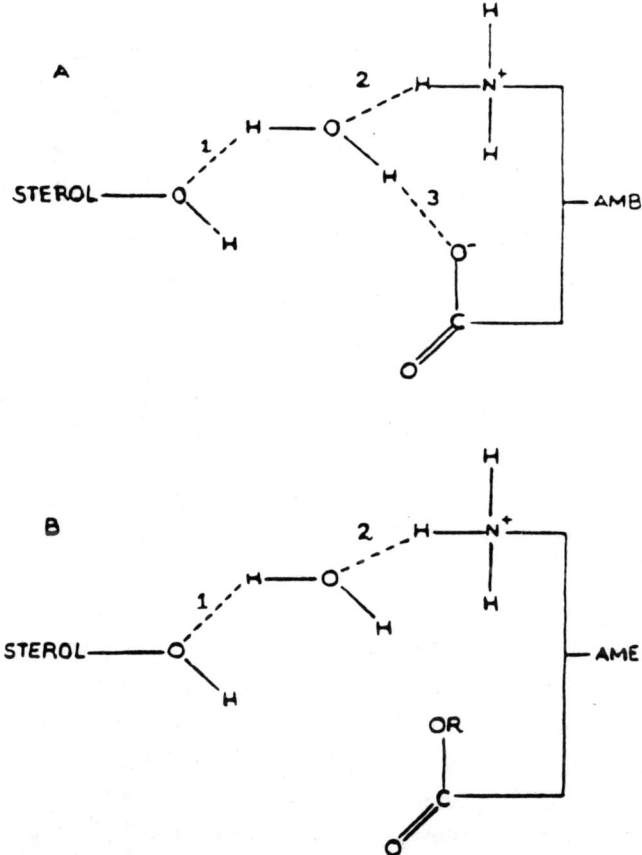

Fig. 9.3: Schematic representation of the H bond formation between the polyene derivatives polar head groups and the sterol OH group in (A) derivatives with free ionizable carboxyl group (B) derivatives with blocked carboxyl group. (Adapted from Cybulska, 1985).

The degree of saturation of phospholipids plays an important role in reducing the toxicity of L-Amp B (saturated acyl chains being less toxic than unsaturated acyl chain). The reduction in the toxicity of L-Amp B in the presence of sterol and saturated phospholipid may be due to an increase in the order and stability of liposomes thus reducing the exchange of phospholipid molecules across the membrane (Szoka, 1991). The reduction in the toxicity of the liposomal formulation allows the use of higher dose of Amp B, thereby improving the therapeutic index. Mehta *et al.* (1984) observed that while free Amp B was toxic to both fungal as well as mammalian cells, the liposomal drug was preferential to the fungal cells without being toxic to the mammalian cells. Janoff *et al.* (1988) have reported a new formulation of high Amp B to lipid mole ratios (consisting of nearly equimolar AmpB-lipid molar ratio). These ribbon like structures were found to significantly reduce the toxicity to mammalian cells but were effective on fungal cells.

Our studies have shown that (1) L-Amp B is less toxic and more effective as compared to fungizone in the control of experimental aspergillosis, (2) cholesterol was found to increase the therapeutic efficacy as well as reduced the toxicity of neutral liposomes; however, in the case of charged lipids, cholesterol had no effect in altering the toxicity or therapeutic efficacy of L-Amp B, (3) mannose grafted liposomes were the most effective in reducing the toxicity as well as in enhancing the therapeutic efficacy of L-Amp B, and (4) altered tissue distribution was observed in infected mice as compared to uninfected animals at various intervals after administration of the liposomal drug (Ahmad *et al.*, 1989, 1990, 1991). Encouraged by these findings, Ahmad and coworkers are conducting human clinical trials of SPC/ Chol: 7:3 liposomal formulation at the K.E.M. Hospital, Bombay, India.

Recently our lab has also been interested in investigating the role of a potent aromatic polyene, hamycin entrapped in liposomes on the *in vivo* toxicity as well as the antifungal potential in the treatment of experimental aspergillosis in mice. Hamycin, isolated from *Streptomyces pimprina* is a potent aromatic polyene belonging to the heptaene group of macrolides. Although hamycin has been shown to be a potent drug, its use has been limited due to its acute toxicity. Our studies have shown that liposomal hamycin is more effective in reducing the toxicity as well as in enhancing the *in vivo* antifungal activity as compared to free hamycin in the control of aspergillosis virulence in mice.

Incorporation of cholesterol into phosphatidic acid (PA) liposomes led to a decrease in the toxicity of hamycin in a dose dependent manner (Moonis *et al.*, 1993a). The LD_{50} (mg/Kg) of hamycin contained in liposomes (Soya phosphatidyl choline/Cholesterol/Phosphatidic acid) (molar ratio 4:5:1) was found 2.8 whereas that in SPC/PA liposomes (molar ratio 9:1) was 0.35. Although the free drug had little or no protective effect on the animals, those administered hamycin incorporated into PA liposomes at an equivalent dose (0.1 mg/Kg) in the absence of cholesterol showed 90% survival after 7 days of therapy. On the other hand, the presence of cholesterol in the carrier PA liposomes at a similar dose led to 60% survival over the same time period. Hamycin incorporated in PA liposomes both in the presence or absence of cholesterol was found to be effective in reducing the fungal load in lung, liver, spleen, and kidney. Similar results were obtained with hamycin incorporated in mannosylated liposomes which also produced less toxicity and enhanced the *in vivo* antifungal activity (Moonis *et al.*, 1993b). The

LD_{50} (mg/Kg) of hamycin contained in SPC/Chol/DPPE-Mannose liposomes (molar ratio 4:5:1) was 2.8 whereas that in SPC/DPPE-Man (molar ratio 9:1) was 1.4. Incorporation of cholesterol into mannosylated liposomes increased the survival rates of infected animals; 70% survival was recorded after 7 days of therapy as well as reduced fungal load in various organs.

The antifungal activity of liposomal hamycin in the presence and absence of cholesterol showed variation in comparison to non aromatic polyenes like Amp B. Colony forming units (CFUS) measured at various time intervals were much lower in certain formulations with liposomal hamycin in comparison to liposomal hamycin containing cholesterol. All these observations suggested that cholesterol did not have any enhancing effect on the *in vivo* antifungal activity of liposomal hamycin. Our *in vivo* observations seem to agree well with the physicochemical studies done on aromatic polyenes which show the same binding affinities for cholesterol- as well as ergosterol-containing vesicles (Witzke & Bittman, 1984). We inferred from our comparative studies with the free liposomal hamycin that reduction in toxicity as well as enhanced *in vivo* antifungal activity of the liposomal drug was probably due to its comparatively faster clearance from the circulation and consequent uptake by the reticuloendothelial system. In addition, our results show that a lower dose of liposomal hamycin is required as compared to non aromatic polyene Amp B in controlling aspergillosis virulence in mice.

To investigate the probable involvement of macrophages on the uptake of the liposomal drug as well as in controlling infection, we studied the effect of elimination of resident macrophages from the liver and spleen by liposomal dichloromethylene diphosphonate (L-DMDP). Liposomal DMDP has been used to deplete macrophages *in vivo* temporarily (Rooijen & Niewmegen, 1984). Administration of L-DMDP by i.v. route selectively and reversibly eliminates resident macrophages from the liver and spleen without affecting alveolar macrophages. It is well documented that macrophages phagocytize the fungal cells during systemic infection. Tissue macrophages play a key role in the control of infectivity against intracellularly growing pathogens, e.g. *Listeria monocytogenes*. Our studies have shown that tissue macrophages and PMN cells play an important role in the control of aspergillosis virulence although *Aspergillus fumigatus* is not an intracellular pathogen.

It can also be inferred that the resident macrophages from liver and spleen seem to be playing an important role in the control of the spread of the fungal disease. This was evident from the fact that when a sublethal dose of fungal spores was injected in animals without macrophage depletion, the mice were able to eliminate the fungal load to a considerable extent and survived even after the 6th day of infection. In contrast, with the same sublethal dose of fungal infection in macrophage depleted mice, the fungal load increased progressively over time and there was 100% mortality after 3 days of infection (Moonis *et al.*, 1993b).

The enhanced toxicity of liposomal Amp B in macrophage depleted mice could possibly be due to the absence of resident macrophages/PMN cells in spleen and liver resulting in an increase in the concentration of the L-Amp B in circulation. The increased circulation time of the liposomal drug in turn may enhance the capability of the drug to interact with sensitive cells and thus to increase its toxicity (Moonis *et al.*, 1993b). The importance of resident macrophages/ PMN cells in the control of the disease process was established by the fact that there was 80% survival of the infected animals which were given L-Amp B therapy after 7 days whereas in macrophage/PMN cells depleted mice there was 100% mortality of infected animals wherein L-Amp B therapy followed the infection. Moreover, the fungal load did not go down in macrophage depleted mice in contrast to the progressive decrease that was seen in animals without macrophage depletion after L-Amp B therapy. In earlier studies, we had observed an alleviating effect of Amp B therapy incorporated in mannosylated liposomes (Ahmad *et al.*, 1991). However, in macrophage depleted mice, Amp B incorporated into mannosylated liposomes could not increase the survival of infected mice which further substantiates our contention that resident macrophages/PMN cells play an important role in the capture of the drug in the diseased tissue enhancing thereby its effectiveness against the pathogen.

The most pronounced effect of macrophage depletion on tissue distribution of liposomal Amp B was in terms of an increase in the Amp B concentration in circulation, a lower concentration in the liver and a higher concentration in lung, spleen and kidney as compared to animals without macrophage depletion. The increase in the concentration of Amp B in the spleen may possibly be due to phagocytic activity of other cells like sinus lining cells of the spleen which have been shown to take over part of the phagocytic function (Claassen & van Rooijen, 1984).

CONCLUSIONS

The observations on the *in vivo* toxicity, therapeutic efficacy and tissue distribution of liposomal Amp B with and without macrophage depletion show that resident macrophages and PMN cells play an important role in the containment of aspergillosis virulence although *Aspergillus fumigatus* is not an intracellular pathogen. Also, resident macrophages/PMN cells play a pivotal role in the clearance of liposomal Amp B from circulation which may be responsible for the decreased toxicity of the drug.

Acknowledgements

B.K.B. is a Bhatnagar Fellow of C.S.I.R., India and honorary Professor,Jawaharlal Nehru Center for Advanced Scientific Research, Bangalore. The award of Senior Research Fellowship to Mona Moonis by CSIR is thankfully acknowledged . We are also grateful to the Department of Biotechnology for financial assistance. Our thanks are also due to Dr. Imran Ahmad for his collaboration in various studies.

References

Ahmad, I., Sarkar, A.K. & Bachhawat, B.K. (1989). Design of liposomes to improve delivery of amphotericin B in the treatment of aspergillosis. Mol. Cell. Biochem. **91**, 85-90.

Ahmad, I., Sarkar, A.K. & Bachhawat, B.K. (1990). Effect of cholesterol in various liposomal compositions on the in vivo toxicity, therapeutic efficacy and tissue distribution of amphotericin B. Biotechnol. Appl. Biochem. **12**, 550-556.

Ahmad, I., Sarkar, A.K. & Bachhawat, B.K. (1991). Mannosylated liposome-mediated delivery of amphotericin B in the control of experimental aspergillosis in Babl/c mice. J. Clin. Biochem. Nutr. **10**, 171-179.

Alving, C.R., Steck, E.A., Hanson, W.L., Loizeaux, P.S., Chapman, W.L. & Waits, V.B. (1978). Improved therapy of experimental leishmaniasis by use of a liposome-encapsulated antimonial drug. Life Sci. **22**, 1021-1026.

Alving, C.R. (1983). Delivery of liposome-associated drugs to macrophages. Pharmacol. Ther. **22**, 407-424.

Andreoli, T.E. (1974). The structure and function of amphotericin B-cholesterol pores in lipid bilayer membranes. Ann. N.Y. Acad. Sci. **235**, 448-468.

Bjorkholm, M., Kallberg, N., Grimfors, G., Eklund, L.H., Eksborg, S., Juneskans, O.T. & Uden, A.M. (1991). Successful treatment of hepatosplenic candidiasis with liposomal amphotericin B preparation. J. Int. Med. **230**, 173-177.

Bolard, J. (1986). How do the polyene macrolide antibiotics affect the cellular membrane properties. Biochem. Biophys. Acta. **864**, 257-304.

Bolard, J. (1991). Mechanism of action of an anti-candida drug: Amphotericin B and its derivatives. In *Candida albicans,* Cellular and Molecular Biology, Edited by R. Prasad. Springer Verlag. N.Y. pp 214-238.

Claassen, E. & Van Rooijen, N. (1984). The effect of elimination of macrophages on the tissue distribution of liposomes containing [^{3}H] methotrexate. Biochem. Biophys. Acta. **802**, 428-434.

Cohen, M.S., Isturiz, R.E., Malech, H.L., Root, R.K., Wilfert, C.M., Gutman, L. & Buckley, R.H. (1981). Fungal infection in chronic granulomatous disease. Am. J. Med. **71**, 59-66.

Cybulska, B., Mazerski, J., Zielinski, J., Ziminski, T. & Borowski, E. (1980). The role of structural factors in the modification of membrane permeability to external ions induced by polyene macrolide antibiotics. Drugs Exptl. Clin. Res. **6**, 449-456.

Cybulska, B., Borowski, E., Prigent, Y. & Gary-Bobo, C.M. (1981). Cation permeability induced by two aromatic heptaenes vacidin A and candicidin D on phospholipid unilamellar vesicles. J. Antibiot. **34**, 884-891.

Cybulska, B., Borowski, E., Prigent, Y. & Gary-Bobo, C.M. (1984). Hemolytic activity of aromatic heptaenes: A group of polyene macrolide antifungal antibiotics. Biochem. Pharmacol. **33**, 41-46.

Cybulska, B., Herve, M., Borowski, E. & Gary-Bobo, C.M. (1985). Effect of the polar head structure of polyene macrolide antifungal antibiotics on the mode of permeabilization of ergosterol - and cholesterol containing lipidic vesicles studies by ^{31}P-NMR. Mol. Pharmacol. **29**, 293-298.

De Kruijff, B., Gerritsen, W.J., Oerlemans, A., Demel, R.A. & Van Deenen, L.L.M. (1974). Polyene antibiotic-sterol interaction in membranes of *Acholeplasma laidlawii* cells and lecithin liposomes. I. specificity of the membrane permeability changes induced by the polyene antibiotics. Biochem. Biophys. Acta. **339**, 30-43.

De Kruijff, B. & Demel, R.A. (1974). Polyene antibiotic-sertol interactions in membranes of *Acholeplasma laidlawii* cells and lecithin liposomes. III. Molecular structure of the polyene antibiotic-cholesterol complexes. Biochem. Biophys. Acta. **339**, 57-70.

Frazer, D.W., Ward, J.I., Ajello, L. & Plikaytis, B.D. (1979). Aspergillosis and other systemic mycoses. J.A.M.A. **242**, 1631-1635.

Gale, E.F. (1984). Mode of action and resistance mechanisms of polyene macrolides. In Macrolide Antibiotics; Chemistry, Biology and Practice. Edited by S. Omura. Academic Press, Orlando, pp 425-455.

Hammond, S.M. (1977). Biological activity of polyene antibiotics. In Progress in Medicinal Chemistry. Edited by G.P. Ellis and G.B. West. Amsterdam, Elsevier/North Holland Biomedical, pp 105-179.

Janoff, A.S., Boni, L.T., Popescu, M.C., Minchey, S.R., Cullis, P.R., Madden, T.D., Taraschi, T., Gruner, S.M., Shyam Sunder, E., Tate, M.W., Mendelsohn, R. & Bonner, D. (1988). Unusual lipid structures selectively reduce the toxicity of amphotericin B. Proc. Natl. Acad. Sci. **85**, 6122-6126.

Kasumov, Kh. M., Mekhtiev, N.K. & Karakozov, S.D. (1981). Potential-dependent formation of single conducting ion channels in lipid bilayers induced by the polyene antibiotic Levorin A_2. Biochem. Biophys. Acta. **372**, 141-153.

Kotler-Brajtburg, J., Medoff, G., Kobayashi, G.S., Boggs, G.S., Schlessinger, D., Pandey, R.C. & Rinehalt, K.L. (1979). Classification of polyene antibiotics according to chemical structure and biological effects. Antimicrob. Agents Chemother. **15**, 716-722.

Kotler-Brajtburg, J., Medoff, G., Kobayashi, G.S. & Elberg, S. (1980). Influence of extracellular K^+ or Mg^{2+} on the stages of the antifungal effects of amphotericin B and filipin. Antimicrob. Agents. Chemother. **18**, 593-597.

Liras, P. & Lampen, J.O. (1974). Sequence of candicidin action on yeast cells. Biochem. Biophys. Acta. **372**, 141-153.

Mazerski, J., Bolard, J. & Borowski, E. (1982). Self association of some polyene macrolide antibiotics in aqueous media. Biochim. Biophys. Acta. **719**, 11-17.

Mechlinski, W., Schaffner, C.P., Ganis, P. & Avitabile, G. (1970). Structure and absolute configuration of the polyene macrolide antibiotic amphotericin B. Tetrahedron. Lett. **44**, 3873-3876.

Medoff, G., Kotler-Brajtburg, J., Kobayashi, G.S. & Bolard, J. (1983). Antifungal agents useful in therapy of systemic fungal infections. Ann. Rev. Pharmacol. Toxicol. **23**, 303-330.

Mehta, R.T., Lopez-Berestein, G., Hopfer, R.L., Mills, K. & Juliano, R.L. (1984). Liposomal amphotericin B is toxic to fungal cells but not to mammalian cells. Biochem. Biophys. Acta. **770**, 230-234.

Mehta, R.T. (1989). Liposomes as drug carriers for polyene antibiotics. Adv. Drug Deliv. Rev. **3**, 283-306.

Moonis, M., Ahmad, I. & Bachhawat, B.K. (1992). Mannosylated liposomes as carriers for hamycin in the treatment of experimental aspergillosis. J. Drug Targeting (In Press).

Moonis, M., Ahmad, I. & Bachhawat, B.K. (1993a). Liposomal hamycin in the control of experimental aspergillosis in mice: Effect of phosphatidic acid with and without cholesterol. J. Antimicrob. Chemother. **31**, 00-00 (In Press).

Moonis, M., Ahmad, I., & Bachhawat, B.K. (1993b). Effect of elimination of phagocytic cells by liposomal dichloromethylene diphosphonate on aspergillosis virulence and toxicity of liposomal amphotericin B in mice (communicated).

New, R.R.C., Chance, M.L. & Health, S. (1981). Antileishmanial activity of amphotericin B and other antifungal agents entrapped in liposomes. J. Antimicrob. Chemother. **8**, 371-381.

Omura, S. & Tanaka, H. (1984). Production, structure and antifungal activity of polyene macrolides. In Macrolide Antibiotics, Chemistry, Biology and Practice, Edited by S. Omura. Academic Press, Orlando, pp 351-424.

Poznasky, M.J. & Juliano, R.L. (1984). Biological approaches to the controlled delivery of drugs: A critical review. Pharmacol. Rev. **36**, 277-336.

Rooijen, N.V. & Niewmegen, R.V. (1984). Elimination of phagocytic cells in the spleen after intravenous injection of liposome - encapsulated dichloromethylene diphosphonate. Cell Tissue Res. **238**, 355-358.

Szoka, F.A. (1991). Liposomal drug delivery. In Membrane Fusion, Edited by J. Wilschut and D. Hoekestra. Marcel Dekker, New York. pp 845-890.

Witzke, N.M. & Bittman, R. (1984). Dissociation kinetics and equilibrium binding properties of polyene antibiotic complexes with phosphatidyl choline/sterol vesicles. Biochemistry **23**, 1668-1674.

10

A cost/benefit analysis of the 3'→5' proofreading exonuclease activity of three DNA polymerases

Navin K. Sinha[1] and Kailash C. Upadhyaya[2]
Waksman Institute of Microbiology
Rutgers-The State University of New Jersey
Piscataway, NJ 08855

ABSTRACT

The efficiency with which 8 different mispairs (T.C., T.G., T.T., A.A., A.C., A.G., G.A., and G.G.) are proofread by the 3'-->5' exonuclease activity of 3 DNA polymerases [*E. Coli* polymerase I (Klenow) and bacteriophage T4 and T7 polymerases] during DNA synthesis *in vitro* has been determined using a sensitive infectivity assay. The cost, as wasteful hydrolysis of correctly inserted deoxynucleotides, during ongoing DNA synthesis has also been measured. The results show that the proofreading function of phage T4 DNA polymerase is approximately 100-fold and 600-fold more cost-efficient than the *E. coli* DNA polymerase I (Klenow) and phage T7 DNA polymerase, respectively. The specificity of proofreading (i.e., the hierarchy of least- to most-edited mispairs) of *E. coli* DNA polymerase I and phage T7 DNA polymerase are quite similar and very different than that of the phage T4 enzyme.

INTRODUCTION

Studies of mutation rates in organisms having genomes of different sizes reveal that there is an optimal mutation load of approximately 3×10^{-3} mutations per genome per doubling (Drake, 1991). Because of this constancy of mutation load, the accuracy of DNA replication has to increase as the genome size increases. For the replication of viral DNA, an error rate of 10^{-6} is sufficient. However, cellular organisms, whose genomes are $10\text{-}10^4$ fold larger than the largest viral genomes, require an accuracy greater than this. For example, in *E. coli* average mutation rate is approximately 10^{-9}. In eukaryotes, whose genomes are 100 to 1,000-fold larger, even greater accuracy is probably needed (at least for the protein information coding, non-repeated sequences). From the study of the replication of genomes of *E. coli* and its bacteriophages, a

---

[1]Supported by grants from the USPHS (GM-24391) and Charles and Johanna Busch Foundation.
[2]Recipient of a Merck Foundation Foreign Scholar award.

fairly detailed understanding of the mechanisms involved in achieving this accuracy has emerged (Kornberg & Baker, 1991; Echols & Goodman, 1991).

Replicative DNA polymerases have a high enough accuracy to satisfy the requirements of viral genomes. DNA polymerases achieve this by a 2-step process. In the first step (deoxynucleotide polymerization), an error rate of approximately 10^{-4} is achieved (Kornberg & Baker, 1991; Echols & Goodman, 1991). Thermodynamic studies of base-pairs in aqueous solutions suggest that the correct base pairs should be approximately 100-fold more stable than incorrect base-pairs due to the energy contributed by 2-3 H-bonds in the correct base-pairs. Non-enzymatic nucleotide polymerization studies have confirmed these predictions (Orgel & Lohrmann, 1974). Though the exact mechanisms involved in obtaining the 100-fold greater accuracy during the nucleotide insertion step in the DNA polymerase catalyzed reaction are not yet fully understood, two mechanisms have been proposed. It is thought that the active site in DNA polymerases is constructed such that only the correct Watson-Crick base pairs fit properly. All other base pairs are strongly discriminated against due to presumed unfavorable stearic hinderances (Topal & Fresco, 1972). It has also been proposed that the active site in the DNA polymerase is relatively free of water. As a result of this hydrophobic environment, the energetic contribution of H-bonds is increased (Petruscka *et al.*, 1986).

DNA polymerases lower the error rate during DNA replication to 10^{-6} or less, by a second scrutiny of the inserted nucleotide using the $3' \rightarrow 5'$ editing exonuclease activity. Use of primer chains having a mismatched nucleotide at their 3'-ends revealed that the unpaired nucleotide is removed by the editing exonuclease before DNA synthesis can occur (Brutlag & Kornberg, 1972; Muzyczka *et al.*, 1972). Mutants of bacteriophage T4 that have lower than normal levels of the editing exonuclease activity have an elevated overall mutation rate ("mutator" mutants) and mutations having a higher than wild type level of the $3' \rightarrow 5'$ exonuclease have a lower mutation rate ("antimutator" mutants). These observations (Muzyczka *et al.*, 1972) indicated that the $3' \rightarrow 5'$ exonuclease activity, that is associated with replicative DNA polymerases, also plays an editing function during DNA replication *in vivo*. In *E. coli*, there is a third correction step (postreplication mismatch repair) which further raises the accuracy of DNA copying by as much as thousand-fold (Schaaper & Dunn, 1987).

Over the past 20-years (since the original observation of Brutlag & Kornberg in 1972), several different approaches have been used to try to decipher the specificity and efficiency of editing of different mismatches during DNA synthesis (Brutlag & Kornberg, 1972; Sinha, 1987; Kuchta *et al.*, 1988). Perhaps, the most sensitive assay is the one developed by Sinha (1987). In this assay deoxy-oligonucleotides containing a mismatched nucleotide at the 3'-end of the primers, hybridized to ØX174 single-stranded viral genome from an amber mutant, are extended using a DNA polymerase *in vitro* (Fig. 10.1). The fraction of the 3'-terminal mismatch surviving the editing function of the DNA synthesis is determined by transfecting the product of the DNA synthesis reaction into *E. coli* spheroplasts and measurement of the frequency of revertant phages produced. This assay can detect as little as 1 mispair in 10^{6} primer termini escaping the editing exonuclease activity of a DNA polymerase (see Sinha, 1987 for further details). The primer and G.G) are shown in Fig. 10.1. Here, we summarize the results of studies measuring the efficiency and specificity of editing of these mispairs by 3 DNA polymerases- E. coli DNA polymerase I (Klenow fragment), and the polymerases of bacteriophages T4 and T7

A

am16 codon (5276-5278)   ØX174 am16 ss DNA

5'   3'

GATGAGATTTAGGCTGGGAAAAGTTACTGTAGCCGACGTTTTGGCGGCGCAACCTGTGAC

MISPAIR			PRIMER
None		3' CGCCGCGTTGGACACACTG 5'	5309-26'
T.C	CTCCGACCCTTTTCAA		5276'C
T.C+9	CTACTCTAACTCCGACCCTTTTCAA		5276'C+9
T.G	GTCCGACCCTTTTCAA		5276'G
T.G+9	CTACTCTAAGTCCGACCCTTTTCAA		5276'G+9
T.T	TTCCGACCCTTTTCAA		5276T
T.T+9	CTACTCTAATTCCGACCCTTTTCAA		5276T+9
A.C	CCCGACCCTTTTCAAT		5277'C
A.C+9	TACTCTAAACCCGACCCTTTTCAAT		5277'C+9
A.G	GCCCGACCCTTTTCAAT		5277'G
A.G+9	TACTCTAAAGCCGACCCTTTTCAAT		5277'G+9
G.A	ACGACCCTTTTCAATG		5278'A
G.A+9	ACTCTAAATACGACCCTTTTCAATG		5278'A+9
G.G	GCGACCCTTTTCAATG		5278'G
G.G+9	ACTCTAAATGCGACCCTTTTCAATG		5278'G+9

B

am3 codon (586-588)   ØX174 am3 ss DNA

5'   3'

GGACTTTGTAGGATACCCTCGCTTTCCTGCTCCTGTTGAGTT

None		3' GGACGAGGACAACTCAA 5'	603-619'
A.A		3' ACCTATGGGAGCGAAA 5'	587'A
A.A+9	CCTGAAACAACCTATGGGAGCGAAA		587'A+9
A.C		CCCTATGGGAGCGAAA	587'C
A.C+9	CCTGAAACACCCTATGGGAGCGAAA		587'C+9
A.G		GCCTATGGGAGCGAAA	587'G
A.G+9	CCTGAAACAGCCTATGGGAGCGAAA		587'G+9

Fig. 10.1: (A) The nucleotide sequence of bacteriophage ØX174 single-stranded DNA near the amber 16 codon (gene B⁻) and the synthetic deoxyoligonucleotide primers used to measure the editing efficiency of the 3'--> 5' exonuclease activity of phage T4 DNA polymerase. All primers with a mismatch at their 3'-OH ends numbered by the position of the mismatched nucleotide. Thus, 5276'C indicates that the primer contains a cytosine on the ØX174 complementary (-) strand at position 5276 of the Sanger *et al.* 1978. Primer 5276'C+9 refers to a primer that is identical to the 5276'C prime except for nine correct nucleotides beyond the mismatched nucleotide. All primers with 3'-terminal mispair are 16-mers, and mismatched+9 primers are 25-mers. These primers allow the measurement of editing of T.C, T.G, T.T, A.C, A.G, G.A and G.G mismatches. The primers named correct and correct+9 are similar to these primers, except that they pair correctly with ØX174 amber mutant DNA template. **(B)** Nucleotide se quence of ØX174 DNA in the vicinity of the amber 3 (gene E⁻) codon and the primers which allow the measurement of the editing specificity and efficiency of A.A, A.C and A.G mispairs at this site.

Table 10.1: Efficiency of editing of mispairs at the φX174 am 16 and am3 sites by the 3'→5' exonuclease activity of bacteriophage T4 DNA polymerase

Primer	Terminal mispair (Template-Primer)	Observed frequency of revertants[*]		Expression of Mismatch[*]	Corrected frequency of Revertants[*]
		-Synthesis	+Synthesis		
amber 16 mutant					
Mock-Primed	None	1.4×10^{-7}	$<3.1 \times 10^{6}$	---	$<3.1 \times 10^{-6}$
5309-5326	None	2.2×10^{-7}	1.0×10^{-6}	---	1.2×10^{-6}
5276'C	T.C.	3.4×10^{-6}	1.4×10^{-5}	0.49	2.7×10^{-7}
5276'G	T.G.	2.0×10^{-5}	8.0×10^{-5}	0.57	1.4×10^{-4}
5276'T	T.T.	3.7×10^{-7}	1.4×10^{-6}	0.59	$<1 \times 10^{-6}$
5277'C	A.C.	2.3×10^{-7}	4.6×10^{-7}	0.23	$<1 \times 10^{-6}$
5277'G	A.G.	5.7×10^{-7}	2.1×10^{-5}	0.62	3.2×10^{-5}
5278'G	G.G.	1.4×10^{-7}	8.6×10^{-7}	0.50	$<1 \times 10^{-6}$
5278'A	G.A.	2.5×10^{-7}	2.0×10^{-6}	0.74	1.3×10^{-6}
amber 3 mutant					
Mock-Primed	None	2.5×10^{-6}	2.5×10^{-6}	---	$\leq 2.5 \times 10^{-6}$
603-619	None	2.0×10^{-6}	2.7×10^{-6}	---	$\leq 2.7 \times 10^{-6}$
587'A					
587'C	A.A.	2.8×10^{-6}	6.5×10^{-6}	0.50	8.0×10^{-6}
587'G	A.C.	6.6×10^{-6}	8.9×10^{-6}	0.24	2.7×10^{-5}
	A.G.	2.2×10^{-6}	1.8×10^{-4}	0.42	4.3×10^{-4}

[*] These values were obtained using infective center assays at 30°C. The values for the expression of the mismatches were obtained by doing DNA synthesis *in vitro* and infectivity assays in parallel using mismatched+9 primers described in Fig. 1. For further details see Sinha (1987).

Results obtained with the bacteriophage T4 DNA polymerase are shown in Table 10.1. This primer chains that permit the measurement of the efficiency of editing of all 8 possible mismatches at an amber codon that can lead to viable revertants (T.C, T.G, T.T, A.C, A.G, A.A, G.A enzyme edits all 8 mismatches with a very high efficiency. The frequency of revertants observed (due to escape from the editing exonuclease function) ranges from 8×10^{-5} (G.T mismatch) to less than 10^{-5} (G.T mismatch) to less than 10^{-6} for A.C and G.C mismatches at the amber 16 mutation site in the ØX174 template DNA. Extension of primers having the same mismatches, but with an additional 9 correct nucleotides past the mismatch, was used to determine the efficiency of expression of these mismatches after transfection into amber su⁻ bacterial cells.

This ranges from 0.24 to 0.74 (Table 10.1) indicating that the mismatches that escaped editing have a fairly high probability (24 to 74%) of yielding a revertant plaque. The actual efficiency of editing of these 8 mispairs can be obtained by correcting the observed frequency of revertants with the probability of expression of that mismatch. The data in Table 1 show that the T.G mismatch at the amber 16 site is edited most poorly (1 in 7,000 mismatched escapes editing) and several mismatches (T.T, A.G, G.G and G.A at the amber 16 site and A.A and A.C at the am3 site) are edited very efficiently (less than 1 in 10^5 mismatches survives the editing step). Based on these results, the specificity of editing of these mispairs by the T4 DNA polymerase is $T_{template} G_{primer} < (A.G, T.C) < (T.T, G.A, G.G, A.C) < A.A$. A comparison of the efficiencies of editing of those mismatches that are common to both the am16 and the am3 site (A.C and A.G), indicates that the efficiency of editing depends not only upon the mispair, but also on the neighbouring nucleotide sequence (beyond the nearest neighbour nucleotides). Results of similar experiments measuring the efficiency and specificity of editing with bacteriophage T7 DNA polymerase are shown in Table 10.2. This enzyme edits these mismatches much less efficiently. The observed frequency of revertants range from 8.6×10^{-2} to 2.0×10^{-4}.

Actual efficiency of editing, after corrections for the expression of these mismatches (Table 10.2), ranges from 0.18 (for A.C mismatch at the am3 site) to 3.6×10^{-4} (for the G.G mispair at the am16 site). These results indicate that the specificity of editing by phage T7 DNA

Table 10.2: Efficiency of editing of mismatches at the ϕX 174 am 16 and am3 sites by the 3' → 5' exonuclease editing activity of phage T7DNA Polymerase

Primer	Terminal Mispair	Reversion Pathway	Observed frequency of Revertants	Expression of Heteroduplex	Corrected frequency of Revertants
		OX174 am16 site			
5276'C	T.C.	TAG -> GAG	3.8×10^{-2}	0.51	7.4×10^{-2}
5276'G	T.G.	-> CAG	4.2×10^{-2}	0.45	9.3×10^{-2}
5276'T	T.T.	-> AAG	9.8×10^{-3}	0.25	3.9×10^{-2}
5277'C	A.C.	-> TGG	2.7×10^{-3}	0.095	2.8×10^{-2}
5277'G	A.G.	-> TCG	6.8×10^{-4}	0.18	3.8×10^{-3}
5278'A	G.A.	-> TAT	2.0×10^{-4}	0.37	5.5×10^{-4}
5278'G	G.G.	-> TAC	5.4×10^{-4}	0.15	3.6×10^{-4}
		OX174 am3 site			
587'A	A.A.	TAG -> TTG	1.4×10^{-3}	0.72	2.0×10^{-3}
587'C	A.C.	-> TGG	8.6×10^{-2}	0.48	1.8×10^{-1}
587'G	A.G.	-> TCG	2.7×10^{-3}	0.50	5.4×10^{-3}

polymerase is (T.G, A.C, T.C and T.T) < (A.A, A.G) < (G.A and G.G). Purine purine mismatches are edited out most efficiently.

The efficiencies of editing by the *E. coli* DNA polymerase I (Klenow) are shown in Table 10. 3. The observed frequencies of revertants ranges from 0.11 (for T.G mispair at the am16 site) to 3.5×10^{-5} (for G.A mispair at the am16 site). The corrected frequencies of escape from the 3' → 5' exonuclease editing function vary from 0.82 (T.G) to 3.5×10^{-4} (G.A). T.G mispair is left essentially unedited (82% of this mispair survives the editing step). In contrast, only about 1 in 1,000 of the A.G and G.A mispairs is left unexcised from the DNA primer terminus. The specificity of editing is (T.G, T.C, A.C) < (T.G, G.G) < (A.G, G.A).

The efficiencies of editing of mispairs by these 3 DNA polymerases are compared in Table 4. The 3'→ 5' exonuclease activities of both the phage T7 DNA polymerase and the *E. coli* DNA

Table 10.3: Efficiency of editing of mismatches at the ϕX 174 am 16 am3 sites by the 3' → 5' exonuclease editing activity of *E. coli* DNA polymerase 1 (Klenow)

Primer	Terminal Mispair	Reversion Pathway	Observed frequency of Revertants	Expression of Heteroduplex	Corrected frequency of Revertants
		OX174 am16 site			
Mock-Primed	None	---	$<5.1 \times 10^{-6}$	---	$<5.1 \times 10^{-6}$
5309-5326	None	---	5.4×10^{-6}	---	$\sim 5 \times 10^{-5}$
5276'C	T.C.	TAG -> GAG	8.4×10^{-2}	0.17	4.9×10^{-1}
5276'G	T.G.	-> CAG	1.1×10^{-1}	0.13	8.2×10^{-1}
5276'T	T.T.	-> AAG	3.2×10^{-2}	0.13	2.5×10^{-2}
5277'C	A.C.	-> TGG	6.5×10^{-3}	0.026	2.5×10^{-1}
5277'G	A.G.	-> TCG	3.9×10^{-5}	0.098	4.0×10^{-4}
	G.A.	-> TAT	3.5×10^{-5}	0.10	3.5×10^{-4}
5278'A 5278'G	G.G.	-> TAC	9.7×10^{-4}	0.088	1.1×10^{-2}
		OX174 am3 site			
Un-primed	None	----	4.8×10^{-6}	---	----
603-619	None	----	1.1×10^{-5}	(0.1-0.2)	$\sim 4.5 \times 10^{-5}$
587'A	A.A.	TAG -> TTG	1.6×10^{-3}	0.12	1.3×10^{-2}
587'C	A.C.	-> TGG	2.3×10^{-2}	0.11	2.1×10^{-1}
587'G	A.G.	-> TCG	3.2×10^{-4}	0.16	2.1×10^{-3}

polymerase I are much less efficient at editing of mismatches than the phage T4 enzyme. All mispairs (except for the $A_{template}$. G_{primer} mispair) are edited, at least 100-fold worse. The greatest difference in the efficiencies of editing is seen for the A.C and the T.T mispairs (greater than 10^4-fold). The specificities of the editing nuclease activities of phage T7 and *E. coli* DNA polymerase I (Klenow) are very similar (Table 10.4). These range from essentially identical efficiencies for the editing of G.A mispair at the am16 site and A.C mispair at the am3 site to as much as 31-fold difference for the editing of G.G mispair at the am16 site. For all, but one, mispair the efficiencies of exonucleolytic editing differ by less than 10-fold for these 2 enzymes. In contrast, not only is the efficiency of editing of mispairs much greater for the phage T4 DNA polymerase, but the specificity is quite different. Only one mispair (A.G. at both the am3 and the am16 sites) has relatively similar efficiency of editing (5 to 120 fold- difference). All other mispairs are edited with very different efficiencies (to as much as 250,000-fold for the A.C mismatch at the am16 site for the phage T4 and *E. coli* pol I (Klenow) (for comparison see Table 10.4).

The great similarity between the editing specificities of the phage T7 and *E. coli* DNA

Table 10.4: The editing specificities of *E. coli* DNA polymerase I (Klenow) and phage T7 and T4 DNA polymerases

φX174 Mutant	Mispair	Unedited Fraction			Relative Efficiency of Editing		
		E.coli	T7	T4	T7/ E.coli	T4/ E.coli	T4/T7
am3	A.A.	1.3×10^{-2}	2×10^{-3}	8.0×10^{-6}	6.5	1,600	250
	A.C.	2.1×10^{-1}	1.8×10^{-1}	2.7×10^{-5}	1.2	7,800	6,670
	A.G.	2.1×10^{-3}	5.4×10^{-3}	4.3×10^{-4}	0.4	5	13
am16	T.C.	4.9×10^{-1}	7.4×10^{-2}	2.7×10^{-5}	6.6	18,150	2,740
	T.G.	8.2×10^{-1}	9.3×10^{-2}	1.4×10^{-4}	8.8	590	66
	T.T.	2.5×10^{-2}	3.9×10^{-2}	$<1 \times 10^{-6}$	6.4	>25,000	>39,000
	A.C.	2.5×10^{-1}	2.8×10^{-2}	$<1 \times 10^{-6}$	8.9	>250,000	>28,000
	A.G.	4×10^{-1}	3.8×10^{-3}	3.2×10^{-5}	0.11	13	120
	G.A.	3.5×10^{-4}	5.5×10^{-4}	1.3×10^{-6}	0.64	270	420
	G.G.	1.1×10^{-2}	3.6×10^{-4}	$<1 \times 10^{-6}$	31	>11,000	360

polymerase I (Klenow fragment) editing exonucleases is not surprising since the amino acid sequences of these enzymes are very similar, but only in the exonuclease portion of the enzymes (Ollis *et al.*, 1985). In contrast, the amino acid sequence of the bacteriophage T4 DNA polymerase shows no homology with the amino acid sequences of T7 DNA polymerase and *E. coli* DNA polymerase I (Spicer *et al.*, 1988). Therefore, it is to be expected that the relative efficiencies with which different mismatches are edited out is dissimilar for T4 DNA polymerase in comparison to the editing specificities of the other 2 polymerases used in this study.

The $3' \to 5'$ editing exonuclease function of DNA polymerases, in addition to excising mispaired nucleotides at the primer termini, also removes a certain fraction of correctly inserted deoxynucleotides, presumably, because these had become unpaired momentarily ("frayed end"), and, as a result the editing exonuclease recognized these as incorrectly paired nucleotides. Bacteriophage T7 DNA polymerase has been reported to have the highest ratio of the $3' \to 5'$ exonuclease activity, relative to the DNA polymerase activity, among all known DNA polymerases (Tabor *et al.*, 1987). In an assay, where the DNA polymerase is used exclusively as a $3' \to 5'$ exonuclease due to the complete absence of the deoxynucleoside triphosphate substrates, T7 and T4 DNA polymerases had approximately 1000-fold and 250-fold greater activities, respectively than *E. coli* DNA polymerase I (Tabor *et al.*, 1987; Kornberg & Baker, 1991).

Perhaps a more meaningful estimate of the ratio between the DNA polymerase and the $3' \to 5'$ exonuclease activities is given by the fraction of the correctly inserted deoxynucleotides that are wastefully excised during ongoing DNA synthesis. In such an experiment, given a high, saturating concentration of the dNTP substrates (so that the DNA polymerase activity is at its maximum and the editing exonuclease activity is at its minimum), approximately 1.2% of the inserted labeled dATP substrate is released as dAMP by *E. coli* DNA polymerase I (Klenow). T4 DNA polymerase wastes approximates 13% of the dATP substrate (data not shown). Bacteriophage T7 DNA polymerase is the most wasteful. It excises approximately 40% of the correctly inserted labeled dAMP (Fig. 10.2).

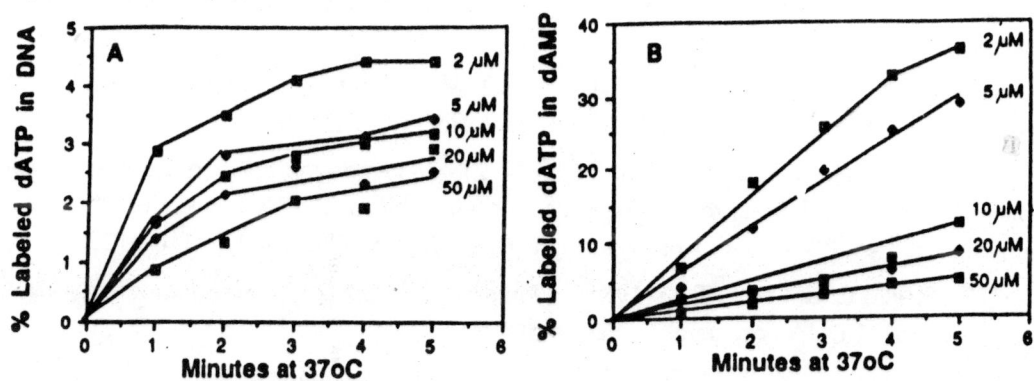

Fig. 10.2: Incorporation and excision of dATP-α-^{32}P by phage T7 DNA polymerase. DNA was synthesized under standard conditions (Sinha, 1987) on ØX174 single-stranded DNA template primed with 603'-619' primer (Fig. 10.1). The concentration of all 4 dNTP substrates together was varied from 2 to 50 μM each. DNA polymerase concentration was 0.1 unit/30 μl reaction. The rates of incorporation of [32]P dATP into DNA and excision as dAMP were simultaneously measured by thin layer chromatography on polyethylene-imine-cellulose coated sheets. (A) Incorporation of dATP into ØX174 DNA. (B) Excision of dAMP after insertion of the labeled deoxynucleotide into DNA.

CONCLUSIONS

Since the phage T4 polymerase is at least 1,000-fold more efficient in editing out most mismatches (Table 10.4) at an approximately 10-fold higher cost in wasted correct deoxynucleotide substrates, this enzyme uses its editing exonuclease activity most efficiently (at least 100-fold more efficiently than the *E. coli* Klenow polymerase). Phage T7 enzyme edits most mismatches about 5-fold more efficiently than the *E. coli* (Klenow) polymerase but at an approximately 30-fold higher cost in wasted deoxynucleotide substrates. Therefore, the editing $3' \rightarrow 5'$ exonuclease activity of phage T7 is about 5-fold more efficient in editing mismatches but at a 6-fold higher cost than the *E. coli* enzyme. It remains to be determined why the phage T7 enzyme is more wasteful than the *E. coli* enzyme despite having a very similar structure of the exonuclease domain.

References

Brutlag, D. & Kornberg, A. (1972). Enzymatic synthesis of deoxyribonucleic acid. XXXVI. A proofreading function for the $3' \rightarrow 5'$ exonuclease activity in deoxyribonucleic acid polymerase. J. Biol. Chem. **247**, 7116-7122.

Drake, J. (1991). A constant rate of spontaneous mutation in DNA-based microbe. Proc. Natl. Acad. Sci. U.S.A. **88**, 7160-7164.

Echols, H. & Goodman, M.F. (1991). Fidelity mechanisms in DNA replication. Ann. Rev. Biochem. **60**, 477-511.

Kornberg, A. & Baker, T. (1991). DNA Replication, Second, Edited by W.H. Freeman and Company, and Francisco.

Kuchta, R.D., Benkovic, P. & Benkovic, S.J. (1988). Kinetic mechanisms whereby DNA polymerase I (Klenow) replicates DNA with high fidelity. Biochemistry **27**, 6716-6725.

Muzyczka, N., Poland, R.L. & Bessman, M.J. (1972). Studies on the biochemical basis of spontaneous mutations. I. A comparison of deoxyribonucleic acid polymerases of mutator, antimutator and wild type strains of bacteriophage T4. J. Biol. Chem. **247**, 7116-7122.

Ollis, D.L., Kline, C. & Steitz, T.A. (1985). Domain of *E. coli* polymerase I showing sequence homology to T7 DNA polymerase. Nature **313**, 818-819.

Orgel, L.E. & Lohrmann, R. (1974). Prebiotic chemistry and nucleic acid replication. Acc. Chem. Res. **1**, 368-377.

Petruscka, J., Sower, L. & Goodman, M.F. (1986). Comparison of nucleotide interactions in water, proteins and vacuum: Model for DNA polymerase fidelity. Proc. Natl. Acad. Sci. U.S.A. **83**, 1559-1562.

Sanger, F., Coulson, A.R., Friedmann, T., Air, G.M., Barrell, B.G., Brown, N.L., Fiddes, J.C., Hutchison, C.A., Slocombe, P.M. & Smith, M. (1978). The nucleotide sequence of bacteriophage ØX174. J. Mol. Biol. **125**, 225-246.

Schaaper, R.M. & Dunn, R.L. (1987). spectra of spontaneous mutations in *Escherichia coli* strains defective in mismatch correction: The nature of *in vivo* DNA replication errors. Proc. Natl. Acad. Sci. U.S.A. **84**, 6220-6224.

Sinha, N. (1987). Specificity and efficiency of editing of mismatches involved in the formation of base-substitution mutations by the 3' → 5' exonuclease activity of phage T4 DNA polymerase. Proc. Natl. Acad. Sci. U.S.A. **84**, 915-919.

Spicer, E.K., Rush, J., Fung, C., Reha-Kranzt, L.J., Karam, J.D. & Konigsberg, W.H. (1988). Primary structure of DNA polymerase: Evolutionary relatedness to eukaryotic and other prokaryotic DNA polymerases. J. Biol. Chem. **263**, 7478-7486.

Tabor, S., Huber, H.E. & Richardson, C.C. (1987). *Escherichia coli* thioredoxin confers processivity on the DNA polymerase activity of the gene 5 protein of bacteriophage T7. J. Biol. Chem. **262**, 16212-16223.

Topal, M.D. & Fresco, J.R. (1972). Complementary base-pairing and the origin of substitution mutations. Nature **263**, 285-289.

11

Genetic instability and DNA rearrangements in Streptomyces

K. Dharmalingam

Genetic Engineering Research Unit
Department of Biotechnology, School of Biological Sciences
Madurai Kamaraj University, Madurai 625 021

ABSTRACT

Genetic instability is ubiquitous in Streptomyces (Huter & Eckhardt, 1988). Eventhough *Streptomyces* genome has about 10% of reiterated sequences, these sequences are stable. In some cases even much larger homologies exist stably in *Streptomyces*. In *S. lividans* TK64 three direct repeats of 1.1 kb each could exist stably. In the other derivative, *S. lividans* TK23 two direct repeats (a single AUD) exist stably. However, a deletion in the region encoding chloramphenicol resistance initiates the instability of *argG* gene, by inducing deletions and amplifications. Depending on the presence or absence of a duplicated AUD, the ADS carrying 5.7 kb repeat units get amplified.

INTRODUCTION

Genetic instability is a rare process in biological systems. However, in *Streptomyces* some of the characteristics are extremely unstable and they mutate at frequencies much higher than 0.1% of spores. This genetically unstable trait was first reported by Beijerinck (1913). Subsequently genetically unstable pigment production and agarase activity were described by Stanier (1942).

Streptomyces are common soil bacteria. They are commercially the most important group, since these produce more than 60% of the known antibiotics (Okami & Hotta, 1988). They also exhibit some of the most uncommon properties. *Streptomyces* are mycelial forms and differentiate to produce spores. This property coupled with the production of several secondary metabolites including antibiotics makes these organisms extremely interesting. The hyphal compartments carry more than one genome unlike the spores which are typically haploid. Surprisingly, the linkage map of the genetically well studied *S. coelicolor* has markers in only two quadrants of the circular map (Hopwood & Kieser, 1990). Fig. 11.1 shows the linkage map of *S. coelicolor*.

Renewed interest in genetic instability began with the demonstration of the presence of reiterated DNA sequences, which existed in multiple tandem copies in genetically unstable strains. The amplification of specific DNA sequences associated with genetic instability was first demonstrated in recombinants derived from the fusion of protoplasts of *S. jumojiensis* and *S. lipmanii* (Robinson *et al.*, 1981). Subsequently the association of DNA amplification and genetic instability was demonstrated in several species of *Streptomyces* (Cullum *et al.*, 1986).

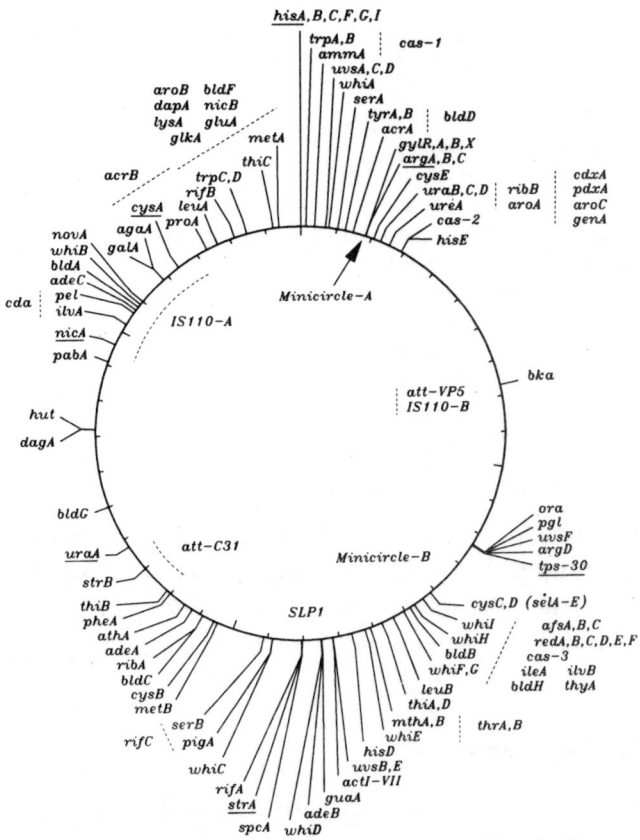

Fig. 11.1: Genetic map of S. coelicolor A3(2)

Genetic length of the map based on the frequency of recombination is 260 centimorgans. The map is divided into 10 intervals and the beginning of the map is at hisA set at 0/360. The distance between the underlined loci on the outside is based on heteroclone analysis and represents probably the actual distance. The distance between other loci is relative. For a complete description see the review by Hopwood and Kieser (1990)

RESULTS

Genetic instability is very common among Streptomyces strains (see Table 11.1). It is clear that this property is not unique to any one particular species. Further, even in natural isolates one could demonstrate genetic instability and DNA amplification (unpublished results). It appears

Table 11.1 *Streptomyces* strains showing genetic instability

S. achromogenes var rubradiris
S. achromogenes var streptozoticus
S. alboniger
S. ambofasciens
S. antibioticus
S. azureus
S. bikiniensis
S. cattleya
S. coelicolor
S. coriofaciens
S. fradiae
S. glaucescens
S. griseus
S. jumojiensis/S. lipmanii hybrid
S. lavendulae
S. lividans
S. reticuli
S. rimosus
S. scabies
S. tendae
S. venezuelae
S. violaceus ruber

that almost all strains of *Streptomyces* might carry genetically unstable characteristics and these may not be laboratory artifacts. However, why this property is unique to *Streptomyces* is still unknown. Once the nature and the role of this process is uncovered we might understand the reason for its prevalence in *Streptomyces* and the evolutionary significance of this paradoxical process.

Genetically unstable characters

There are several genetically unstable characteristics listed in Table 11.2. It is apparent that most of the affected characteristics are secondary metabolic functions. Besides, the loss of primary metabolic or essential functions will lead to cell death and hence it is natural to recover

Table 11.2:*Streptomyces* characters showing genetic instability

A - factor determinant (*afsA*)
Agarase (*dag*)
Aerial mycelium formation
Arginine succinate synthase (*argG*)
Chloramphenicol resistance (*cml*)
Tyrosinase (*melC*)
Streptomycin sensitivity
Oxytetracycline resistance
Tylosin resistance (*tyr*)
Tylosine biosynthesis
Alpha amylase inhibitor (tendamistat)

mutations only in non-essential genes. In any case, genetic instability affects only certain genes since in Streptomyces the frequency of spontaneous mutations in other genes is about 10^{-6} per cell generation, like in other bacteria.

Genetic instability and DNA rearrangements

The mutants which show genetic instability carry additional DNA rearrangements (see Robinson *et al.*, 1981; Cullum *et al.*, 1986). Prominent rearrangements include amplification and deletion of specific DNA sequences. Additional rearrangements like inversions were also observed in those cases where this process was carefully analysed. In all the cases studied, genetic instability is accompanied by DNA amplification and or deletion. In most cases the unstable gene itself is lost due to the deletion of DNA, since these unstable mutations do not revert to wild type. In some cases (e.g., arginine auxotrophy) deletions were physically demonstrated by Southern analysis using appropriate probes (Altenbuchner & Cullum, 1984, 1987; Betzler *et al.*, 1987).

Genetic instability in *S. lividans*

One of the well studied cases of genetic instability in *S. lividans* is arginine auxotrophy which is due to *arg*G mutation affecting the synthesis of arginine succinate synthase (Cullum *et al.*, 1986). It has been shown that *arg*G instability occurs by a two step process. In the first step the wild type cells become Cml^S at a high frequency and these Cml^S mutants segregate to *arg*G also at higher frequencies. In our experiments the Cml^S mutants arose at a frequency of 0.3% of the spores, and the Cml^S mutants gave rise to *arg* derivatives at a frequency of about 4% spores. However, the frequency of formation of Cml^S from *Cmlr* and *arg* from arg^+ (in *Cmls* derivatives) varied considerably from one isolate to another (Mathumathi *et al.*, 1990). It has been shown that Cml^S *arg* mutants carry amplification of specific DNA sequences. Based on Southern hybridizations using the cloned amplified DNA sequence as probe, the organisation of the amplified DNA in the Cml^S *arg* and its parents was deduced (Cullum *et al.*, 1989; Schrempf *et al.*, 1989). It was first proposed by Hershberger and Fishman (1985) that the multiple copies of the amplifiable DNA sequences (ADS) and the repeating units have one copy of the direct repeat each. The amplifiable unit of DNA (AUD) in this case is a 4.4 kb DNA sequence flanked by 1.2 kb right and left repeats. Therefore, the AUD is 6.8 kb in length in *S. lividans* TK23 (Fig. 11.2A). Interestingly *S. lividans* TK64 carries a duplication of AUD sequences (Fig. 11.2B, 11.2C).

Chloramphenicol sensitivity in *S. lividans* was shown to be reversible as well (Freeman *et al.*, 1977). However, it was demonstrated clearly that the Cml^S mutation which arose as a prelude to *arg*G mutations are actually deletions (Flett & Cullum, 1987). Southern analysis revealed deletions of upto 40 kb in Cml^S mutants. These mutants gave rise to *arg*G mutants and these were also shown to be deletion by Altenbuchner and Cullum (1987) and Betzler *et al.* (1987). These deletions could be upto 500 kb. Surprisingly, one of the deletion end points in *arg*G is always within the ADS and the other at different locations (see Cullum *et al.*, 1989; Schrempf *et al.*,

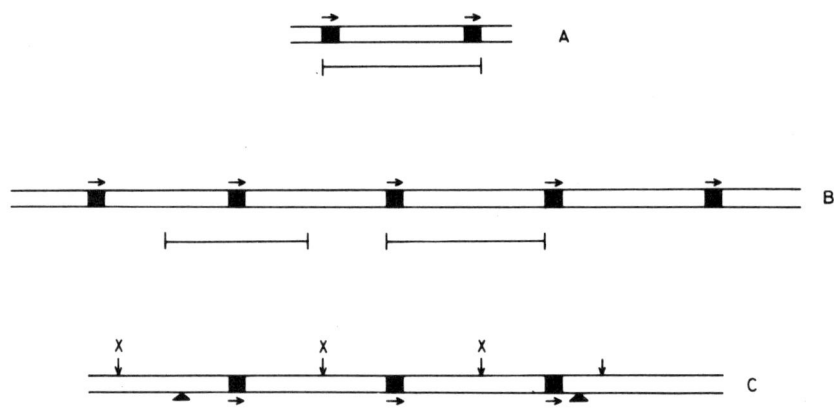

Fig. 11.2: *Structure of AUD and ADS in S. lividans*

The amplifiable DNA sequences (ADS) are drawn according to the proposal of Hershberger. Right and left repeats (1.2 kb each) flanking the central 4.4 kb region is seen in *S. lividans* TK23 (A). This is the amplifiable unit of DNA (AUD) with a length of 6.8 kb. In *S. lividans* TK64 the AUD is duplicated (C). The total ADS with the repeating unit length of 5.7 kb is shown in B.

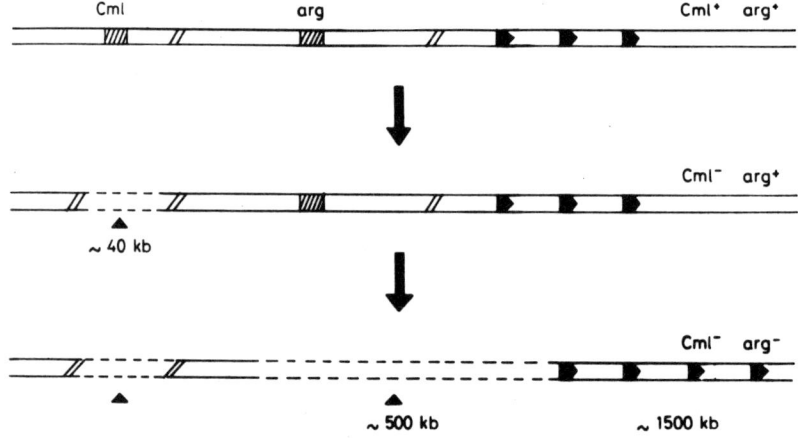

Fig. 11.3: *DNA arrangement at the genetically unstable argG loci*

Initial deletion occurs at the *Cml^s* region covering a span of 40 kb. This is followed by another deletion at the *argG* region and one deletion end point is always within the ADS. The ADS could carry 300 copies of the repeating units of 5.7 kb length.

1989; Mathumathi *et al.*, 1990). The *argG* gene is about 30 kb away from the ADS. Fig. 11.3 illustrates the organisation of the chromosomal DNA in *Cml^s argG* mutants which suffered the amplification and deletion events. Recent genetic mapping data positioned both *Cml^s* and *argG* at a location clockwise from ura-6 marker on the *S. lividans* chromosome. These experiments also imply the *Cml^r* and *argG* might be linked (Betzler *et al.*, 1987). However, this conclusion needs further confirmation.

Stability of variants carrying amplified DNA

As described already there are several different amplified DNA sequences in *S. lividans*. The 6.8 kb AUD always gets amplified in *Cml*[s] *argG* mutants. This ADS is stable for several generations and repeated storage also does not affect the amplified DNA (Betzler *et al.*, 1987; Mathumathi *et al.*, 1990). In *S. fradiae* also the amplified DNA was shown to be stable (Mathumathi *et al.*, 1990).

Induced deletion of amplified DNA

We have observed accidently that the ADS could be deleted if the strains of *S. fradiae* and *S. lividans* carrying ADS are treated with subinhibitory concentrations of antibiotics. Further analysis was carried out to explore this process. Since degeneration of antibiotic producer strains could be mediated by the ADS, removal of potentially amplifiable sequence could be industrially important (Table 11.3).

Table 11.3: *Streptomyces* strains showing amplified DNA and DNA deletions

Strain	Presence of amplified DNA	Occurrence of deletion
S. achromogenes var rubradiris	+	+
S. achromogenes var streptozoticus	+	
S. alboniger		+
S. ambofasciens	+	-
S. antibioticus	+	-
S. azureus		+
S. bikiniensis		-
S. cattleya		-
S. coelicolor	+	-
S. coriofaciens	+	?
S. fradiae	+	+
S. glaucescens	+	+
S. griseus	-	+
S. jumojiensis/S. lipmanii hybrid	+	
S. lavendulae	-	+
S. lividans	+	+
S. reticuli	+	+
S. rimosus	+	+
S. scabies	+	+
S. tendae	+	
S. venezuelae	-	+
S. violaceus ruber	-	+

The effect of antibiotic treatment was examined using two different approaches. In the first set of experiments *Cml*[s] *argG* spores (which carry ADS) were propagated on minimal agar plates containing subinhibitory concentrations of antibiotics. Under these conditions none of the physiological processes like sporulation or growth were affected. The spores prepared using this method were inoculated into liquid medium and grown for 48 hours without any added antibiotics. Chromosomal DNA was isolated from these cultures and cleaved with XhoI.

Southern analysis using the 15 kb duplicated AUD of *S. lividans* TK64 as probe showed a 5.7 kb DNA as well as the two junction fragments of 21 kb and 15 kb hybridizing with the probe.

Similarly, XhoI digests of $Cml^s\ arg^+$ chromosomal DNA also showed an identical hybridization pattern, indicating that the duplicated AUD was unaffected. As already described by Cullum and his coworkers, (1989) the Cml^s mutation was a deletion occurring elsewhere in the chromosome. The 15 kb and 21 kb fragments are presumably right and left junction fragments. When the total DNA isolated from mycelia grown from antibiotics pretreated spores was cleaved with XhoI and analysed by hybridization using the 15 kb probe neither the 5.7 kb amplified sequence nor the 15 kb and 21 kb junction fragments were seen. The new end points created due to this deletion are being analysed now.

Another surprising finding was that the antibiotic treatment of $Cml^s\ arg^+$ parents which carry

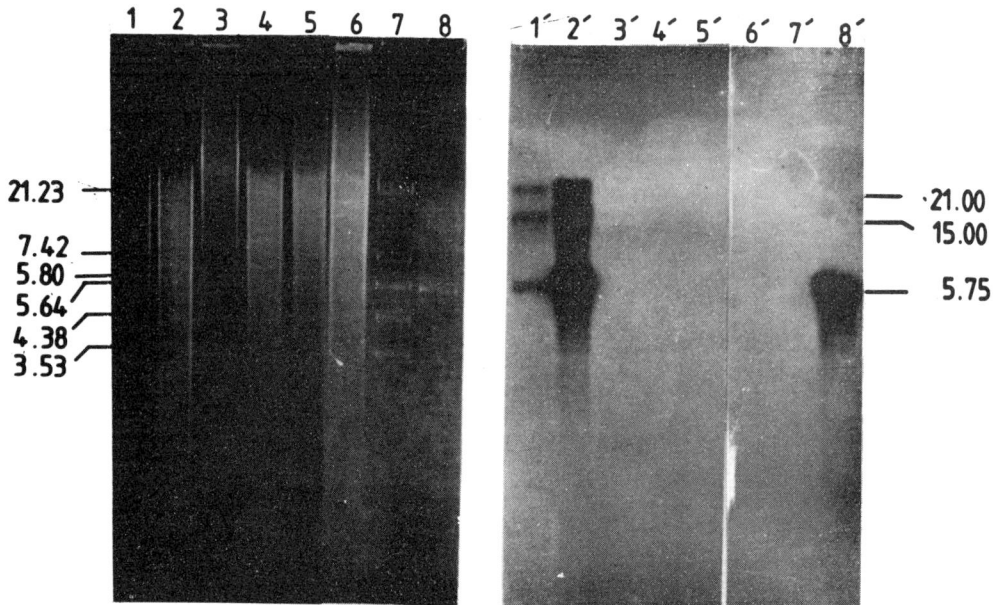

Fig. 11.4: *Antibiotic induced deletion of ADS in S. lividans*

Total chromosomal DNA from *S. lividans* and its derivatives was isolated and cleaved with XhoI. DNA fragments were separated on 0.7% agarose gel and Southern transferred to nitrocellulose and hybridized with the 15 kb probe. Tracks 1,2 and 8 are $Cml^r\ arg^+$; $Cml^s\ arg^+$ and $Cml^s\ arg$, respectively. In track 8 ($Cml^s\ arg$) only half the amount of total DNA was loaded. Due to the decreased amount of loaded DNA the junction fragments are not seen in track 8. Track 3, $Cml^s\ arg$ derivative isolated in the presence of antibiotics. Track 4,5,6-$Cml^s\ arg$ spores treated with chloramphenicol, thiostrepton, and kanamycin, respectively. Tracks 1'-8' show the autoradiogram of the above sample.

only the duplicated AUD led to the deletion of this parental duplicated AUD as well (Fig. 11.4 track 3 and 3'). Further there is a significant reduction in the frequency of Cml^s arg mutants, from Cml^s parents if the mutants were scored in the presence of antibiotics. This observation is significant since one could delete the potentially amplifiable sequences and thereby avoid the problems associated with the genetic instability in industrial producer strains.

In the second approach, the antibiotic treatment was limited to the duration of the mycelial growth of untreated spores. In these experiments antibiotics spectinomycin or streptomycin was added immediately after inoculating the spores of the untreated Cml^s arg strain into liquid media. Spectinomycin treatment, even at 0.5 µg/ml, completely abolished the amplified sequence within 48 hr of mycelial growth (Fig. 11.5). Eventhough the total amount of DNA in tracks 4 and 5 appeared to be greater, hybridization with the probe (tracks 3' and 4') did not reveal any amplified DNA. This indeed strengthens the above conclusion. Interestingly, when total DNA isolated from cells grown at different concentrations of streptomycin was digested with *Bgl* II and hybridized with the 15 kb probe, there was a progressive decrease in the amount

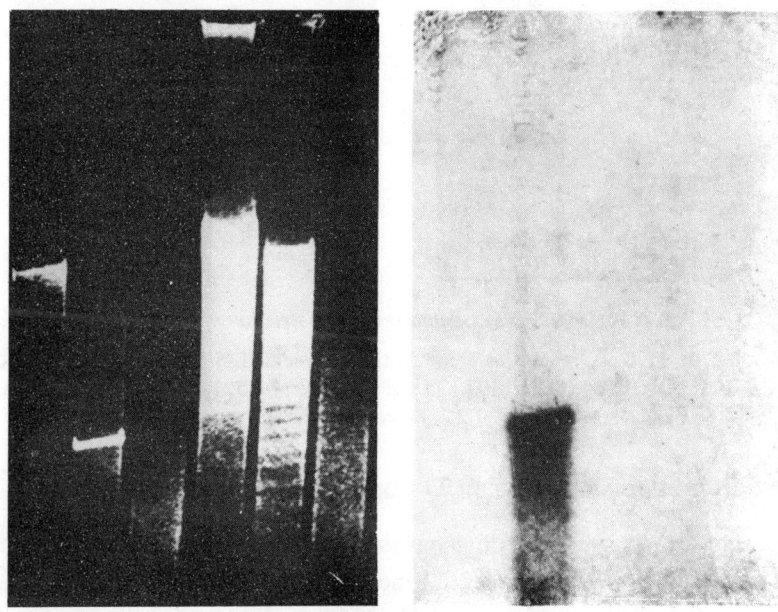

Fig. 11. 5:*Spectinomycin treatment induces deletion of ADS in mycelia*

Total chromosomal DNA from different *S. lividans* strains was digested with BglII, and electrophoresed in 0.7% agarose gel. DNA was transferred to nitrocellulose filters and hybridized with ^{32}P-labelled 15 kb probe and exposed to X-ray film.

Track 1 - Cml^r arg^+, track 2 - Cml^s arg, tracks 3,4,5,6 - Cml^s arg mycelia grown in the presence of 0.5, 1.0, 2.0 and 10.0 µg/ml spectinomycin, respectively. Tracks 1'-6-show the autoradiogram of the above samples. Total DNA loaded was 1 µg per track.

of the 5.7 kb sequence as can be seen in the ethidium-bromide-stained total chromosomal DNA digest (Fig. 6, tracks 3 and 4). Southern hybridization using the 15 kb probe also showed a progressive decrease in the signal when the autoradiogram was analysed. During photographic printing the light band in track 4' did not show up and there was no signal in track 5 even in the autoradiogram. However, in the cells grown in 10 µl/ml streptomycin, both the 5.7 kb sequence and the junction fragments were completely lost (Fig. 11.6, track 5 and 5').

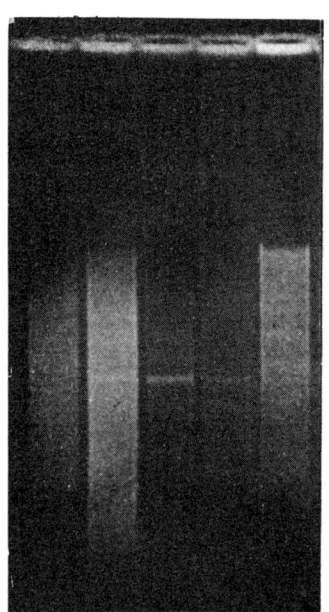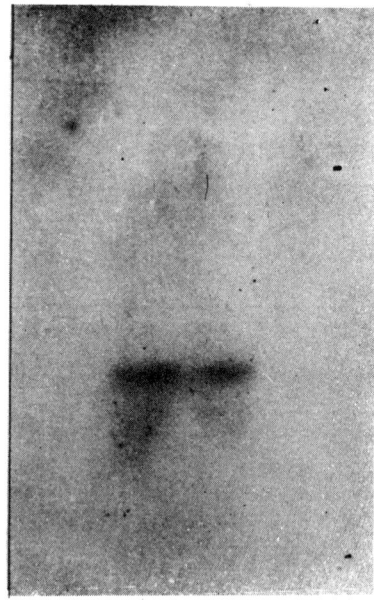

Fig. 11.6:_Streptomycin treatment induces deletion of ADS in mycelia_

The procedures followed were essentially the same as those described in Fig.11. 5. Track 1 - *Cml^r arg^+*, track 2 *Cml^s* arg, tracks 3,4,5 - *Cml^s arg* mycelia grown in the presence of 0.5, 1.0, and 10.0 µg/ml streptomycin, respectively. Tracks 1'-5' show the autoradiogram of the above samples.

Induced deletion and reduction of AUD copy number in *S. fradiae*

In order to understand whether this antibiotic induced elimination of ADS is a general phenomenon, we tested the effect of antibiotics on a new amplification system in *S. fradiae* discovered by us (Mathumathi, unpublished Ph.D. Thesis).

Fig. 11.7 track 3 shows that the chromosomal DNA of *S. fradiae* DM14 (*cml^S*) derivative digested with *Ava*I, has the amplification of a 5 kb DNA unlike DM02, the *cml^r* parent (track 2). This fragment was eluted and cloned (Fig. 11.8) on pBR322 (pDM32). The DNA was used as probe to detect the amplified DNA in spectinomycin and streptomycin treated cultures. Results in Fig. 11.7 clearly show that the copy number of the ADS is reduced due to antibiotic treatment.

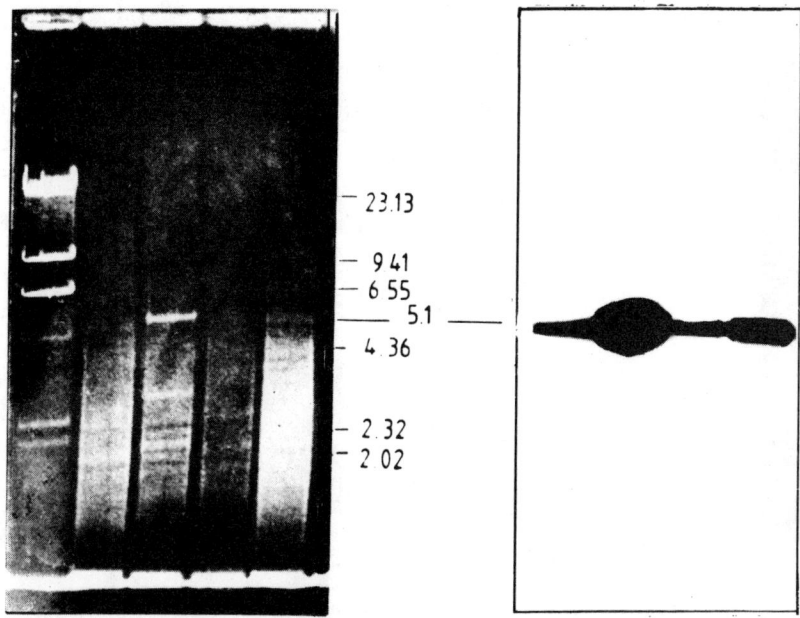

Fig. 11.7: *Antibiotic induced reduction in copy number of the AUD in S. fradiae*

Total chromosomal DNA from different *S. fradiae* strains were digested with AvaI and electrophoresed in 0.7% agarose gel. Southern hybridization was done using ^{32}P-labelled pDM32 (see Fig. 11.8). Track 1 - lambda DNA HindIII, digested track 2 - DM02 (*Cml^r*), track 3 - DM14, (*Cml^s*), tracks 4,5 - DM14 (*Cml^s*) mycelia grown in the presence of 2 μg/ml spectinomycin and 10 μg/ml streptomycin, respectively. Tracks 2'5' show the autoradiogram of the above samples. Each track was loaded with 3 μg chromosomal DNA.

Induced elimination of ADS is a two step process

It is clear from the above results that antibiotics first induce reduction in AUD copy number. Perhaps the recombinases induced by the antibiotic first recognize the large homology available in the direct repeats (Fig. 11.9). Subsequently, either the cells continue to have this reduced copy number or they undergo further deletions. The second stage perhaps uses the limited homology outside the ADS which lead to the loss of the initial right and left junctions as well.

CONCLUSIONS

We have demonstrated that deamplification and deletion could be induced. This demonstration opens up new ways to stabilize the strains and one could also now understand the mechanism of amplification. Deletion and introduction of such large tracts of apparently non-coding sequences without any lethal effect is reflected in the absence of genes in the two quadrants of *Streptomyces* genome (Fig. 11.1). These silent regions could be the regions where these events occur.

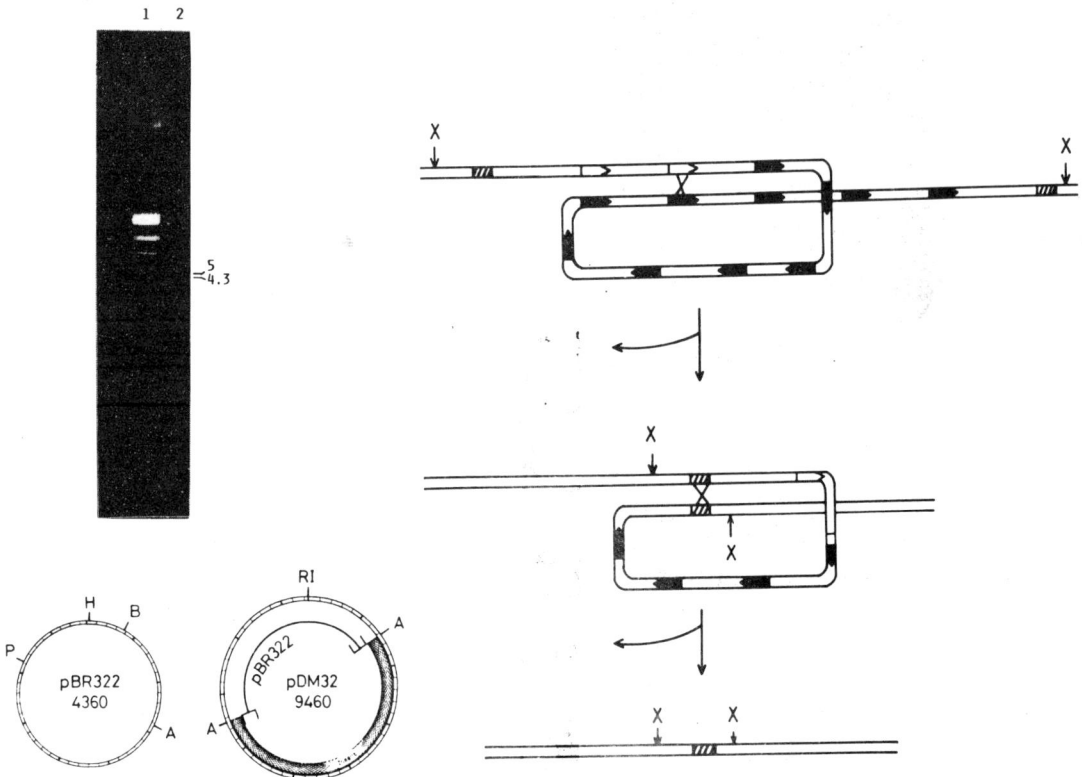

Fig. 11.8:*Construction of pDM32 carrying the ADS from S. fradiae*

Total chromosomal DNA from DM14 (*Cml^s*) was cleaved with *Aval* and the fragments separated on 0.7% agarose gel. The 5.0 Kb amplified DNA was eluted and cloned on the pBR322 at the *Aval* site.

Fig. 11.9:*Model for antibiotic induced deletion of the ADS*

The second stage of recombination presumably uses the limited homology available outside the ADS. (See text for details)

References

Altenbuchner, J. & Cullum, J. (1984). DNA amplification and an unstable arginine gene in *Streptomyces lividans* 66, Mol. Gen. Genet. **195**, 134-138.

Altenbuchner, J. & Cullum, J. (1987). Structure of an amplifiable sequence in *Streptomyces lividans*. Mol. Gen. Genet. **201**, 192-197.

Beijerinck, M.W. (1913). Ueber schroter and Cohn's Lakmusmicroccus. Folia Microbiol. **2**, 185-200.

Betzler, M.P., Dyson, P. & Schrempf, H. (1987). Relationship of an unstable arg gene to a 5.7

Kb amplifiable DNA sequence in *Streptomyces lividans* 66. J. Bacteriol. **169**, 4804-4810.

Cullum, J., Flett, F. & Piendl, W. (1989). Genetic instability, deletions, and DNA amplification in *Streptomyces species*. In Genetics and Molecular Biology of Industrial Microorganisms, Edited by C.L. Hershberger, S.W. Queener and G. Hageman. American Society for Microbiology, Washington, D.C. pp 127-132.

Cullum, J., Altenbuchner, J. Flett, F. & Piendl, W. (1986). DNA amplification and genetic instability in *Streptomyces*. Biotechnol. Genet. Eng. Rev. **4**, 59-78.

Flett, F. & Cullum J. (1987). DNA deletions in spontaneous chloramphenicol - sensitive mutants of *Streptomyces coelicolor* A3(2) and *S. lividans* 66. Mol. Gen. Genet. **207**, 499-502.

Freeman, R.F., Bibb, M.J. & Hopwood, D.A. (1977). Chloramphenicol-acetyltransferase - independent chloramphenicol resistance in *Streptomyces coelicolor* A3(2). J. Gen. Microbiol. **98**, 453-465.

Hershberger, C.L. & Fishman, S.W. (1985). Amplified DNA structure and significance. Microbiology 1985, Edited by I. Leive. American Society for Microbiology, Washington, D.C. pp 427-430.

Hopwood, D.A. & Kieser, T. (1990). The *Streptomyces* genome. In The Bacterial Genome, Edited by K. Drlica and M. Riley. American Society for Microbiology, Washington, D.C. pp 147-162.

Huter, R. & Eckhardt, T. (1988). Genetic manipulation. In Actinomycetes in Biotechnology, Edited by M. Goodfellow, S.T. Williams and M. Mordarski. Academic Press, Inc. (London) Ltd., London. pp 89-184.

Mathumathi, R., Kumaravel, S. & Dharmalingam, K. (1990). Deamplification and deletion of amplified DNA in *Streptomyces lividans* and *S. fradiae*. Appl. Microbiol. Biotechnol. **33**, 291-295.

Okami, Y. & Hotta, K. (1988). Search and discovery of new antibiotics. In Actinomycetes in Biotechnology, Edited by M. Goodfellow, S.T. Williams and M. Mordarski. Academic Press. Inc. (London), Ltd. London. pp 33-67.

Robinson, M., Lewis, E. & Napier, E. (1981). Occurrence of reiterated DNA sequences in strains of *Streptomyces* produced by an interspecific protoplast fusion. Mol. Gen. Genet. **182**, 336-340.

Schrempf, H., Kessler, A., Bronneke, V., Dittrich, W. & Betzler, M.P. (1989). Genetic instability in *Strpetomyces*. In Genetics and Molecular Biology of Industrial Microorganisms, Edited by C.L. Hershberger, S.W. Queener and G. Hageman. American Society for Microbiology, Washington, D.C. pp 133-140.

Stanier, R.V. (1942). Agar decomposing strains of the *Actinomyces coelicolor* species group. J. Bacteriol. **44**, 555-570.

12

Nitrogen fixing enterobacter: NIF-genetics and risk assessment for releases

Walter Klingmüller

Department of Genetics, University of Bayreuth
W08580 Bayreuth, Germany

ABSTRACT

Nitrogen-fixing *Enterobacter agglomerans* occur in the rhizosphere of cereals. The genes for nitrogen fixation (nif-genes) are located on large indigenous plasmids of 100-200 kb. Two of these, pEA3 and pEA9, were characterized by genetical methods.

Plasmid pEA3 (110 kb) contains the nif-genes (about 20 in a contiguous cluster) similar in order to *Klebsiella*. This similarity even holds for genes V, W, T, and Z, still unsufficiently studied. However, the two nif-genes coding for electron transport proteins, nifF and nifJ, are linked, located on one side of the nif-gene group and transcribed together in the order of function of the proteins. Because this gene arrangement is particularly simple and plausible (in contrast to all other known nitrogen-fixers), it is suggested that the nif-region on these plasmids of *Enterobacter* represents a cornerstone in bacterial nif-gene group development.

Plasmid pEA9 (200 kb) also contains the nif-genes. This plasmid was labelled with either Tn5 or Tn1725. It was then used in matings of *Enterobacter* donors with cured, plasmid-free *Enterobacter* recipients. On plates with complete medium, transfer rates of 10^{-4} per donor, for optimized conditions were obtained. To collect data on possible uncontrolled gene spread, for planned releases of such bacteria, additional matings were done with soil as substrate. The soil was from a plot into which an actual release is being planned. Under standard conditions, sterilized soil gave low rates of plasmid transfer (10^{-6} per donor) but non-sterilized soil gave none. Adding complete medium or sucrose to non-sterilized soil elicited strong cell propagation, together with plasmid transfer (optimum after incubation for 1 day: 10^{-4} exconjugants per donor). No transfer could be registered in the presence of wheat seedling roots for periods up to 3 months, nor in the presence of cattle manure. However, mashed sugarbeets, added at 0.4 g per 50 g soil, elicited distinct plasmid transfer. These experiments are designed and continued towards a first test release of N_2-fixing *Enterobacter* under the new Gene Law in Germany.

INTRODUCTION

During recent years bacterial nitrogen-fixation has been studied in many laboratories and with increasing intensity. Not only biochemical methods, but also molecular genetic techniques have been applied to such research. The best known nitrogen-fixing bacterium is still *Klebsiella pneumoniae*, which is free living and related to *E. coli*. It was, therefore, easily amenable to all relevant methods developed for the latter.

The genus *Rhizobium*, also nitrogen-fixing, has been exploited for a long time to improve yields of leguminous plants. In contrast to *Klebsiella*, Rhizobia are symbiotic bacteria, i.e. they infect the roots of host plants and initiate nodules in which the bacteria fix nitrogen, the energy for which is provided by plant photosynthesis. In return the bacteria contribute ammonium to the plant. Data on the genetics of nitrogen-fixation in Rhizobia are now accumulating with remarkable speed.

In countries, where nutrition is mainly based on cereals (e.g. maize, wheat and rice), other nitrogen-fixing bacteria have attracted considerable interest. Azospirilla, in tropical and sub-tropical climates, occur in high numbers in the rhizosphere, on the root surface and within the roots of various C_4 and C_3 cereals and grasses (Klingmüller, 1988a, Skinner *et al.*, 1989). Recent genetic studies were directed towards the collection of *nif* structural mutants, regulatory mutants, and mutants probably defective in general nitrogen control; in addition, resistance mutants and mutants with alterations in a number of physiological or metabolic properties were obtained (Singh & Klingmüller, 1986a; Abdel-Salem & Klingmüller, 1987). Such mutants are required for elucidating the molecular basis of plant-bacterial interaction, and for devising strategies for obtaining beneficial effects of inoculations.

In temperate climates, other bacteria have to be found with the potential to assist agriculture via biological nitrogen-fixation. An ideal bacterium should not only be able to fix nitrogen and to live in close association with plant roots (e.g. wheat) in order to utilize the exudates from the roots as an energy source, but also form a substantial proportion of the root microflora in such climates. We have taken up this problem, identified such bacteria as *Enterobacter agglomerans* strains, and studied them extensively (Kleeberger *et al.*, 1983; Singh *et al.*, 1983, 1988; Klingmüller, 1985,1988a,b; Singh & Klingmüller, 1986b; Kreutzer *et al.*, 1989; Klingmüller *et al.*, 1989). In the course of these studies, we have identified the complete set of nitrogen fixation genes on large indigenous plasmids of these bacteria, and analyzed their nif-genes at the DNA level and searched the comparative development of nif-gene groups in nitrogen-fixing bacteria. We have also tried to devise scenarios of future releases of such nitrogen-fixing bacteria for agriculture.

he presumptive risks involved in deliberate releases of genetically altered bacteria are a topic of much public concern. Assessment of these risks is hampered by insufficient experimental work. We have investigated bacterial survival and plasmid transfer in soil environments for *Pseudomonas* and *Escherichia coli* strains (Schilf & Klingmüller, 1983) and also *Rhizobium leguminosarum* (Döhler & Klingmüller, 1988), the results showed low risk. To corroborate that work, in particular for plasmid transfer, as an example of gene spread, we later investigated nitrogen fixing *Enterobacter* strains. Nif-plasmid pEA3 has been described in detail previously. Nif-plasmid pEA9 is similar, apart from its larger size (ca. 200 kb, Steibl, 1989) and its self-transmissibility (Klingmüller *et al.*, 1989). The latter was used for our experiments.

Initially, matings with *Enterobacter* were done on plates with complete medium (Klingmüller *et al.*, 1989). The rate of plasmid transfer was in the range of 10^{-4} exconjugants per donor. To assess the possible risk of uncontrolled plasmid spread if such nitrogen-fixing *Enterobacter* are released into the environment, we then began to study survival and plasmid spread for such strains in soil. In sterilized agricultural soil, plasmid transfer does occur, however in non-steril-ized soil, no exconjugants could be detected (Klingmüller *et al.*, 1990; Kiingmüller, 1991a,b).

In recent studies, we have extended these investigations, measuring the influence of several biological and ecological parameters, amongst them two different transposon labels, bulk soil versus wheat rhizosphere, and several amendments of possible agricultural significance, on bac-terial survival and on transfer of nif plasmid pEA9 in the soil. The data obtained are being used to support an application under the new German Gene Law for permission to run a test release with such genetically engineered bacteria.

In the following report, a short review of the earlier results together with the presentation of recent, yet unpublished findings is given. It is hoped not only to stimulate further research into the hereditary development of the complete set of nitrogen-fixation genes, but also into the problems of risk assessment for future bacterial releases as inoculants.

RESULTS

a) Taxonomical survey

We conducted a taxonomical survey screening for bacteria in the innermost rhizosphere of wheat and barley roots isolated from fields in the vicinity of Bayreuth, Germany. The frequen-cies of bacteria of different groups were registered.

A detailed account of the results obtained has been published (Kleeberger *et al.*, 1983). As can be seen from Table 12.1, the largest class of bacteria in the innermost rhizosphere of both barley and wheat was *Pseudomonas* (52%). The next largest class (23%), isolated from the in-

Table 12.1: Bacteria in rhizosphere of cereals (percent of total). Roots were either carefully washed and surface sterilized by H_2O_2-treatment, or only washed, before homogenizing them and spreading aliquots onto nutrient agar plates to obtain individual colonies. These were then classified.

	barley		wheat	
	washed	washed and H_2O_2 treated	washed	washed and H_2O_2 treated
	(n=61)	(n=45)	(n=28)	(n=62)
Pseudomonas spp.	34	53	45	52
Enterobacteriaceae	-	16	4	23[*]
Achromobacter spp.	-	18	-	8
Corynebacterium spp.	61	13	50	15
Bacillus spp.	5	5	-	2

[*] *E. agglomerans* 14, *E. cloacae* 5, *S. liquefaciens* 4.

nermost rhizosphere of wheat, was that of Enterobacteriaceae, containing the very homogenous group of *Enterobacter agglomerans* as the largest component, amounting to 14%. These bacteria are related to *E. coli* as well as to *Klebsiella*. It is, therefore, not surprising that some of them fix nitrogen.

b) Analysis of nif-genes in *Enterobacter agglomerans*

Altogether 50 strains of *E. agglomerans* were isolated, five of which were found to fix nitrogen. These strains were analyzed at the molecular level, seeking the location, distribution, and structural composition of their nitrogen-fixation genes (Singh *et al.*, 1983, 1988; Kreutzer, 1986, 1989; Kreutzer *et al.*, 1989). Recalling that the *Klebsiella* nif-genes form a cluster of about 25 kb on the chromosome but that certain *Rhizobium* species have nif-genes located on plasmids, we decided first to screen the *E. agglomerans* plasmids. The investigation led to the identification of large indigenous plasmids with molecular weights between 73 and 100 Md (Table 12.2).

Table 12.2: Designations and molecular weight of plasmids from nif-positive *E. agglomerans* strains

E. agglomerans strains	No. of plasmids	Plasmid designation	Molecular weight $(\times 10^6) \pm$ S.E.M.)
243	2	pEA1 pEA2	73±4 100±1
333	1	pEA3	77±2
334	1	pEA4	77±2
335	1	pEA5	100±1
339	1	pEA9	100±1

The molecular weight of plasmids were determined on the basis of relative mobilities in agarose gels using plasmid standards of known molecular weight. There were five molecular weight determinations performed for each plasmid.

Two of the five nif[+] strains, viz. nos. 335 and 339, could be cured of their plasmids by heat treatment. Then, not only did they loose the respective plasmid, but also their nitrogen-fixing capability. This was a first indication that at least some (if not all) of the nif-genes were localized on these plasmids. The plasmid DNA was separated from the chromosomal DNA in agarose gels, and the *Klebsiella* nif KDH gene region was used as a radioactive probe, which was found to hybridize with a homologous region on one plasmid in each strain, confirming the presence of nif KDH.

Further investigations focussed on plasmid pEA3 from strain 333. This plasmid, from more recent investigations, is 110 kb in size (Kreutzer, 1986; Singh *et al.*, 1988). By preparing a cosmid gene bank with overlapping fragments of this plasmid, subcloning, restriction mapping and DNA-hybridization against suitable nif-gene probes from *Klebsiella*, a physical and genetic map of the plasmid was obtained. The plasmid contains a 23 kb cluster of nif-genes which shows homology (by Southern hybridization and heteroduplex analysis) not only to the KDH genes, but all the other nif-genes of *Klebsiella pneumoniae* M5a1. Apart from *nif* J and F, which are located on the right end of the pEA3 *nif*-gene cluster (near *nif* QB) (Kreutzer et al. in press), all the other nif-genes on pEA3 including nif *T*, W, and Z (Dippe, 1990) are organized in the same manner as in *K. pneumoniae*. Preliminary evidence (Steibl, 1989) points to the possibility that the same holds for others of the remaining four nif-plasmids from *E. agglomerans*, in particular, pEA9. A *Bam*HI restriction map of pEA3 and a detailed restriction map of the 23 kb nif-region on pEA3 is presented in Figure 12.1.

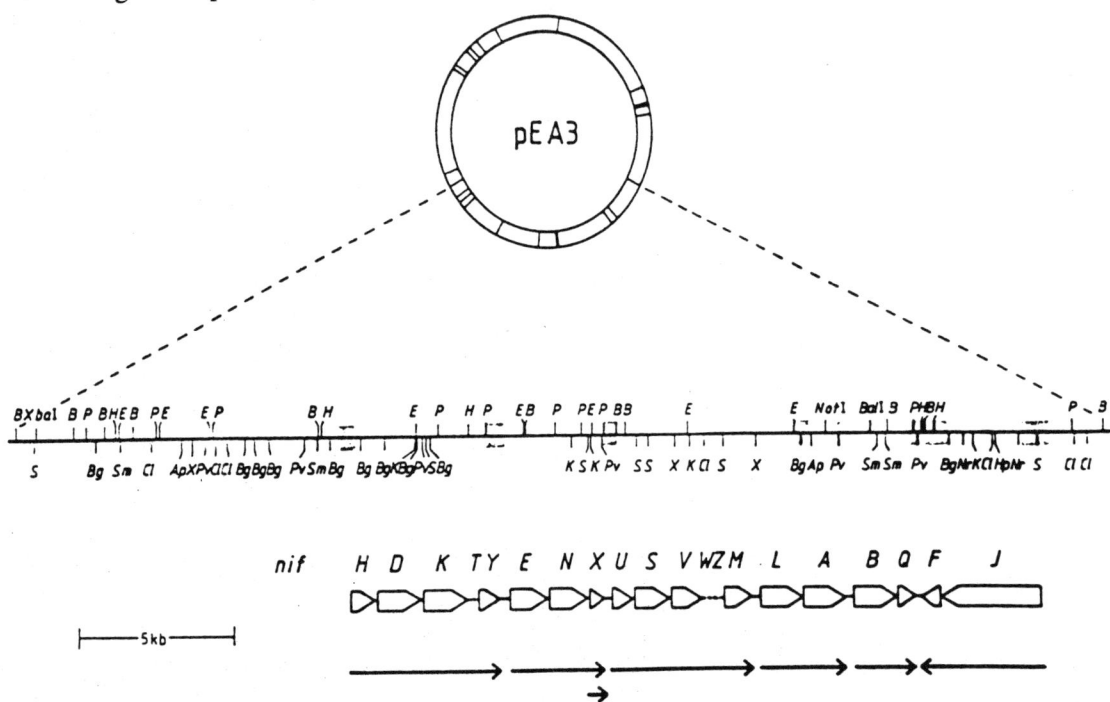

Fig. 12.1: Organization of nif-gene group in *E. agglomerans* 33 (pEA3). Top: *Bam*HI restriction map of pEA3. Center: Restriction map of insert in cosmid clone peaMS2-2. Those regions sequenced are dashed. Bottom: Position and orientation of nif-genes and preliminary arrangement in operons (From Kreutzer, 1989).

To check whether the nif information in this region is complete, i.e. sufficient for nitrogen fixation, we have tried to transfer it as entire functional unit into *E. coli* and to look for nitrogen fixation of such cells. The cosmid vector had a high copy number which very likely would interfere with nitrogen fixation in *E. coli* (Riedel *et al.*, 1983). Therefore, the cosmid clones as such were inappropriate. Instead, low copy number vectors had to be cloned into those constructs and be used in combination with a polA⁻ *E. coli* strain, to block the colE1-origin.

One such construct was #27-1 (Figure 12.2). By partially digesting this with *Bam*HI and allowing self ligation, construct # 49 was obtained, which had only the RK-origin. The former construct, transformed into *E. coli* polA⁻, and the latter, even in *E. coli* HB101, made the recipients nif-positive (acetylene reducing). This demonstrates that the genes we are studying are not only DNA-fragments, homologous to nif-genes on the molecular level, but are functional nif-genes themselves, and that this gene group contains all the genetic information required to fix nitrogen (Klingmüller, 1988b).

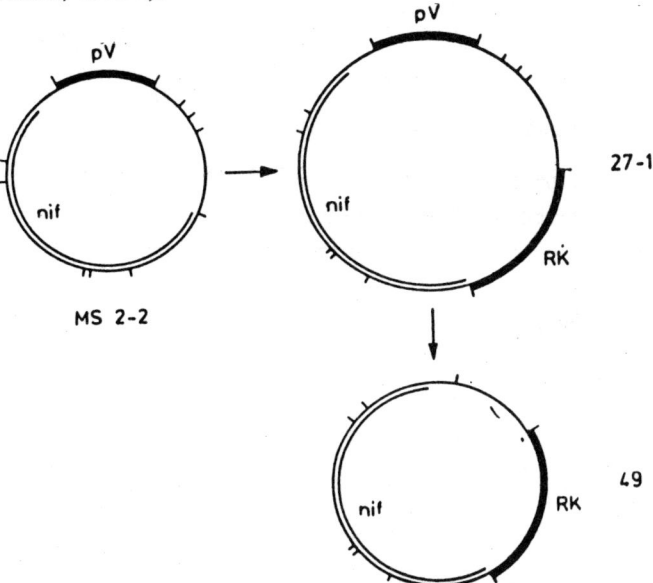

Fig. 12.2: Nif plasmid construction with broad host range vectors. Left: Cosmid peaMS2-2 from a pEA3 - *Bam*HI gene bank. pV: cosmid pV34, derived from pHC79, with Cbʳ gene and colE1-replicon; nif: nif-region of pEA3; *Bam*HI restriction sites are indicated.

Upper right: One of the hybrid plasmids, #27-1, obtained from the former cosmid, after partially digesting it with *Bam*HI and ligating the fragments into the single *Bam*HI site of vector plasmid pRK415. The latter carries the *tet*ʳ genes. In hybrid plasmid #27-1, pRK415 is closely linked to the pEA nif-gene group.

Lower right: One of the smaller derivatives of plasmid 27-1, viz. #49, obtained after partially digesting 27-1 with *Bam*HI, self ligation and selection for Tetʳ, Cbˢ. In hybrid plasmid #49, cosmid pV34 and some additional material of the former plasmid is lost.

Recent studies on the nucleotide sequence of electron transport protein genes J and F, and on their transcription, as probed by transcriptional fusions with promoter probe vectors containing the *lac* z gene, have shown that these 2 genes in *E. agglomerans*, form one single operon. This operon is located on the right part of the nif-gene group beyond *nif*Q, and transcribed from a nif-promoter, in the opposite direction of all other nif-genes of the group (Fig. 12.3).

Comparing all N₂-fixing bacterial groups studied so far, the nif-gene group in *E. agglomerans* hence offers the best ordered example of nif-genes known. It can be considered operationally as an archetype in nif-gene group evolution, from which all other nif gene groupings have originated. Alternatively, the *Enterobacter* nif-gene group can be considered as a final stage,

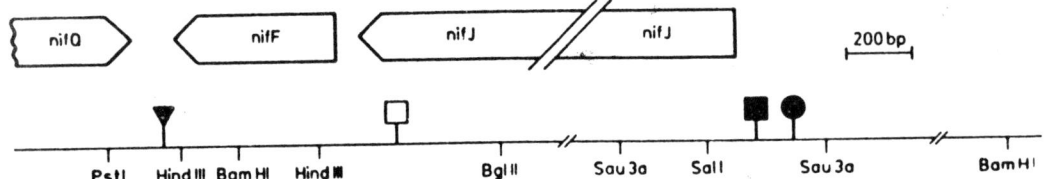

Fig. 12.3: Restriction map of the right end of the nif-gene group in *E. agglomerans* 333. The positions of the genes *nifQ, nifF* and *nifJ* as revealed by DNA-sequencing are indicated by open arrows above the map. The positions of the *nifJ* promoter and promoter-like motif are indicated by filled and open boxes, respectively. The *nifJ* upstream activating sequence is represented by a filled circle, and the putative transcriptional terminator by a filled triangle (From Kreutzer, Dayananda, and Klingmüller, 1990, in press).

into which all other nif-gene groupings will ultimately develop. The possible sorting of nitrogen-fixing bacteria according to this principle is exemplified in Fig. 12.4.

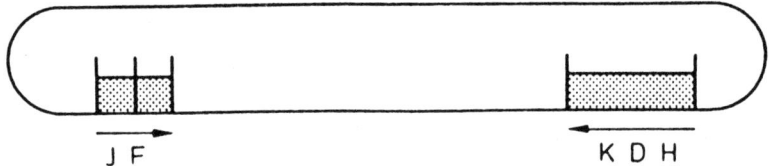

E. agglomerans. pEA3, plasmid, well ordered
Rhizobium meliloti, plasmid, complex
Bradyrhizobium japonicum, chromosome, compled
Azotobacter vinelandii, chromosome, additional groups
Azospirillum, chromosome, A gene separate
Anabaena, chromosome, rearrangement cycle
klebsiella. chromosome, F gene interspersed

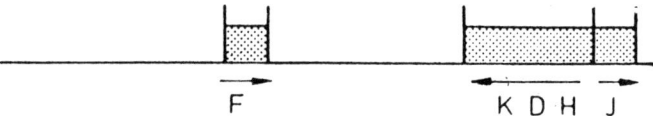

Fig. 12.4: A sorting of nitrogen-fixing bacteria, based on the organization of their nif-region(s). Top: *E. agglomerans*, nif-plasmid pEA3, with (indicated as blocks) structural *nif*-genes HDK (right) and electron transfer genes J and F (left). In between (not marked) the complete set of all other nif-genes. The arrows give the direction of transcription.

Bottom: *Klebsiella*, nif-region on chromosome, with genes indicated as above. The electron transfer genes are dispersed. Of them *nifF* in the midst of the (not marked) complete set of all other *nif*-genes. Center: Other cases of *nif*-gene arrangement in N_2-fixing bacteria, listed from *Rhizobium* to *Anabaena*, according to plasmid or chromosomal location and to chaotic or ordered arrangement.

c) Survival of *E. agglomerans* and plasmid transfer in soil

In a first series of experiments, we studied survival and plasmid transfer in homologous matings, as influenced by the plasmid label used, which was either Tn1725 or Tn5-Mob (Klingmüller, 1991 a,b). In principle, the two types of matings gave similar results, both in survival data and in plasmid transfer rates. Most important is that although the plasmid transfer is possible on Luria Broth (LB) plates and in sterile soil, it is extremely rare in natural soil, if the cells are offered in saline solution (0.85% NaCl). In all following experiments, the Tn5-Mob labelled plasmid was used exclusively to be able to check later for any helper effect of RP4-like plasmids from the environment on transfer of plasmids with related Mob-site.

Table 12.3: Plasmid transfer frequencies in natural soil. 1 day incubation at 22 °C, with different additives, at 4 ml per 50 g*

Experiment#	1	2	3
saline	0	0	0
LB	1×10^{-5}	2×10^{-4}	7×10^{-5}
sucrose 0,75%	$2,4 \times 10^{-5}$	$2,0 \times 10^{-5}$	$2,1 \times 10^{-5}$
manure	0	$7 \times 10^{-8**}$	0
beat mash	2×10^{-6}	1×10^{-6}	1×10^{-6}

* For details of method see explanation to Figure 12.1

** 1 colony

The influence of LB compared to saline, and of bentonit, sucrose, manure, and mashed sugar beets was then assessed in more detail. To check LB seemed important, since rich media will probably be used to grow cells if releases are undertaken. Adding bentonit (10%) to unsterile soil and inoculating the mixture with cells suspended in saline did not give any exconjugants. Manure was equally ineffective. Sucrose, however, gave high transfer rates (Table 12.3), as did LB (Fig. 12.5). Sugar beet mash was similarly effective. With LB (Fig. 12.5), besides of exconjugants, drastic cell propagation was registered. The highest titers of donors, of recipients, and also of exconjugants were obtained after one day. Later, the titer of all three classes of cells declined. In contrast, with the saline cell suspensions, no exconjugants were obtained, and the titer of donors and recipients remained nearly constant.

d) Varying several ecological parameters

Non-sterilized soil was inoculated with the mating mix, in 4 ml LB per 50 g soil sample, and incubated for 3 days at temperatures from 18 to 26°C. The titers of donor and recipient were not influenced, but a slight temperature optimum for plasmid-transfer was found around 20°C. LB plates were buffered with 2x Soerensen buffer, in the pH range of 5.0 to 8.0. Then the mating mix was added on top and incubated at 22°C for 3 days. The mixture was processed and plated onto selective plates as before. Growth of both donor and recipient seemed independent of pH. Plasmid transfer was optimal around pH 5.5, but zero for pH 8.0.

Non-sterile soil of graded humidity was incubated. After adding the mating mix, it contained 10.6 to 24.0% water (-7.6 to -3.6 bar water potential). The samples were incubated for 3 days with LB instead of saline. Dry soil (10.6% final RH) gave distinctly higher numbers of exconjugants and better survival of both donor and recipient, than moist soil.

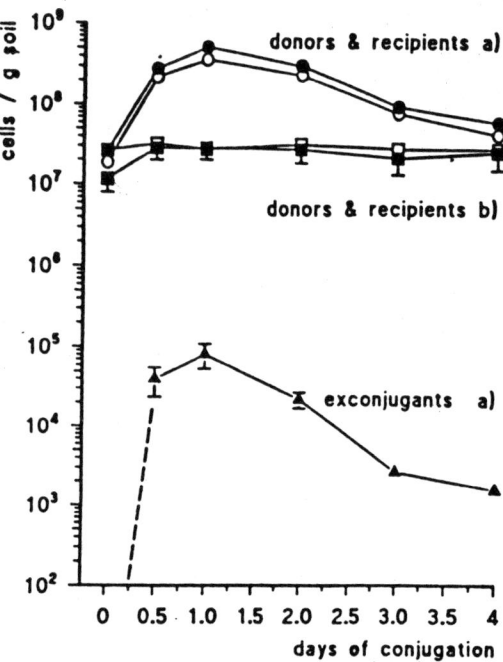

Fig. 12.5: Survival (donor and recipient cells per g soil) and plasmid pEA9 transfer (exconjugants per g soil) in homologous matings of *E. agglomerans* 339 in non-sterilized soil (50 g samples in Erlenmeyer flasks) for increasing mating periods with (a) and without (b) addition of Luria Broth. Mating conditions: 22°C, pH 5.2, humidity 15.5%. The symbols (open symbols: recipients) give the means and 1 x S.E.M. from 6 samples each (2 independent experiments with 3 replicates each). The limit of detection without Luria Broth (b) was 6 exconjugants per 10^8 surviving donors, the number of exconjugants with Luria Broth (a) was zero for zero time conjugation (indicated by the broken line).

Pregerminated wheat seedlings were put on top of the inoculated non-sterilized soil (bacteria in saline) in the flasks (7 seedlings/flask) and grown for upto 5 weeks. At two weeks, the roots were already forming a dense mat in the soil. The soil was then resuspended as before (free soil) and samples plated. In addition, the remaining roots were washed carefully, homogenized with quartz sand, and the material (rhizosphere homogenate) plated as before. Up to that time, and in contrast to 3 days incubation, there was a drastic decrease in the survival of both donor and recipient cells in the flasks without plants. This decrease was less dramatic if wheat roots were present (shielding effect). There were no exconjugants in the free, non-sterilized soil, neither with nor without plants, and none in the root material, not even after 5 week of incubation, or after 3 months, if plants were grown in pots in a growth chamber (data not shown). Hence exudates, if present, in contrast to LB, sucrose or sugar beet mash, are not sufficient to bring transfer rates up to the level of detection.

Further data have been published (Klingmüller, *et al.* 1990) on the influence of inoculation density, of soil packing, on the composition of the soils used and the absence of toxic factors in them, as well as on host range of the plasmids used. All these are laboratory or growth chamber studies.

e) Prospective test release with meso-cosms

To carry the analysis even further, a test release using meso-cosms with semi-containment is being scheduled. These meso-cosms are PVC tubes (14 cm diameter) that are closed at the bottom by a stainless steel sieve and a membrane filter (Fig. 12.6). Moisture and ions can pass freely, but inoculated bacteria are retained. Putting such meso-cosms, filled with inoculated soil, in the field, we want to see the survival of donors and recipients, and whether plasmid transfer is obtained or not, under such "close to natural" conditions. In the first stages of those releases, only saline washed inoculants and soil without the addition of higher plants will be used. Plasmid transfer will be checked to added homologous recipients, but also to other indigenous soil bacteria. This restricted schedule will hopefully enable the BGA-Berlin, acting under the new German Gene Law, to grant permission, which would get things going in Germany. Later, more complex situations will be checked under semi-containment, and if justified, the containments will be gradually relieved.

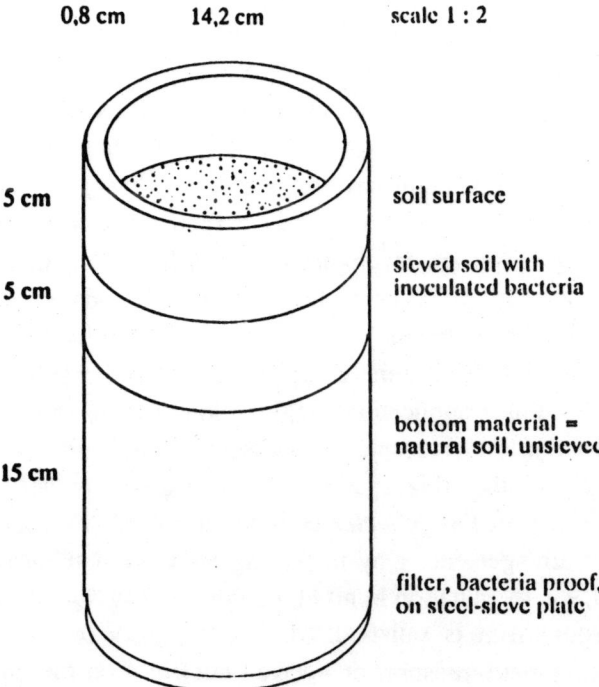

Fig. 12.6: Sketch of meso-cosm for field experiments.

DISCUSSION

Bacterial nitrogen fixation is a property with potential to satisfy the nitrogen requirement of agricultural crops. For cereals, attention has to be focussed on strains from their rhizosphere. Such strains include, among others, *Enterobacter agglomerans*.

In the above report, the genetic basis of nitrogen-fixation in such strains has been described. The nif-plasmids they harbour and the nif-gene group on them have been studied.

The location of the nif-gene group on plasmids is different from *Klebsiella pneumoniae* and several other Enterobacteriaceae where the nif-genes are on the chromosome (e.g. *Klebsiella oxytoca*: Wang, *et al.* 1985; *Enterobacter cloacae*: Zhu *et al.*, 1986). It is, however, similar to *Rhizobium leguminosarum* and *R.meliloti*. Yet, in these Rhizobia, the nif-genes, although on plasmids, are dispersed in several subclusters, interrupted by *fix* genes, *nod* genes and by other material. This arrangement is different from *E. agglomerans*, where they form one contiguous, tightly packed cluster, similar to *Klebsiella*. However, in *E. agglomerans*, they surpass *Klebsiella* by their particularly well ordered arrangement: nif-genes J and F, which have similar function but are dispersed in *Klebsiella* (Arnold, *et al.* 1988), are located in tandem arrangement on one side of the group, transcribed together as one operon.

For environmental considerations, it is remarkable, that some of the *Enterobacter* plasmids, carrying the nif-gene group, are self transmissible. Their spread to other Enterobacteriaceae may be a factor in the fluctuation of numbers of N_2-fixing bacteria in the rhizosphere of cereals, observed in different locations (Lindberg & Granhall, 1984, Jagnow, 1984, 1988; Haahtela, 1985; Väisänen *et al.*, 1985) and environmental situations.

The compact and well ordered organization of the nif-genes in these *Enterobacter* strains and their positioning on transmissible plasmids can be taken as an indication of a cornerstone in nif-gene evolution and dissemination. This system can be envisaged as the one plausible archetype, from which all other N_2-fixing systems have originated and diverged. Alternatively, it can be envisaged operationally as the ultimate step in the arrangement of these genes towards which all other N_2-fixing bacterial systems are developing. It will be interesting in this context to learn more about the organization of the nif-genes in primitive and in highly advanced bacterial groups.

Apart from elucidating further the molecular basis of nitrogen fixation in *E. agglomerans* and its correlation and connection to nif-gene group evolution, further work on these strains seems required with emphasis on their application, e.g, Develop methods to increase the absolute number of nif$^+$ *Enterobacter* in the rhizosphere. Develop methods to spread the nif information from nif$^+$ *Enterobacter* to other rhizosphere bacteria that do not yet have it. Find ways to amplify the nif-gene group in *Enterobacter* cells to an optimal number, e.g. by providing the gene group in tandem arrangement or by increasing the copy number of the plasmids. Make nif-genes derepressed, e.g. by mutation in nifAL operon, so that N_2-reduction does not stop once the bacterium's N-requirement is satisfied. Make cells leaky for NH_4 or amino acids synthesized, e.g. by making them transport or uptake deficient (*amt$^-$*) so that the plant roots have easier access to the N_2 fixed. Integrate nif plasmid genes into the recipient's chromosome to avoid uncontrolled spread of the nif information in a release situation. Some of these points are closely connected to the biological safety issue.

Our studies on risk assessment document that the conditions in natural soil are strongly opposed to plasmid transfer. This is in line with reports of others (Trevors, *et al.* 1987), and our own earlier findings (Schilf & Klingmüller, 1983, Döhler & Klingmüller, 1988). One reason for this could be the action of protozoa (Casida, 1989). Although, they may contribute to the disappearance of the inoculated bacteria, it is unlikely that protozoa would discriminate and preferentially feed on exconjugants. Another, more attractive reason could be the competition of indigenous bacteria with the inoculated ones. However, it could be documented (Klingmüller, 1991 a), that the titer of culturable indigenous bacteria in our natural soil samples was much less than that of the inoculated bacteria. Further, upon addition of LB, the titer of both groups of bacteria increased, and in particular that of the indigenous group. But these were just the conditions where exconjugants were obtained. This shows that competition is not the cause for low rates of plasmid transfer. It can be concluded that the lack of plasmid transfer in natural soil is primarily due to energy limitation. The occurrence of exconjugants in sterilized soil can then be explained by the assumption that by killing the indigenous microbes, energy containing substrates become available to the inoculated bacteria.

The finding that adding LB to the sample elicited high propagation and plasmid transfer raises the question whether LB grown material as inoculant bears the risk of uncontrolled gene spread. However, calculating the total liquid required, if a farmer were to apply the future inoculant as undiluted LB suspension at a rate of 4 ml per 50 g soil, on the basis of an equal distribution of the inoculant in the upper 10 cm layer of the field, it can be shown that per hectare 80,000 l LB would be needed, which - apart from labour - means costs of approximately 100,000.- US$ alone for the substrate. Hence not only cheaper substrates, e.g. molasses, but also smaller amounts of the inoculant, as in routine *Rhizobium* inoculations, would have to be used. Since in our experiments a dilution of as little as 1/100 of the LB cancelled the stimulating effect (Klingmüller *et al.*, 1990), the problem of plasmid transfer induced by LB supplementation would not arise in agricultural practice. A different situation would hold, if fields where sugar beets had grown before, were inoculated with *Enterobacter*. Here, locally, from decaying beets, the sucrose concentration may reach values that permit cell propagation and gene spread. Such situations would, therefore, have to be avoided.

Apart from energy depletion, the possible transfer of nif plasmid pEA9 is restricted by its narrow host range, hence barely significant in nature as a means of gene spread. It is known from other studies that the chances of gene spread by chromosomal transfer, by transduction or by transformation are even smaller than those of plasmid transfer, hence negligible here. Our data, indicating only low chances of gene spread for our bacterial system in natural soil, add to the findings of others with other bacterial systems. Together they tend to throw some doubt upon fanciful speculations that might invoke more thrilling scenarios.

In evaluating our risk assessment data, it should be kept in mind that these experiments deal with a very special case, e.g. nitrogen fixing *Enterobacter agglomerans*.

CONCLUSIONS

1. Bacterial survival and plasmid transfer data obtained with sterile soil are not representative of results in non-sterile soil, or of the release situation. Studies on the latter should, therefore, be increased.

2. Bacterial propagation and plasmid transfer in non-sterile soil, upon inoculation, can in general barely be detected, unless energy sources are supplied. The particular source of energy is unimportant. This holds for all Gram-negative soil bacterial systems studied so far for homologous and heterologous transfer. The possible exception of plasmids of RP4 type is still controversial.

3. Further studies are needed, with other bacteria and plasmids, with and without higher plants, in particular under true release situations, on the presumptive risks of such releases and their environmental impact.

Acknowledgements

I am grateful to Mrs. Christine Fentner, Marion Steinlein and Ruth Hösl for excellent technical assistance, Dipl. agr. Ch. Rappold and E. Evguenieva helped with some of the experiments. Support for this work was provided by grants from the Bundesministerium für Forschung und Technologie, Bonn, the European Community, Brussels and the Bayer, Staatsministerium Für Landesentwicklung und Umweltfragen, München.

References

Abdel-Salam, M.S. and Klingmüller, W. (1987). Mol. Gen. Genet. **210,** 165-170.

Arnold, W., Rump, A., Klipp, W., Priefer, U. and Pühler, A. (1988). J. Molec. Biol. **203,** 715-738.

Casida, L.E. Jr. (1989). Appl. Environ. Microbiol. **55**, 1857-1859.

Dippe, R. (1990). Diplom-thesis, University of Bayreuth.

Döhler, K. and Klingmüller, W. (1988) . In Risk Assessment for Deliberate Releases. Edited by W. Klingmüller. Springer-Verlag, Heidelberg, pp. 18-28.

Haahtela, K. (1985). FEMS Microbiology Ecology. **33**, 211-214.

Jagnow, G. (1984). Kali-Briefe (Büntehof) **17**, 341-350.

Jagnow, G. (1988). In Nitrogen fixation: Hundred years after. Edited by H. Bothe, F.J. de Bruijn and W.E. Newton. Gustav Fischer, Stuttgart - New York, p 795.

Kleeberger, A., Castorph, H. and Klingmüller, W. (1983). Arch. Microbiol. **136,**306-311.

Klingmüller, W. (1985). In Trends in Molecular Genetics. Edited by U. Sinha and W. Klingmüller. Spectrum Publishing House Patna and New Delhi, pp 37-61.

Klingmüller, W. (Ed) (1988a). Azospirillum IV: Genetics, Physiology, Ecology. Springer-Verlag Berlin - Heidelberg, New York, Tokyo.

Klingmüller, W. (1988b). In: Proceedings of the International Symposium on Plant Biotechnology, Gyeongsang National University, Chinju, South Korea, pp 41-56.

Klingmüller, W. (1991a). FEMS Microbiology Ecology, (In press).

Klingmüller, W. (1991b). In Proceedings of Kiawah Island Conference. Edited by D.R. MacKenzie and S. Henry. United States Department of Agriculture, (In Press).

Klingmüller, W., Dally, A., Fentner, C. and Steinlein, M. (1990). In Bacterial genetics in natural environments. Edited by J.C. Fry and M.J. DayChapman and Hall, London, pp 133-151.

Klingmüller, W., Herterich, S. and Min, B.W. (1989). In Nitrogen fixation with non-legumes. Edited by F.A. Skinner, R.M. Boddey and I. Fendrik. Kluwer Academic Publishers, Dordrecht, pp 172-178.

Kreutzer, R. (1986). Diplom-thesis, University of Bayreuth.

Kreutzer, R. (1989). Ph.D. thesis, University of Bayreuth.

Kreutzer, R., Singh, M. and Klingmüller, W. (1989). Gene **78**, 101-109.

Lindberg, T. and Granhall, U. (1984). Appl. Environ. Microbiol. **48**, 683-689.

Riedel, G.E., Brown, S.E. and Ausubel, F.M. (1983). J. Bacteriol. **153**, 45-56.

Schilf, W. and Klingmüller, W. (1983). Recombinant DNA Technical Bulletin, **6**, 101-102.

Singh, M. and Klingmüller, W. (1985) In: Sinha, U. and Klingmüller, W. (Eds) Trends in Molecular Genetics, Spectrum Publishing House Patna and New Delhi, pp 15-35.

Singh, M. and Klingmüller, W. (1986a). Mol. Gen. Genet. **202**, 136-142.

Singh, M. and Klingmüller, W. (1986b). Plant and Soil. **90**, 235-242.

Singh, M., Kleeberger, A. and Klingmüller, W. (1983). Mol. Gen. Genet. **190**, 373-378.

Singh, M., Kreutzer, R., Acker, G. and Klingmüller, W. (1988). Plasmid **19**, 1-12.

Skinner, F.A., Boddey, R.M. and Fendrik, I. (Eds) (1989). Kluwer Academic Publishers, Dordrecht, Boston, London.

Steibl, H.-D. (1989). Diplom-thesis, University of Bayreuth.

Trevors, J.T., Barkar, T. and Bourquin, A.W. (1987). Can. J. Microbiol. **33**, 191-198.

Väisänen, O., Haahtela, K., Bask, L., Kari, K., Salkinoja-Salonen, M. and Sundman, V. (1985). Arch. Microbiol. **141**, 123-127.

Wang, P.-L., Koh, S.K., Chung, K.-S., Uozumi, T. and Beppu, T. (1985). Agric. Biol. Chem. 49, 1469-1477.

Zhu, J.-B., Li, Z.-G., Wang, L.-W., Shen, S.-S. and Shen, S.-C. (1986). J. Bacteriol. **166**, 357-359.

13

Molecular Biology of Pituitary Gonadotropins

S.K. Jain

Department of Biochemistry, Hamdard University
Hamdard Nagar, New Delhi-110062

ABSTRACT

The neuro-endocrine axis regulates a number of metabolic pathways through the production of many hormones by hypothalamus and pituitary and other neuronal factors. Gonadotropins regulate the sexual development and reproductive functions. These are a group of complex glycoprotein hormones, each having two subunits, α and β, coded by separate genes present on different chromosomes. The expression of genes for α and β subunits of LH and FSH is mutually co-ordinated, the precise mechanism of which is not yet fully understood.

The cDNA and genomic clones for α and β subunits of LH and FSH have been isolated and characterized; the complexity of the related coding genes have also been analysed. The genes for the two subunits have varying degree of homology and it is predicted that the α and β subunit genes have diversified from a common ancestral gene during the course of evolution. There is a strong inter-species homology in the α and β subunits of LH and FSH, though some restriction site based polymorphisms are seen.

The molecular cloning and characterization of full length cDNAs for α and β subunits of sheep LH and FSH is reported. These clones have been expressed in mammalian cell expression vectors and the synthesis of immuno-reactive and bioactive hormone by rDNA technology has been achieved. The analysis of r-hormone reveals that this is similar to natural hormone in its size and is recognized by specific antibodies. It is secreted in dimer form and seems to have correct glycosylation. Further, the r-hormone is recognized by FSH specific receptors and causes the maturation of oocytes in a hormone dependent *in vitro* system. Furthermore, the bioactivity of r-hormone is comparable to natural hormone. The implications of these studies have been discussed.

INTRODUCTION

Pituitary secretes a number of hormones which have diversified actions and regulate a variety of physiological processes. Some of these hormones act directly on their target tissues triggering a chain of events resulting in various metabolic changes. Another group of hormones, known as tropic hormones, act on other endocrine glands and control the function of these glands. These hormones modulate the synthesis and secretion of hormones produced by the target glands. The function of pituitary itself, on the other hand, is regulated by the secretion of hormones from hypothalamus and by some neuronal factors produced in brain cells. A fine balance exists between various factors. This mutually inter-dependent regulatory mechanism or the neuro-endocrine axis, plays the key role in synchronization of various physiological functions. The regulation of this balance is very precise and any disturbance here is amplified many fold in its manifestation at the target cell level resulting in a number of adverse effects. Though a great deal of studies aimed at understanding the regulation of gene expression for various pituitary hormones have been reported, our knowledge in this field is not complete. In last few years the advances in genetic engineering technology have not only helped our understanding immensely, but also offered an alternate route for the production of gonadotropins in relative abundance.

Structure and Function of Gonadotropins

Gonadotropins control the development and regulation of sexual functions and reproductive processes. The group consists of three interrelated hormones, the luteinizing hormone (LH), the follicle stimulating hormone (FSH) and the chorionic gonadotropin (CG). The LH and FSH are produced by the gonadotrope cells of anterior pituitary while the CG is produced by placenta. Each of these hormones consists of two dissimilar but related subunits, the α and the β. The α-subunit is common to all these hormones and also to thyroid stimulating hormone (TSH). The β-subunit, on the other hand, differs from each other and confers the biological specificity to the hormones. Both the subunits are highly glycosylated and the carbohydrate component is necessary for the biological activity of the hormone. Deglycosylated form loses the hormonal activity but maintains immuno-reactivity. The natural hormone is a dimer of the two subunits in equimolar ratio. It is well established that only the dimer form of the hormone is biologically active and neither of the two subunits present in free form, has the hormonal activity (Archer, 1980; Pierce & Parsons, 1981; Chin, 1985; Jain, 1994).

The synthesis of LH and FSH is regulated by a common gonadotropin releasing hormone (GnRH) produced by hypothalamus. A number of other neuronal factors are also involved in regulation of the synthesis of these hormones. Inhibin, a small polypeptide found in brain inhibits FSH synthesis while another protein factor, the FSH releasing peptide (FRP) stimulates the dimer formation and secretion of FSH. The LH and FSH, both act on primary gonads and influence the synthesis of estradrol and progesterone in females and testosterone in male. The circulating levels of sex steroids on the other hand, exert a negative feed back on the synthesis of gonadotropins. A delicate balance is maintained between the synthesis of various hormones and the factors. Any disturbance in this system upsets the whole reproductive process (Archer, 1980; Forage *et al.*, 1986; Ling *et al.*, 1986; Mayo *et al.*, 1986; Vale *et al.*, 1986).

The primary function of the gonadotropins is to modulate the gonadal activity, and hence the name gonadotropins. The LH and FSH act synergistically in many of their actions. In females, these are necessary for the maturation of ovarian follicles and oocytes and for the release of mature ovum. The LH is necessary for the ovulation, and a LH surge occurs just before ovulation. It is also responsible for the formation of corpus luteum and for its maintenance. The FSH is the primary factor for the induction of ovarian growth and causes the maturation of oocytes in females. In males, FSH regulates spermatogenesis. The sertoli cells in seminiferous tubules are the target for its action. It also stimulates the synthesis of androgen binding protein (ABP) and thus facilitates the testosterone in carrying out its functions. It has also been implicated as a fertility regulator in males. An antifertility vaccine based on anti-FSH is undergoing clinical trials currently (Moudgal *et al.*, 1988). The diagrammatic representation of neuro-endocrine axis and the role of gonadotropins has been illustrated in Fig. 13.1.

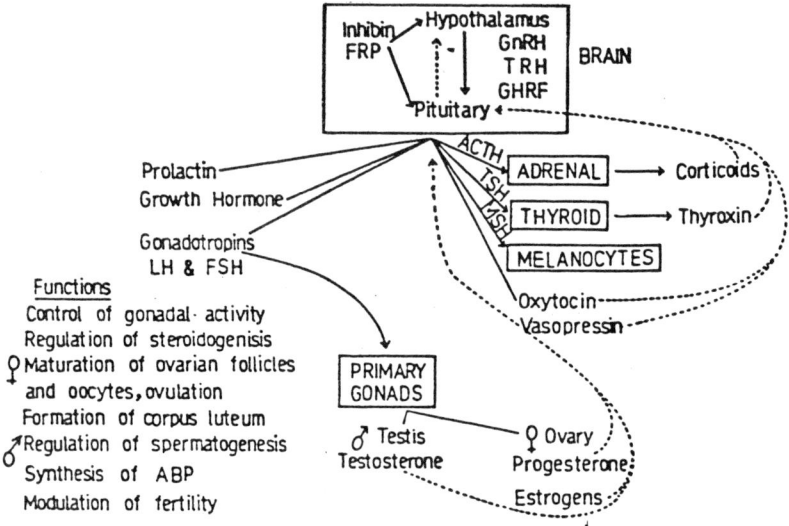

Fig. 13.1 The Neuro-endocrine axis.

In all the species, the α and β subunits of LH and FSH are relatively small, having a molecular weight of about 10-15 KD. The α-subunit common to both LH and FSH, is 90-96 amino acid long. It is a secretory protein and has a signal peptide of 24-25 amino acids in its precursor form. The LHß is larger in all the species, about 135-145 amino acid long including the signal peptide of 20 amino acids. The mature FSHß is about 110 amino acid in size and has a signal peptide of 19 amino acids. The two subunits get annealed post-translationally and are held together by a number of interchain S-S bridges (Kornfeld & Kornfeld, 1980; Goverman *et al.*, 1982; Chin, 1985).

There is a high degree of inter-species cross-reactivity at both the protein and gene level of both α as well as β subunits. Also, there is some degree of homology between the α and β subunit of the same hormone and also between the ß-subunits of different hormones. The different subunits are coded by separate but related genes which are present on different chromosomes but their expression is mutually coordinated. This homology raises the important question about the evolution of subunit genes, and the studies reveal that the gene for common β subunit and the genes for β subunits of different hormones have evolved from a single ancestral gene. The diversion of α and β subunit genes probably occurred earlier than the diversion of β-subunit genes of LH, FSH, and CG during the course of evolution (Chin, 1985). The mRNA for α-subunit is present in much higher abundance than the LHβ or FSHβ mRNAs. Further, α-subunit is synthesized in much higher amounts than what can be accounted for dimer formation. As a result, certain amount of free α-subunit is always present in cells which does not participate in biological function of the hormones. The importance of this free α-subunit pool is not well understood (Vaitukaitis *et al.*, 1976; Parsons *et al.*, 1983; Chin, 1985). It, therefore, is of great interest and relevance to understand the regulatory mechanism of the expression of these genes. The gene cloning techniques offer a very powerful tool. Besides, it also provides an alternate route for the production of these hormones independent of physiological system. The cDNAs and genes coding for α and β subunits of LH and FSH have been cloned from a number of species and analysed. My laboratory has a long standing and continuous interest in the molecular biology of pituitary hormones. We have constructed a number of pituitary cDNA and genomic libraries and have isolated full length cDNA clones for the α and β subunits of ovine LH and FSH, analysed the genes and mRNAs coding for these and have expressed them in mammalian cell expression system.

Molecular cloning of sheep gonadotropin cDNAs

The cDNAs and genes for the common α-subunit of gonadotropins from mouse (Chin *et al.*, 1981), rat (Godine *et al.*, 1980), cow, sheep (Jain *et al.*, 1987) and human (Fiddes &Goodman, 1979) have been cloned. Similarly the gene clones for LHβ of cow (Maurer, 1985), rat (Chin *et al.*, 1983), sheep (Jain et al., 1987) and human (Talmadge *et al.*, 1984) and the FSHβ genes for human (Watkins *et al.*, 1987), cow (Esch *et al.*, 1986), pig, sheep (Mountford *et al.*, 1989) and rat (Maurer, 1987) have been isolated. We have already isolated full length cDNA clones for sheep, α and β-subunits of LH and FSH and analysed their genomic complexities (Jain *et al.*, 1987, 1990; Garg & Jain, 1992a, b). The strategies for the cloning of sheep gonadotropin subunit genes are shown in Fig. 13.2. The first step to achieve the cloning of genes for the α-subunit, LHβ and FSHβ is to construct the gene libraries. The cDNA library of sheep pituitary was constructed by the method of Gubler and Hoffman (1983) with certain modifications (Jain *et al.*, 1987). The total cellular RNA was prepared from fresh pituitaries by homogenization in guanidinium isothiocyanate followed by CsCl density gradient centrifugation (Ullrich *et al.*, 1977). The mRNA was isolated by affinity chromatography over oligo-dT-cellulose columns (Aviv & Leder, 1972). It was quantitated by its absorbance at 260nm, the size was analysed on agarose gels, and the biological activity was assessed by *in vitro* translation using rabbit reticulocyte lysate. The first strand of cDNA was synthesized from the mRNA using oligo-dT primer and MLV reverse transcriptase; and second strand was synthesized with RNase H and *E. coli* DNA polymerase I added together. The ds:cDNA showed a heterogenous size distribution of 0.3-2.5 kb with 0.5-1.5 kb being the predominant class. This is in agreement with the size distribution of majority of eukaryotic mRNAs. A 20-25 base long poly(dC) tail was added to

3'-end of the cDNA with terminal deoxynucleotidyl transferase (Jameson *et al.*, 1972) and the tailed cDNA was annealed to *Pst*I digested, poly(dG) tailed pBR322 (New England Biolabs, Boston). The open circular DNA was used for transforming *E. coli* cells. Cloning of cDNA at the *Pst*I site of the plasmid resulted in inactivation of β-lactamase gene of pBR322 (Sambrook *et al.*, 1988), so that the recombinants could be selected for their sensitivity to ampicillin and resistance to tetracycline. cDNA was also cloned in λgt10 arms by *Eco*RI linkers. Representative libraries of about 10^5 independent clones were thus obtained and screened with radiolabelled heterologous probes (courtesy Dr. W.W. Chin, Harvard Medical School, Boston) by colony hybridization of plasmid library and by plaque lifting of phage library (Sambrook *et al.*, 1988). Essentially full length cDNA clones for the α-subunit, LHβ and FSHβ have been obtained. The clones were purified and their identities were established by following criterian. (I) The specific hybridization with respective rat cDNAs, (II) Hybrid selection of specific mRNAs on Northern blots, (III) The restriction analysis of the inserts, and (IV) By the partial sequencing and its comparison with cDNA and deduced amino acid sequence of rat hormonal subunits (Jain *et al.*, 1987, 1990, Garg & Jain, 1992a).

Table: 13.1 Properties of sheep gonadotroptin hormone clones

S.N.	Property	α-subunit	LHβ	FSHβ
1.	Abundance	high	moderate	very low
2.	Size of cDNA	0.6 kb	0.65 kb	1.5 kb
3.	mRNA	single mRNA, size 0.75 kb	single mRNA, size 0.7 kb	single mRNA, size 1.7 kb
4.	Genes	single gene	single gene	multiple genes
5.	Restriction analysis			
	HindIII	0 sites	0 sites	0 sites
	EcoRI	0 sites	0 sites	0 sites
	Bam HI	0 sites	0 sites	0 site
	BglII	0 sites	0 sites	0 sites
	PstI	2 sites	0 sites	2 sites
	NcoI	1 site	0 sites	0 sites
	AluI	3 sites	3 sites.	4 sites
6.	Homology with other subunit genes	very little	very little	very little
7.	Inter-species homology	high homology with cow, moderate homology with rat and human	very high homology with cow, moderate homology with rat and human	99% homology with cow, moderate homology with rat and human

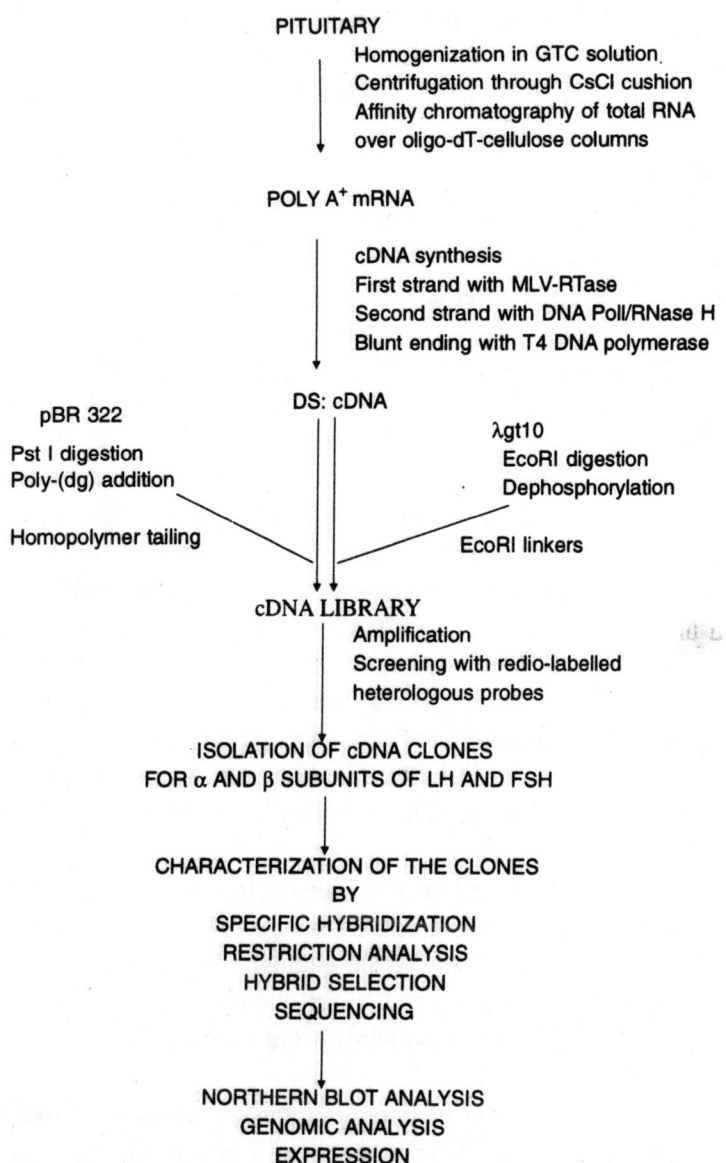

PITUITARY

Homogenization in GTC solution.
Centrifugation through CsCl cushion
Affinity chromatography of total RNA
over oligo-dT-cellulose columns

POLY A⁺ mRNA

cDNA synthesis
First strand with MLV-RTase
Second strand with DNA PoII/RNase H
Blunt ending with T4 DNA polymerase

DS: cDNA

pBR 322

Pst I digestion
Poly-(dg) addition

Homopolymer tailing

λgt10
EcoRI digestion
Dephosphorylation

EcoRI linkers

cDNA LIBRARY
Amplification
Screening with redio-labelled
heterologous probes

ISOLATION OF cDNA CLONES
FOR α AND β SUBUNITS OF LH AND FSH

CHARACTERIZATION OF THE CLONES
BY
SPECIFIC HYBRIDIZATION
RESTRICTION ANALYSIS
HYBRID SELECTION
SEQUENCING

NORTHERN BLOT ANALYSIS
GENOMIC ANALYSIS
EXPRESSION

Fig. 13.2: Diagrammatic representation of strategy for cloning of pituitary hormone genes

For the analysis of genes, the large molecular weight genomic DNA was isolated from the sheep lymphocytes purified by repeated extractions with phenol, phenol/CHCl₃, CHCl₃/isoamyl alcohol and precipitations with ethanol and analysed on gels. It was digested with different restriction enzymes in 10-15 fold molar excess to complete digestion and was fractionated on agarose gels, transferred to nitrocellulose by the method of Southern (1975) and hybridized with the nick-translated (Rigby *et al.*, 1977) probes. Similarly, formamide denatured mRNA was fractionated on formaldehyde containing agarose gels (Sambrook *et al.*, 1988) and transferred to nylon membranes (Thomas, 1983). The Northern blots were analysed by molecular hybridization with the probes (Ausubel *et al.*, 1987; Sambrook *et al.*, 1988).

The α-subunit Genes

The analysis of α-subunit cDNAs from human (Fiddes & Goodman, 1979; Talmadge *et al.*, 1984), rat (Godine *et al.*, 1980, 1982; Chin *et al.*, 1983), mouse (Chin *et al.*, 1981), pig, cow (Chin *et al.*, 1981) and sheep (Jain *et al.*, 1987) show that it is one of the most predominant species of mRNA in all these cases. The relative abundance of α-subunit cDNA in sheep pituitary libraries is about 0.3%. The sizes of cDNA inserts are reported to be 0.6-0.8 kb, and in sheep it is approximately 0.6 kb in size. The insert was excised from the pBR322 by *Pst*I digestion, showed three fragments suggesting the presence of two internal PstI sites (Fig. 13.3A). It differs from human and rat cDNAs in this respect which have one (human) and none (rat) *Pst*I sites, respectively (Jain *et al.*, 1987, 1990; Garg *et al.*, 1991). There are, thus, restriction site based differences amongst different species and the polymorphisms are located both in the coding and non-coding regions of the gene. The detailed restriction analysis of sheep α-subunit cDNA revealed that it does not have any sites for *Hind*III, *Eco*RI, *Bam*HI and *Bgl*II but has a number of sites for *Alu*I and *Hae*III and one site each for *Nco*I and *Pvu*II. The *Nco*I site is immediately upstream of the initiation codon and can be conveniently used for the removal of 5'-non coding region. The restriction analysis is in agreement with α-subunit cDNAs of other species with minor differences (Fig. 13.3B). The sequence analysis shows that the cDNA has a single open reading frame coding for 120 amino acids followed by the termination codon. No other unique feature was discernible.

The analysis of the RNA showed that the sheep α-subunit is coded by a single mRNA species. The size of this mRNA is about 750 bases. The hybridization of the Northern blot with the cDNA insert gave a sharp band on the autoradiogram and showed no heterogenity in size (Jain *et al.*, 1987). In all the species studied so far, the presence of only one mRNA of similar size coding for α-subunit of pituitary gonadotropin has been reported (Chin, 1985). The analysis of mRNA is shown in Figure 13.3C.

The detailed analysis of sheep α-glycoprotein gene revealed that α-subunit is coded by a single gene (Garg & Jain, 1992a) present within a 5 kb *Eco*RI fragment (Fig. 13.3D). The presence of a single band coding for the α-subunit has been reported for human, cow and rat (Chin, 1985). The α-subunit gene is relatively large measuring 13.7 kb in cow and 9.4 kb in human. In comparison, the sheep gene seems to be relatively small. In all the species the gene contains four exons and three introns. The sizes of the introns are variable while those of exons are relatively constant. The positions of intron-exon junction are conserved to a great extent. In cow the first intron is about 10 kb, second is 1.1 kb and the third is 0.6 kb in size while in human these are 6.4 kb, 1.7 kb and 0.4 kb, respectively. The first exon has the promoter region and the bulk of 5'-noncoding region, the coding region starts in the second exon and the last

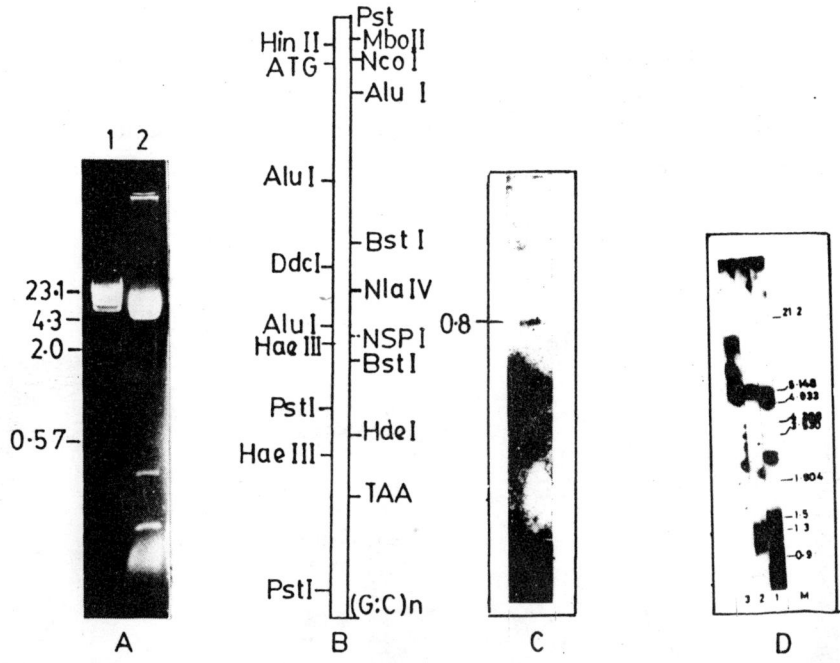

Fig. 13.3: The analysis of α-subunit genes.

Panel A   The cDNA clone. Lane 1, *Hind*III digested DNA run as marker; lane 2, the α-subunit cDNA clone digested with *Pst*I, releasing the insert in three fragments.

Panel B   The restriction map of α-subunit cDNA showing the relative positions of different restriction enzyme sites.

Panel C   The autoradiogram of Northern blot of total pituitary RNA, hybridized with α-probe. Only one mRNA of 0.75 kb gets hybridized.

Panel D   The genes coding for α-subunit. Sheep genomic DNA was digested with *Pvu*II, *Hind*III and *Bam*HI (lanes 1, 2 and 3 respectively) and the Southern blots were hybridized with α-probe. The sizes of genomic bands

exon contains part of the coding region, the termination codon and the 3'-noncoding region along with the polyadenylation signal (Jain, 1991). In human the α-subunit gene is present on chromosome 6 and in mouse it is present on chromosome 4 (Nayler *et al.*, 1983; Chin, 1985).

Genes for LHß

The LHβ cDNA for human (Talmadge *et al.*, 1984), rat (Chin *et al.*, 1983), cow (Maurer, 1985) and sheep (Jain *et al.*, 1987, 1990) have been cloned. The occurrence of LHβ mRNA is relatively low, in sheep it is present in an abundance of about 0.05%. The analysis of sheep LHβ cDNA and genes is shown in Fig. 13.4. The cDNA insert is 0.65 kb (Fig. 13.4A) and has a single open reading frame coding for 141 amino acids. The restriction analysis demonstrates that it has no internal *Pst*I site. There are no sites for *Bam*HI, *Bgl*II, *Eco*RI, *Hind*III, *Nco*I, *Hinf*I and

*Xho*II, but a number of sites for *Hae*III, *Ava*II and *Alu*I are present (Fig. 13.4B). The sequence analysis shows a very high (more than 95% homology) between sheep and cow LHβ while about 80% homology between sheep and rat LHβ cDNAs. There is no cross reactivity between cDNAs for α-subunit and LHβ. The sequence analysis also reveals an unusually small 5' noncoding region a unique feature of LHβ mRNA. In rat it is only 7 nucleotide long (Jain *et al.*, 1987, 1992).

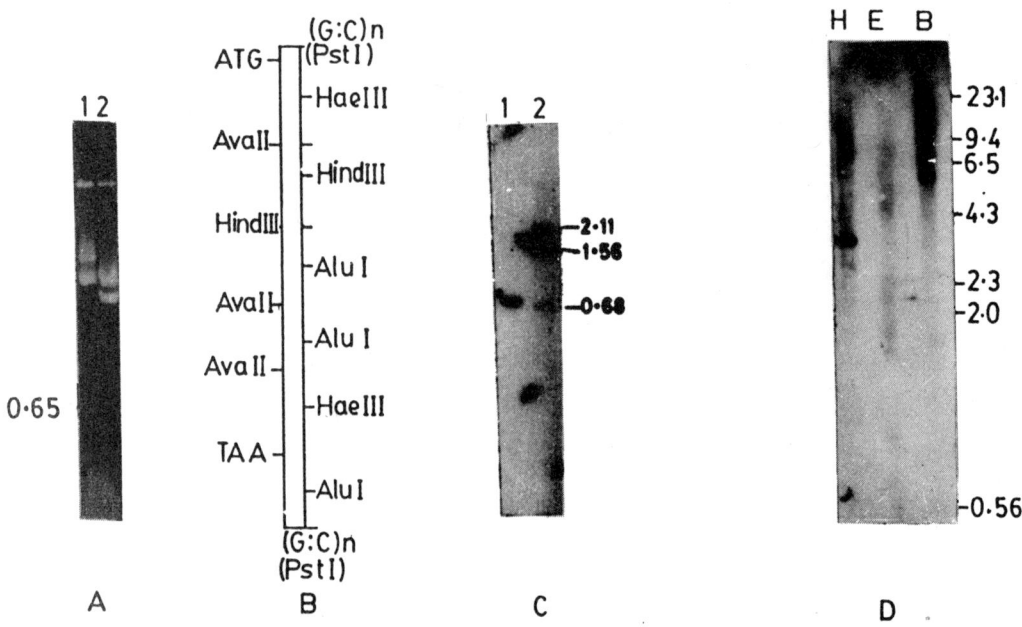

Fig. 13.4: Sheep LHß genes and mRNA.

Panel A   The cDNA clone in plasmid pBR322. Lane 1, the total plasmid DNA (undigested) and lane 2, *Pst*I digestion of the plasmid DNA showing the insert band of 0.65 kb.

Panel B   The restriction map.

Panel C   The Northern blot analysis. A single mRNA of 0.7 kb gets hybridized (lane 1). Lane 2 shows the radio-labelled size marker (*Rsa*I digested pBR322) run in the same gel.

Panel D   Genes coding for LHβ. *Hind*III, *Eco*RI and *Bam*HI (lanes H, E and B respectively) digested genomic DNA was analysed in a Southern blot. A single gene fragment gets hybridized in each case.

The LHβ mRNAs from a number of species have been analysed. In each case, a single mRNA codes for it. In sheep, the mRNA is about 650 bp and gives a sharp band on autoradiograms (Fig. 13.4C). The LHβ is coded by a single gene in all the species. However, there is a very high degree of cross reactivity between LH and CG genes in human. In most of the early studies for human, LHβ and hCG genes were analysed together and 8 genes coding for hLHβ/hCGβ family were reported (Boornstein *et al.*, 1982; Policastro *et al.*, 1983; Talmadge *et al.*, 1984). The later studies using differential hybridization revealed that only one out of these genes codes for LHβ, others are for hCGβ. The LHβ gene is very small; contains two introns and three exons. While in human the total size of the gene is only 1.4 kb, in rat it is about 1 kb. The sizes of the introns are 350 and 250 bp, respectively in human and 240 and 225 bp, respectively in rat. Our results with sheep have shown that it is coded by a single gene present within a 3 kb *Hind*III fragment (Figure 13.4D). We also found a number of related bands hybridizing with LHβ probe with relatively lower intensity which may probably represent the CGβ genes. The LHβ/CGβ genes of human are present on chromosome 19 and that of mouse LHβ gene on chromosome 7 (Garg & Jain, 1992a).

FSHß genes

The FSH cDNAs for rat (Maurer, 1987), human (Watkins *et al.*, 1987), cow (Esch *et al.*, 1986; Maurer & Beck, 1987) and sheep (Mountford *et al.*, 1989; Garg and Jain, 1992a) are cloned. The sheep and cow cDNAs have 99% homology. The sheep cDNA insert is about 1.6 kb long and codes for a protein of 129 amino acids (Fig. 13.5A). It has two internal *Pst*I sites and a number of sites for *Ava*II and *Hinf*I. There is one site each for *Pvu*II, *Bgl*II and *Xho*II and no sites for *Eco*RI, *Bam*HI, *Hind*III, *Nco*I, *Sal*I and *Xho*I (Fig. 13.5B). The sequence analysis reveals a unique feature that FSHβ mRNA has a very long 3'-non coding region of about 1.1 kb (Garg & Jain, 1992b; Jain *et al.*, 1992).

In sheep, FSHß is coded by a single mRNA which gives a relatively broad band on the gel. The size of this mRNA is about 1.7 kb. The mRNA is polyadenylated. Similar sizes for the FSHβ mRNAs of human and cow have also been reported. Northern blot analysis of sheep FSHβ mRNA is illustrated in Figure 13.5C. The FSHβ genes are relatively less characterized. In human, it is coded by a single gene which is present within a 3.7 kb *Eco*RI fragment. The gene is present on chromosome 11. However, our studies with sheep FSHß reveal that it may be coded by a multigene family consisted of atleast 3 genes (Figure 13.5D). These genes require further analysis to get an insight into subgenomic organization and size of these genes. The discrepancy between human and sheep genes needs further analysis (Garg & Jain, 1992b).

The detailed characteristics of cDNA clones for α and β subunits of sheep LH and FSH are summarised in Table 1 and their genomic complexity is given in Table 13.2.

Expression of cloned genes

Both LH and FSH are glycoprotein in nature and require post-translational modifications for their synthesis. As bacteria do not have necessary machinery for these modifications, it is obligatory to use one of the different eukaryotic expression systems available for this purpose. Yeast cells provide the simplest eukaryotic expression system as they are easy to grow and give good yields of the foreign gene products. However, it has been found that very often the glycosylation of mammalian proteins produced in yeast cells is defective and complex biomolecules such as pituitary hormones may not be synthesized in bioactive form. Other systems such as the vaccinia virus system, are efficient for expression and are highly useful. However, vaccinia is a cytoplasmic virus and is unsuitable for stable expression. Baculo virus/insect cell system has a lot of potential. It has a great advantage over other systems as insect cells are easy to grow and unlike mammalian cells do not require the expensive ingredients like FCS. Further, the system can express a foreign gene at a very high levels and has been in usage for a number of genes. However, it is still at the exploratory stage and some preliminary studies have suggested that the system is incapable of correct sialic acid addition. We have, therefore, used vaccinia virus system for the transient expression and the mammalian cell expression system for stable expression of these genes. The general strategy of expression is shown in Figure 6.

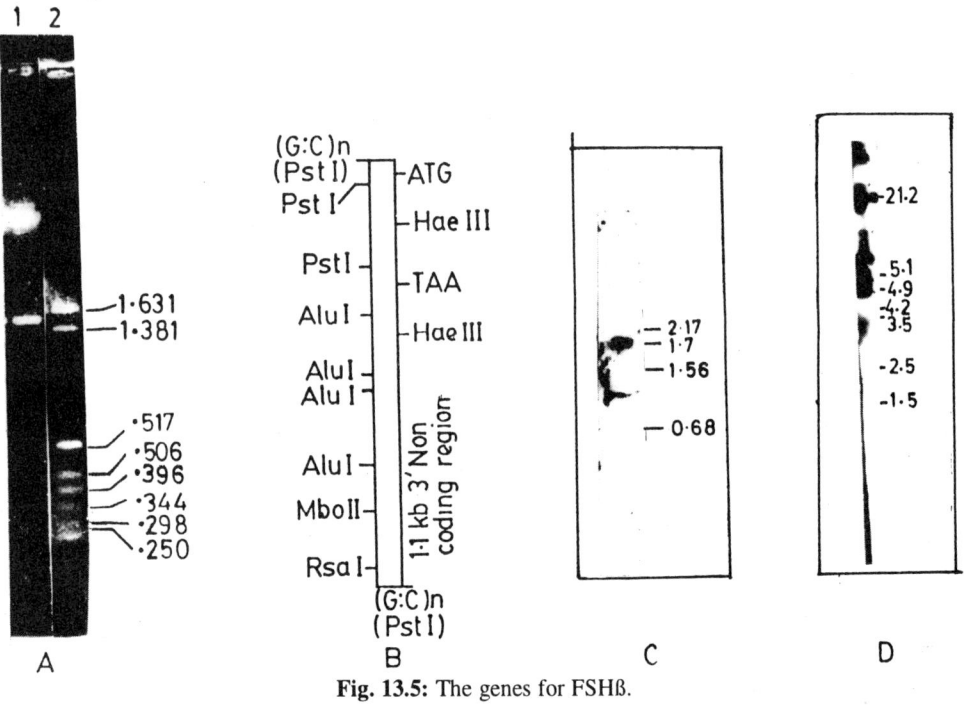

Fig. 13.5: The genes for FSHß.

Panel A   The size of cDNA clone. Lane 1, *Pst*I digested cDNA clone showing the vector and two insert bands of 1.4 and 0.25 kb, lane 2, *Hinf*I/ *Pst*I digested pBR322 DNA run as size marker.

Panel B   The restriction map of FSHß cDNA insert.

Panel C   Northern blot analysis showing a single mRNA of 1.7 kb codes for sheep FSHß.

Panel D   Genes coding for FSHß. A Southern blot of *Bam*HI digested DNA was hybridized when a number of genomic bands got hybridized.

The first strategy was to clone a 900 bp fragment of α-subunit cDNA in pBR322 containing the entire coding region in a vaccinia virus vector under TK promoter. The infection of a wild

Table: 13.2 Genomic complexity of sheep gonadotropin hormones of sheep

S.N.	Hormonal	No. of Subunit	Restriction Genes used for Digestion	No. of Enzyme in cDNA	No. of Bands Sites		Size of Major – Minor Bands in kb
1.	Common α-subunit	1					
a.			HindIII	0	1	2	5.0
b.			Bam HI	0	2	1	6.0 5.0
c.			PvuII	1	2	1	5.0 4.8
2.	LHß	1					
a.			HindIII	0	1	1	3.0
b.			PstI	0	2	0	6.5 5.8
c.			Bam HI	0	1	0	5.4
3.	FSHß	3					
a.			HindIII	0	3	1	3.2 2.7 1.4
b.			PstI	2	5	0	3.2 2.5 2.0 1.3 0.9
c.			Bam HI	0	5	0	25.0 20.0 5.6 5.0 4.8

type virus (attenuated) with this plasmid led to the integration of the α-subunit gene into viral genome. The transfection of fibroblast cells with the recombinant virus resulted in expression of α-subunit gene. The hormone produced by this route was immuno-reactive and was capable of combining with hCGß to form a bioactive dimer, stimulating the steroidogenesis in leydig cells (Lall *et al.*, 1988).

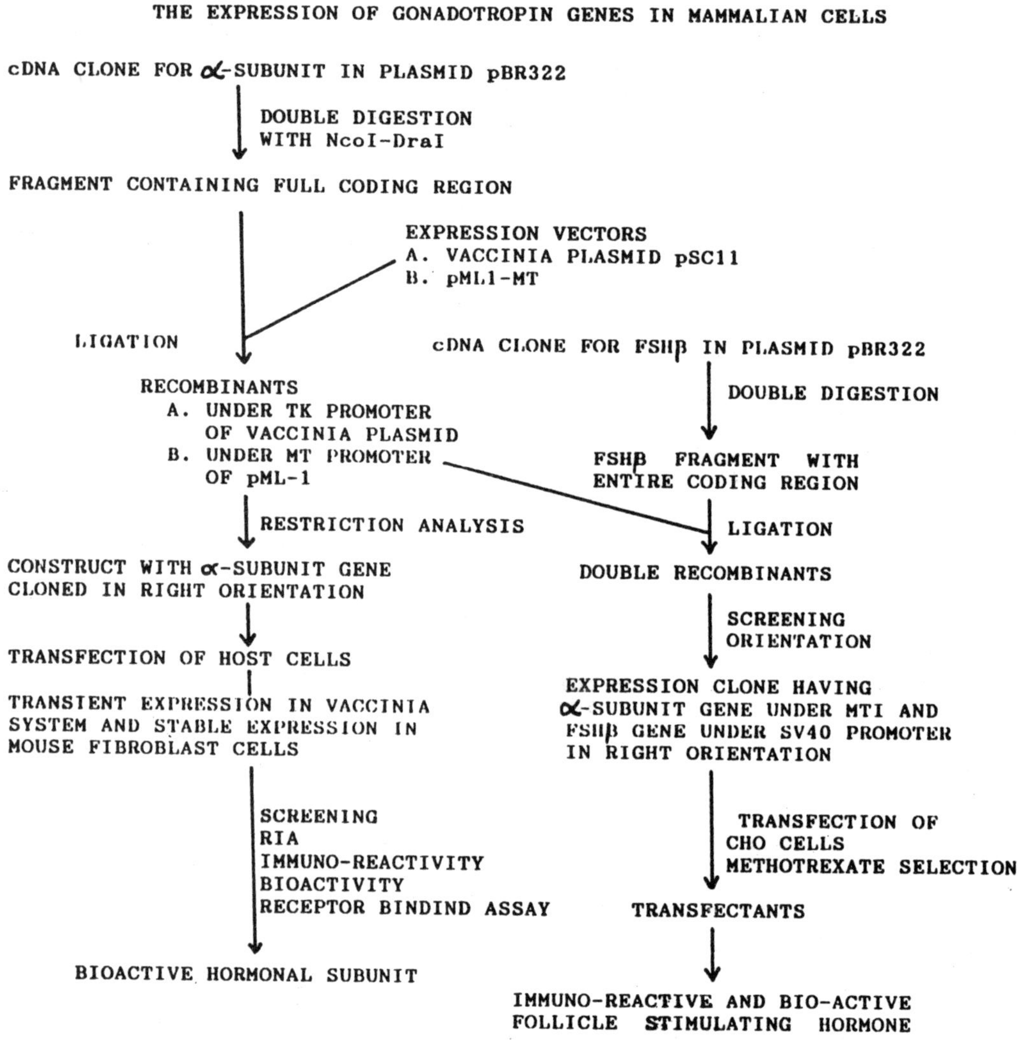

THE EXPRESSION OF GONADOTROPIN GENES IN MAMMALIAN CELLS

cDNA CLONE FOR α-SUBUNIT IN PLASMID pBR322

↓ DOUBLE DIGESTION
WITH NcoI-DraI

FRAGMENT CONTAINING FULL CODING REGION

EXPRESSION VECTORS
A. VACCINIA PLASMID pSC11
B. pML1-MT

LIGATION

cDNA CLONE FOR FSHβ IN PLASMID pBR322

RECOMBINANTS
A. UNDER TK PROMOTER
OF VACCINIA PLASMID
B. UNDER MT PROMOTER
OF pML-1

DOUBLE DIGESTION

FSHβ FRAGMENT WITH
ENTIRE CODING REGION

RESTRICTION ANALYSIS

LIGATION

CONSTRUCT WITH α-SUBUNIT GENE
CLONED IN RIGHT ORIENTATION

DOUBLE RECOMBINANTS

SCREENING
ORIENTATION

TRANSFECTION OF HOST CELLS

TRANSIENT EXPRESSION IN VACCINIA
SYSTEM AND STABLE EXPRESSION IN
MOUSE FIBROBLAST CELLS

EXPRESSION CLONE HAVING
α-SUBUNIT GENE UNDER MTI AND
FSHβ GENE UNDER SV40 PROMOTER
IN RIGHT ORIENTATION

SCREENING
RIA
IMMUNO-REACTIVITY
BIOACTIVITY
RECEPTOR BINDIND ASSAY

TRANSFECTION OF
CHO CELLS
METHOTREXATE SELECTION

TRANSFECTANTS

BIOACTIVE HORMONAL SUBUNIT

IMMUNO-REACTIVE AND BIO-ACTIVE
FOLLICLE STIMULATING HORMONE

Fig. 13.6: The strategy for the expression of the cloned genes for α and ß subunits of FSH.

Having achieved the transient expression, we used fibroblast cells for the stable expression of the α-subunit. By a series of manipulations, the α-subunit fragment was subcloned in a pML-1 derived plasmid under the influence of MT promoter. A mutated mouse fibroblast cell line deficient in TK was transfected with this construct alongwith a TK plasmid by $Ca_3(PO_4)_2$ co-precipitation method (Graham & Vander berEb, 1973). The transfectants were selected on HAT medium and the clones were purified to yield single cells. The cells were grown under the constant selection pressure of HAT. The analysis revealed that the transfected cells produced α-subunit, which was secreted into the culture medium. The immuno-cytology and immuno-fluorescence studies revealed that the hormonal subunit is present in the entire cytoplasmic region of the cells without any preferential accumulation in any subcellular organelle. The RIA studies revealed that the subunit is immuno-active and competes with the natural α-subunit for binding to specific antibodies (Garg *et al.*, 1991; Jain *et al.*, 1992). These results are summarized in Table 13.3.

Table: 13.3 Production of gonadotropin hormone α-subunit by rDNA route

1.	The hormonal subunit is produced by mouse fibroblast cells under the influence of MT promoter
2.	The genomic blot analysis reveals that the transfected cells have the cloned gene integrated into their genome.
3.	The RNA blots show that the sheep α-subunit gene is transcribed by the transfected cell and a-subunit mRNA is synthesized by these cells.
4.	The immuno-cytology and immuno-fluorescence studies reveal that immuno-reactive α-subunit is produced by the cells which is present in the entire cytoplasmic region of these cells without any preferential accumulation in any of the subcellular organelles.
5.	The RIA analysis of the culture supernatant reveal that the hormonal subunit is secreted by the cells into the culture medium.
6.	The bioassay shows that the a-subunit produced by rDNA route is able to conjugate wwith hCGβ to form a dimer which enhances the steroidogenesis in leydig cells

Since the bioactive hormone acts in a dimer form only, it was pertinent to express both the subunits together in the same cell to simulate the natural process. The validity of this approach further enhanced by a report that the α-subunit of hCG does not express efficiently in a SV40 based vector when cos cells were used as the host. The level of expression of the cloned hormone genes increased several folds when the host cells were co-transfected with two separate plasmids containing α and β subunits, respectively (Reddi *et al.*, 1985). Therefore, in the next series of experiments it was decided to transfect the cells with genes for both the subunits. A complex plasmid was constructed by a series of cloning steps which contained the coding regions of both α and β subunits of FSH cloned in the same plasmid, each under an independent promoter. The plasmid also contained a DHFR gene cloned under yet another promoter, and served as a convenient and amplifiable marker. A mutated CHO cell line, deficient in DHFR gene, was transfected with this construct and the transfectants were selected for methotrexate (MTX) resistance. Only the recombinant cells survived MTX treatment and formed colonies. These were grown, purified and the single cell clones were used for further studies.

The analyses of nucleic acids from the transfected cells revealed that these cells have the genes for both α and β subunits integrated into the host genome. The genes are transcribed and translated and FSH protein is produced. The hormone is distributed in the entire cytoplasmic region of the cells. The RIA analysis showed that the cells secrete the hormone into the culture medium. The secreted hormone competes with natural hormone for binding to total FSH (dimer form) specific antibodies. These observations, thus present an indirect but definitive evidence that the two subunits get annealed and the dimer form of the hormone is secreted as under the natural process (Jain *et al.*, 1992).

The analysis of r-FSH vis-a-vis natural FSH by PAGE and by immuno-precipitation of ^{35}S labelled rFSH revealed that the molecular sizes of the two preparations are similar. Further, the rFSH was able to bind to FSH specific receptors and displace the natural FSH in a competitive receptor binding assay. The final proof for the authenticity of rFSH come by its bioactivity. It was found that in a FSH dependent oocyte maturation system, the culture medium isolated from the transfected cells caused the maturation of oocytes. Further, the oocytes thus matured maintained their biological rhythm and can be fertilized in an IVF system. Furthermore, the biological activity of rFSH was found to be comparable to natural FSH (Jain, 1994). Since glycosylation is necessary for the biological activity of the hormone, these experiments also prove that the rFSH is post-translationally modified and undergoes glycosylation. This also suggests that correct dimer formation occurs and rFSH is similar to natural FSH in all respects (Jain, 1994, Jain *et al.*, 1992). These results are summarized in Table 13.4.

Table: 13.4 Properties of FSH produced by genetic engineering route

1. The hormone is produced by CHO cells transfected by a plasmid having both α and ß subunit genes under separate promoters.

2. The hormone is localised in the entire cytoplasmic region of the cells and is also secreted into the culture medium.

3. Both the subunits are synthesized by the cells and these get annealed to form the dimer.

4. The size of r-hormone is similar to the natural hormone.

5. The r-hormone is immuno-reactive and is recognized by the dimer specific antibodies.

6. The r-hormone is recognized by the FSH specific receptor and it competes with natural hormone for binding to these receptors.

7. The r-hormone causes the maturation of oocytes *in vitro* in a FSH dependent system.

8. The r-hormone does not have any toxic effects on cells. The oocytes maturated with rFSH maintain their biological rhythm and get fertilized in an IVF system.

9. The bioactivity of r-hormone is comparable to natural FSH.

CONCLUSIONS

The biosynthesis of gonadotropins is regulated in a very complex manner. Though the genes for α and β subunits of LH and FSH are localized on different chromosomes, these are expressed in a mutually coordinated manner. Further, the GnRH affects the transcription of these genes. The regulation is both at the transcription and the translation level. We have isolated the full length clones for α and β subunits of LH and FSH. These cDNAs and genomic clones for these subunits are being used to further understand these regulatory mechanisms.

A number of foreign genes have been expressed in various heterologous systems and many biologically important macromolecules have been synthesized by rDNA technology. However, the synthesis of complex glycoprotein molecules like gonadotropin hormones has met with only limited success. We have been able to achieve the synthesis of immuno-reactive and bioactive FSH by genetic engineering route. However, enhancement in the levels of expression is needed before rFSH can be produced economically. The rFSH produced by CHO cells has similar size as the natural hormone. Its biological activity is comparable to that of natural hormone. The comparison of physico-chemical properties of rFSH vis-a-vis its natural counterpart is being further studied. A number of toxicological and clinical studies are needed before this can be put into therapeutic and other biological use.

Acknowledgements

The author wishes to thank Dr. Pinki Garg for her contribution in many of these studies. Typing of the manuscript by Mr. Mueed and illustrations by Ms. Indu Mehra are acknowledged.

References

Archer, R. (1980). Molecular evolution of biologically active polypeptides. Proc. R. Soc. Lond. (Biol) 200, 21-43.

Ausubel, F.M., Brent, R., Kingston, R.E., Moore, D.D., Seidman, J.G., Smith, J.A. & Strihl, K. (1987). Current Protocols in Molecular Biology. John Wiley and Sons, London.

Aviv, H. & Leder, P. (1972). Purification of biologically active globin mRNA by chromatography on oligo-dT-cellulose. Proc. Natl. Acad. Sci. (USA) **69**, 1408-1412.

Boornstein, W.R., Vamvakopoulos, N.C. & Fiddes, J.C. (1982). Human chorionic gonadotropin β-subunit is encoded by atleast eight genes arranged in tandem and inverted pairs. Nature (London) **300**, 419-422.

Chin, W.W., Kronenberg, H.M., Dee, P.C., Maloof, F. & Harberner, J.F. (1981). Nucleotide sequence of the mRNA encoding the pre--subunit of mouse thyrotropin. Proc. Natl. Acad. Sci (USA) **78**, 5329-5333.

Chin, W.W., Godine, J.E., Klein, D.R., Chnag, A.S., Tan, L.K. & Habener, J.F. (1983). Nucleotide sequence of the cDNA encoding the precursor of the ß-subunit of rat lutropin. Proc. Natl. Acad. Sci (USA) **80**, 4649-4653.

Chin, W.W., Maizel, J.V. & Habener, J.F. (1983). Differences in sizes of human compared to murine α-subunits of glycoprotein hormones arises by a four-codon gene deletion or insertion. Endocrinology **112**, 482-485.

Chin, W.W. (1985). Organization and expression of glycoprotein hormone genes. In The Pituitary Gland. Edited by H. Imura. Raven Press New York. pp 103-125.

Esch, F.S., Mason, A.J., Cooksey, K., Mercado, M. & Shimasaki, S. (1986). Cloning and DNA sequence analysis the cDNA for the precursor of the beta chain of bovine follicle stimulating hormone. Proc. Natl. Acad. Sci (USA) **83**, 6618-6621.

Fiddes, J.C. & Goodman, H.M. (1979). Isolation, cloning and sequence analysis of the cDNA for the -subunit of human chorionic gonadotropin. Nature **281**, 351-355.

Fiddes, J.C. & Goodman, H.M. (1981). The gene encoding the common alpha subunit of the four human glycoprotein hormones. J. Mol. Appl. Genet. **1**, 3-18.

Forage, R.G., Ring, J.M., Brown, R.W., McInerney, B.V., Cobon, G.S., Gregson, R.P., Robertson, D.M., Morgan, F.J., Hearn, M.T.W., Findlay, J.K., Wettenhall, R.E.H., Burger, H.C. & DeKretser, D.M. (1986). Cloning and sequence analysis of cDNA species coding for the two subunits of inhibin from bovine follicular fluid. Proc. Natl. Acad. Sci (USA) **83**, 3091-3095.

Garg, P., Reddi, P.P., Ali, A. & Jain, S.K. (1991). Expression of a cDNA clone for alpha subunit of ovine luteinizing hormone in a mammalian cell system. Med. Sci. Res. **19**, 633-637.

Garg, P., Ali, A. & Jain, S.K. (1993). Cloning and expression of α-oLH. In Frontiers in Modern Biology, Edited by A. Ali. (In press).

Garg, P. & Jain, S.K. (1992a). Molecular cloning of a cDNA for β-subunit of sheep follicle stimulating hormone. Med. Sci. Res. **20**, 163-165.

Garg, P. & Jain, S.K. (1992b). Complexity of gonadotropin hormone genes. Med. Sci. Res. **20**, 337-339.

Godine, J.E., Chin, W.W. & Habener, J.F. (1980). Luteinizing and follicle stimulating hormones: Cell free translation of mRNAs coding for subunit precursors. J. Biol. Chem. **255**, 8780-8783. Godine, J.E., Chin, W.W. & Habener, J.F. (1982). α-subunit of rat pituitary glycoprotein hormones: Primary structure of the precursor determined from the nucleotide sequence of the cloned cDNAs. J. Biol. Chem. **257**, 8368-8371.

Goverman, I.M., Parson, T.F. & Pierce, J.G. (1982). Enzymatic deglycosylation of the subunits of chorionic gonadotropin: Effects on formation of tertiary structure and biological activity. J. Biol. Chem. **257**, 15059-15064.

Graham, F.L. & Vander berEb, A.J. (1973). A new technique for the assay of infectivity of human adenovirus 5 DNA. Virology **52**, 456-467.

Gubler, U. & Hoffman, B.J. (1983). A simple and very efficient method for generating cDNA libraries. Gene **25**, 263-268.

Hussa, R.O. (1980). Biosynthesis of human chorionic gonadotropin. Endocrinol. Res. **1,** 268-294.

Jameson, J.L., Chin, W.W., Hollenberg, A.N., Chang, A.S. & Habener, J.F. (1972). The gene encoding the ß-subunit of rat luteinizing hormone: Analysis of gene structure and evolution of nucleotide sequence. J. Biol. Chem. **259**, 15474-15480.

Jain, S.K., Chin, W.W. & Talwar, G.P. (1987). Isolation and characterization of cDNA clones for alpha and beta subunits of ovine LH. J. Biosci. **12**, 349-357.

Jain, S.K., Garg, P. & Kaur, R. (1990). Gonadotropin hormone genes: Strategies for cloning and expression. In Microbial Gene Technology, Edited by H. Polasa. S. A. Publisher, New Delhi. pp 93.

Jain, S.K., Garg, P., Totey, S.M., Singh, G.P. & Guron, C.S. (1992). Construction of genetically engineered cell lines for the expression of pituitary hormone genes. Int. J. Toxicol. Occup. Environ. Hlth. **1**, 53-59.

Jain, S.K. (1991). Genetic engineering: Principals, applications and production of pituitary hormones by recombinant DNA route. In Biotechnology: Human Welfare, Edited by P.P. Reddi and A. Jyothi. Osmania University Press, Hyderabad. pp 37-54.

Jain, S.K. (1994). Follicle stimulating hormone: Genes and proteins. In Frontiers in Modern Biology, Edited by A. Ali (in press)

Kornfeld, R. & Kornfeld, S. (1980). Structure of glycoproteins and their oligosaccharide units. In Glycoproteins and Proteoglycons, Edited by W. J. Lennarz. Plenum Press, New York. pp 1-34. Lall, L., Srinivasan, J., Rao, L.V., Jain, S.K., Talwar, G.P. & Chakrabarti, S. (1988). Recombinant vaccinia virus expresses immunoactive alpha subunit of ovine LH which associates with B-hCG to generate bioactive dimer. Ind. J. Biochem. Biophys. **25**, 510-514.

Ling, N., Ying, S.Y., Ueno, N., Shimasaki, S., Esch, F., Hotta, M. & Guillemin, R. (1986). Pituitary FSH is released by a heterodimer of the ß-subunits from the two forms of inhibin. Nature (London) **321**, 779-782.

Maurer, R.A. (1985). Analysis of several bovine lutropin ß-subunit cDNAs reveal heterogeniety in nucleotide sequence. J. Biol. Chem. **260**, 4684-4687.

Maurer, R.A. (1987). Molecular cloning and nucleotide sequencing of complementary disoxy nucleic acid for the beta subunit of rat follicle stimulating hormone. Molec. Endocrinol. **10**, 717-729.

Maurer, R.A. & Beck, A. (1987). Isolation and nucleotide sequence analysis of a cloned cDNA encoding the ß-subunit of bovine follicle stimulating hormone. DNA **5**, 363-367.

Mayo, K.E., Cerelli, G.M., Spiess, J., Rivier, J., Rosenfeld, M.G., Evans, R.M. & Vale, W. (1986). Inhibin A-subunit cDNAs from porcine ovary and human placenta. Proc. Natl. Acad. Sci. USA **83**, 5849-5853.

Moudgal, N.R., Murthi, G.S., Ravindranath, N., Rao, A. & Prasad, M.R.N. (1988). Development of a contraceptive vaccine for use by the human male: Results of a feasibility study carried out in adult male bonnet monkeys. In Progress in Vaccinology, Edited by G.P. Talwar. Springer Verlag Press. vol 1, pp 253-258.

Mountford, P.S., Bello, P.A., Brandon, M.R. & Adams, T.E. (1989). Cloning and DNA sequence analysis of the cDNA for precursor of ovine follicle stimulating hormone. Nucl. Acid Res. **17**, 6319.

Nayler, S.I., Chin, W.W., Goodman, H.M., Lalley, P.A., Grzeschik, K.H. & Sakaguch, A.Y. (1983). Chromosomal assignment of genes encoding the and ß subunits of glycoprotein hormones in man and mouse. Somatic Cell Genet. **9**, 757-770.

Nilson, J.H., Thomason, A.R., Cserbak, M.T., Moncman, C.L. & Woychik, R.P. (1983). Nucleotide sequence of a cDNA for the common -subunit of the bovine pituitary glycoprotein hormones. J. Biol. Chem. **258**, 4679-4682.

Parsons, T.F., Bloomfeld, G.A. & Pierce, J.G. (1983). Purification of an alternative form of α-subunit of glycoprotein hormones from bovine pituitaries and identification of its o-linked oligosaccharides. J. Biol. Chem. **258**, 240-244.

Parsons, T.F. & Pierce, J.G. (1983). Free -like material from bovine pituitaries: Restoration of the ability to reassociate with native LHB. Fed. Proc. **42**, 1799.

Pierce, J.G. & Parsons, T.F. (1981). Glycoprotein hormones: Structure and functions. Ann. Rev. Biochem. **50**, 465-495.

Policastro, P., Ovitt, C.D., Hoshina, M., Fukuoka, H., Boothby, M.R. & Boim, I. (1983). The ß-subunit of human chorionic gonadotropin is encoded by multiple genes. J. Biol. Chem. **258**, 11492-11499.

Reddi, V.B., Beck, A.K., Garramore, A.J., Vellucci, V., Lustbader, J. & Bernstine, E.G. (1985). Expression of hCG in monkey cells using a single SV 40 vector. Proc. Natl. Acad Sci. (USA) **82**, 3644-3648.

Rigby, P.W.J., Dieckmann, M., Rhodes, C. & Berg, P. (1977). Labelling of DNA to high specific activity in vitro by nick translation by DNA polymerase I. J. Mol. Biol. **113**, 237-249.

Sambrook, J., Fritsch, E.F. & Maniatis, T. (1988). Molecular Cloning: A Laboratory Manual. Cold Spring Harbour Laboratory Press.

Southern, E. (1975). Detection of specific sequences amongst DNA fragments separated by gel electrophoresis. J. Mol. Biol. **98**, 503-517.

Talmadge, K., Vamvakopoulos, N.C. & Fiddes, J.C. (1984). Evolution of the genes for the β-subunits of human chorionic gonadotropin and luteinizing hormone. Nature **307**, 37-40.

Thomas, P.S. (1983). Hybridization of denatured RNA transferred or dotted to nitrocellulose paper. Methods in Enzymol. **100**, 255-267.

Ullrich, A., Shine, J., Chirgwin, J., Pictel, R., Tischer, E., Rutter, W.J. & Goodman, H.M. (1977). Rat insulin genes: Construction of plasmid containing the coding sequences. Science **196**: 1313-1319.

Vaitukaitis, J.L., Ross, G.T., Braustein, G.D. & Rayfold, P.L. (1976). Gonadotropins and their subunits: Basic and clinical studies. Rec. Progr. Horm. Res. **32**, 289-331.

Vale, W., Riever, J., Vaughan, J., McClintock, R., Corrigan, A., Woo, W., Karr, D. & Spess, J. (1986). Purification and characterization of a FSH releasing protein from porcine ovarian follicular fluid. Nature, **321**, 776-779.

Watkins, P.C., Eddy, R., Beck, A.K., Vellucci, V., Laverone, B., Tanzi, R.E., Gusella, J.F. & Shaws, T.B. (1987). Assignment of human gene for the beta subunit of follicle stimulating hormone to chromosome 11. Cytogenet. Cell Genet. **40**, 773-779.

14

Mechanism of cyanobacterial nitrate reduction

S.N. Bagchi

Department of Biological Sciences, R.D. University
Jabalpur 482001, India

ABSTRACT

Since the nitrogen of nitrate has formal charge of +5, an eight electron reduction is required to reach to formal charge of -3, typical of NH_3. This precedes the combination of the inorganic nitrogen with carbon skeleton, arising from photosynthetic CO_2 - fixation, to form nitrogenous compounds whose redox value is similar to ammonia. The assimilatory reduction of nitrate to ammonia in cyanobacteria, as in other nitrate utilizing organisms, is sequentially catalyzed by the enzymatic activities of fer-redoxin - dependent nitrate reductase (NR) and nitrite reductase (NiR). First, NO_3 is reduced to NO_2 in a two electron reduction catalyzed by NR and resulting NO_2 then to NH_3 by a six electron reduction mediated by NiR (Vega *et al.*, 1980; Guerrero *et al.*, 1981; Mendez *et al.*, 1981).

INTRODUCTION

NR activity is the focal regulatory point controlling the net nitrogen input in various organisms (Beevers & Hageman, 1969). Significant positive correlation exists between this enzyme activity and the nitrogen status of an organism (Srivastava, 1980).

NR from the unicellular cyanobacterium *Anacystis nidulans* has been purified to homogeneity and characterized (Candau, 1979) as a single protein with a molecular weight of 75,000 having only one polypeptide chain and exhibiting a Km value of 0.7mM for nitrate. Further, NR was purified and exclusively characterized from a filamentous cyanobacterium *Plectonema boryanum* (Ida & Mikami, 1983; Mikami & Ida, 1986). The *Plectonema* enzyme is a single polypeptide with a molecular weight of 85,000 and contains 0.95 atoms of molybdenum, four iron and four acid labile sulfur atoms per molecule. The apparent Km for NO_3 was nearly 0.75mM. NR from all sources share a common molybdenum-cofactor, which can reconstitute apoprotein fractions of cofactor-free molybdoenzymes such as xanthine oxidase, sulfite oxidase, as well as nitrate reductase (Hewitt, 1975; Ketchum & Swarin, 1973; Lee *et al.*, 1974; Singh *et al.*, 1978 a,b). This property of molybdenum cofactor could easily be tested in a cell-free system containing the cofactor and a source of cofactor-free nitrate reductase apoprotein of *Neurospora crassa* nit^{-1} mutant (Muller & Grafe, 1978; Fernandez & Cardenase, 1981; Miller & Amy, 1983) or *Nostoc muscorum* W-R mutant (Bagchi *et al.*, 1985b). The cofactor activity could be defined by its capacity to reconstitute mutant nitrate reductase in a complementation assay (Muller & Grafe, 1978; Lee *et al.*, 1974). In cyanobacteria, molybdenum-cofactor was characterized only in *Nostoc muscorum* (Bagchi *et al.*, 1987a). Cofactor was found in the soluble fraction of cell-free preparations and could be distributed between two pools of different molecular properties. These are composed of a protein-associated (a protein with the molecular weight of 30,000 and a sedimentation coefficient at 20°C of 2.5) and a protein-free fraction.

RESULTS

Photosynthetic regulation of cyanobacterial nitrate reduction

Photoautotrophically growing cyanobacteria are expected to derive the assimilatory power for various vital reactions from active photosynthesis. NO_3 metabolism is not an exception, and the first evidence for a close and stoichiometric relation was produced between the reduction of nitrate to ammonia and photosynthetic water splitting to O_2 by using the illuminated thylakoid preparations of *A. nidulans* (Candau *et al.*, 1976). Such preparations contained both the processes (PS II reaction and NO_3 reduction) quite intact but lacked CO_2 fixation ability. One molecule of nitrate reduced caused an evolution of two molecules of O_2. Further experimentation on *Anabaena* 7119's (Ortega *et al.*, 1976) revealed a positive correlation between photosynthetic O_2 evolution and NO_3 reduction to nitrite. These workers developed the first test systems, in which ferredoxin received electrons from $NADPH_2$ instead of H_2O via a coupled enzyme $NADP^+$ ferredoxin reductase (FNR). Reduced ferredoxin eventually donated electrons to NR (Manzano *et al.*, 1976).

Consequently, by using whole cells of *A. nidulans* (Flores *et al.*, 1983b), a close coupling of the rates of NO_3 utilization and of photosplitting of water (observed as O_2 evolution) was established. Care was taken to minimize reductant wastage by not allowing ammonia to further metabolize by using a GS inhibitor MSX. A large variety of photosynthetic inhibitors namely DCMU (PS II inhibitor), abruptly ceased O_2 evolution accompanied with concomitant block of NO_3 entry, suggesting a tight coupling of the processes. Apparently cyanobacterial NO_3 reduction is more close to PS II reaction than the conventional CO_2 fixation. NO_3 can be considered as Hill reagent (Losada, 1975/76). Some indirect evidences, such as nitrate and nitrite induced quenching of chlorophyll fluorescence of *Anabaena* 7119 and *Nostoc* 6719 further confirmed a photosynthetic nature of cyanobacterial NO_3 reduction (Serrano *et al.*, 1981, 1982). It is now firmly established that the energy requirement of NO_3 intake (Flores *et al.*, 1983b) and reductants demand of its reduction (Candau *et al.*, 1976) are met by photophosphorylation and coupled electron transport from water (PS II reaction).

In green plants and algae, it has been suggested that light effects on nitrate reduction may involve participation of phytochrome and/or blue-light absorbing pigments (Hewitt *et al.*, 1976).

Thus the question arose whether light directly interacts with the cyanobacterial enzymes. This possibility was first checked in *A. nidulans* (Flores *et al.*, 1983b) in which darkness caused inactivation of nitrate utilization but had hardly any influence on the cellular NR activity. Eventually, Tischner and Schmidt (1984) elaborately studied such effects using another unicellular form *Synechococcus leopoliensis*. This elegant work revealed that NiR but not NR is light stimulatory via participation of ferredoxin-ferredoxin thioredoxin oxidoreductase-thioredoxin complex within the cells. In a cell-free system NiR activity could be reductively activated several folds by including dithiothreitol (DTT) whose action was to reduce thioredoxin of the extract. Several enzymes that include glutamine synthetase and of carbon metabolism are light stimulatory (Tischner & Schmidt, 1984). This brings out a possibility that photosynthetic regulation of NO_3 utilization at least in part involves enzyme activation.

Most of the cellular experiments were conducted in the presence of MSX to avoid errors caused by NH_4 metabolism. These treated cells would not continue NO_3 incorporation for longer period as synthesis of organic nitrogen from a combination of ammonia (resulting from NO_3 - reduction) and photosynthate (resulting from CO_2 - fixation) was not possible. Similar situation could arise if CO_2 - fixation was rendered inhibitory. Therefore, for continuous nitrate assimilation a continuous supply of desired carbon skeleton has to be ensured. A positive effect of CO_2 supply (CO_2 fixation) on nitrate utilization rates was made evident by employing *A. nidulans* cells (Flores, 1982). Besides other control, stable CO_2-fixation directly depends on assimilatory power generated during the light reactions of photosynthesis. One can, therefore, assume that the photosynthesis operates NO_3 utilization indirectly by supplying the necessary carbon-skeleton. One very basic metabolic arrangement by any organism is a right balance between carbon and nitrogen utilization. CO_2-enriched photosynthetic cells would utilize nitrate at an optimum rate. Once CO_2 is excluded or limited the net nitrate entry is immediately checked in order to maintain the C, N balance to a minimum possible value just enough to support the cells to survive. In fact, it has been observed with green algae (Azuara & Aparicio, 1984) and unicellular cyanobacterium *Synechococcus* (Kramer & Schmidt, 1989) that under CO_2 limitation (a condition that normally prevails in the atmosphere) the photoreduction of nitrate followed an excretion of the resulting nitrite. This would be a mechanism to balance the C, N ratio and to minimize the wastage of reductants (NO_2-reduction to NH_4 requires 6 electrons). Since assimilation of NO_3^- consumes photosynthetically generated assimilatory power, a competition can take place between this process and CO_2- fixation for reducing power and/or ATP. This interaction, still a matter of controversy in photosynthetic organisms, was tested in *A. nidulans* (Romero & Lara, 1987) by exposing the cells to excess or limited light (by changing the intensity) that caused a stimulation of photosynthetic O_2 evolution of nitrogen limited and CO_2-sufficient cells. Under light limiting conditions, CO_2-fixation was depressed by NO_3, whereas at photon fluxes saturating condition for CO_2 fixation, the addition of NO_3 had no negative effects on the process. At low light intensity, a strong competition between NO_3 utilization and CO_2 fixation existed which relaxed with increasing light intensity. Therefore, if other factors are not involved, an optimal NO_3 utilization would be operative under saturating carbon status and light intensity.

In all, photosynthesis of a cyanobacterial cell operates and regulates NO_3 utilization of four distinct levels namely: (a) ATP for nitrite transport (Flores *et al.*, 1983b), (b) reductants for reduction to ammonia (Candau *et al.*, 1976), (c) fixed photosynthate for ammonium assimilation (Flores *et al.*, 1983c), and (d) enzyme modification (Tischner & Schmidt, 1984).

The regulatory behaviour of NO_3 assimilation in some bloom-forming cyanobacteria is, however, exceptional. Planktonic *Oscillatoria redeki* performed nitrate intake even in the darkness (Foy & Smith, 1980). With *Phormidium uncinatum* as a candidate, sustained nitrate uptake and reduction to nitrite was experienced even by photosynthetically impaired cells (Bagchi *et al.*, 1989). Reserve glycogen synthesized during photosynthetic regime seems to support the process by dissimilation via oxidative pentose pathway. For the first time chemo-and photoheterotrophic modes of nitrate assimilation was also proposed in this organism (Bagchi *et al.*, 1990), trained to utilise glucose in darkness. Electrons from oxidative pentose pathway were channelled through ferredoxin NADP-oxidoreductase to nitrate reduction. Since the habitats in which these species are successful are rich in organic matter and are barely exposed to bright light, such alteration in mechanisms driving nitrate assimilation is expected in nature (see Foy & Smith, 1980).

Inorganic and organic nitrogen control of nitrate metabolism

Phototrophic cyanobacteria are capable of growing at the cost of a large variety of inorganic (NO_3, NH_4, NO_2) and organic (some amino acids and their amides) nitrogen sources. Although the metabolic pathways for quite a few assimilable nitrogen are common, mainly channelling through the GS-GOGAT and subsequent transaminases mediated reactions, a strict competition exists among the nutrients. This is predominantly at the level of expression of the structural genes responsible to operate the corresponding pathway. Ammonia, for example, will not let the cells utilize molecular nitrogen, nitrate or nitrite. Among cyanobacteria tested, ammonium inhibition of nitrate reduction was observed, with *Anabaena cylindrica* (Hattori, 1962; Ohmori & Hattori, 1970), *Anacystis nidulans, Anabaena* sp. PCC 7119 and *Nostoc* sp. PCC 6719 (Herrero *et al.*, 1981), *Agmenellum quadruplicatum* (Stevens & Van Baalen, 1974), *Anabaena variabilis* and *Synechocystis* sp. (Herrero *et al.*, 1985), *Nostoc muscorum* (Bagchi & Singh, 1984) *Anabaena cycadeae* (Bagchi *et al.*, 1985a) and *Calothrix* sp. 7101 (Martin-Nieto *et al.*, 1989.

The nature of ammonium inhibition on the process of nitrate metabolism was quite ambiguous as it was difficult to distinguish the effects caused by NH_4 *per se* and by product(s) of its metabolism vias GS - GOGAT pathway, that is also operative under such conditions. This problem was partly overcome by introducing a GS inhibitor, MSX, so that the effects caused by the products(s) were avoided. It was concluded on the basis of employing this inhibitor that ammonia *per se* was not an inhibitor molecule of cyanobacterial nitrogenase (Stewart & Rowell, 1975) and thus its metabolism was required. One could easily demarcate the so called ammonium effects on selected metabolic pathways simply by using this or analogous inhibitors. When MSX was added to nitrate assimilating *A. nidulans* cells, almost 85 to 90% of the NO_3

reduced was found released in the external medium in the form of NH_4 due to strong and rapid GS inactivation (Ramos *et al.*, 1982b). Investigations were conducted to ascertain the exact molecule responsible for NH_4 - promoted inhibition of cyanobacterial NO_3 - reduction. Initial work (Herrero *et al.*, 1981) revealed that MSX could effectively reverse the ammonium effect on NO_3 reduction tested in *A. nidulans, Anabaena* sp. 7119, and *Nostoc* sp. 6719, indicating a role of ammonium metabolism product(s) in inhibition. Further studies on other cyanobacteria, namely *A. variabilis,* and *Synechocystis* sp. (Herrero *et al.*, 1985) confirmed above findings. A rapid increase in cellular glutamine/glutamic acid ratio upon NH_4 addition and its reversal upon MSX addition suggest glutamine to be the putative inhibitor (Flores *et al.*, 1980). Glutamine alone certainly can not inhibit NO_3 utilization in *Anacystis* cells as azaserine an inhibitor of glutamine amide transferases would accumulate glutamine inside the cells, but not inhibit nitrate incorporation (Flores *et al.*, 1983c). Some other nitrogenous metabolite(s) derived from glutamine as amino N donor must be involved, together with glutamine, for the characteristic inhibition. Asparagine might well represent such metabolite (Romero *et al.*, 1985). In a particular case, the regulatory effects have been ascribed to a combination of glutamine and asparagine (Cook & Anthony, 1978). A report from Singh *et al.* (1983) claiming that MSX not only inhibits GS activity but also causes a strong inactivation of ammonium transport system of *Anabaena cycadeae* created some confusion on the understanding of the nature of real inactivator responsible for NH_4 inhibition of NO_3 utilization. Moreover, identical reports on bacterial (Kleiner & Castroph, 1982) and some cyanobacterial (Bergman, 1984) systems further confirmed above proposition. Anomaly of this characteristic MSX effect indicated NH_4 itself to be the actual inhibitor rather than a product of its metabolism. This was based on an elaborate study on a glutamine auxotroph, unable to assimilate NH_4 due to genetic loss of GS in *A. cycadeae* (Singh *et al.*, 1983). Bagchi *et al.* (1985a) proposed NH_4 to be one of the potent inhibitors of NO_3 assimilating enzymes at least in this cyanobacterium.

Expression of NR in a variety of cyanobacteria (Herrero *et al.*, 1981; Bagchi & Singh, 1984) required *de novo* protein synthesis, once ammonia was omitted from the medium and conditionally nitrate was included. This, together with the observation that ammonia does not inactivate cell-free enzyme activity indicates that the inhibition by ammonia is basically of repression type. Further, as observed with *N. muscorum* (Bagchi *et al.*, 1985b), ammonium mediated repression of NR involved only the apoprotein moiety of the enzyme while the molybdenum cofactor was relaxed from this control.

Once repressor molecule is withdrawn, development of NR requires presence of nitrate, the authentic inducer, in the external medium as observed with *A. cylindrica* (Hattori, 1962; Ohmori & Hattori, 1970), *Anabaena* 7119 and *Nostoc* 6719 (Herrero *et al.*, 1981), *Agmenellum quadruplicatum* (Stevens & Van Baalen, 1974), *A. cycadeae* (Bagchi *et al.*, 1985), and *Calothrix* sp. 7601 (Martin -Nieto *et al.*, 1989). Filamentous, heterocystous cyanobacteria in absence of any combined nitrogen source would fix atmospheric nitrogen leading to the generation of intracellular ammonia, which in turn can, at least in part, retard nitrate and nitrite reductase activities.

Therefore, just elimination of external ammonia may not be enough to release the NH_4-repression. To avoid this problem, *A. variabilis* mutants unable to fix atmospheric nitrogen (Nif⁻) were used in place of the wild-type. Low levels of nitrate and nitrite reductases without exogenous combined nitrogen (nitrogen starved conditions for mutants and N_2-fixing condition for the wild-type) suggested that these enzymes are not derepressed in the absence of ammonia (Martin-Nieto *et al.*, 1989).

On the contrary, expression of nitrate reductase did not require nitrate as tested in *A. nidulans* (Herrero *et al.*, 1981) and *N. muscorum* (Bagchi *et al.*, 1985b) suggesting derepressible nature of the enzymes.

Avissar (1985) reported an interesting result with *A. variabilis* NR expression, that is the basal enzyme levels of NH_4 and combined nitrogen-free cultures were already high enough to support, at its peak, nitrate reduction. Absence of an active nitrate transport system under these conditions restricted the entry of NO_3. With NO_3 in the medium, first the NO_3 transport was activated (required no new protein synthesis) and subsequently NR was induced (required *de novo* protein synthesis) to a level much higher than what was required. NO_3 uptake system was also found NO_3-activating in *N. muscorum* (Bagchi & Singh, 1984). Relatively little work is done as far as regulation of NiR in cyanobacteria is concerned. Like NR, NiR was universally ammonia repressible (Herrero & Guerrero, 1986). While nitrate or nitrite was required, in absence of ammonia, for this enzyme of filamentous diazotrophic cyanobacteria to express (Ohmori & Hattori, 1970; Martin-Nieto *et al.*, 1989), no such inducer was necessary for *A. nidulans* (Herrero & Guerrero, 1986), and *Phormidium laminosum* enzymes (Arizmendi *et al.*, 1987).

Results of similar studies on *P. uncinatum* (Palod *et al.*, 1990) indicate distinctive modes of ammonium-repression on nitrate and nitrite reductases. Ammonia by itself and through assimilation exerts repression control on nitrite and nitrate reductase, respectively, which in absence of ammonia are otherwise derepressed.

Control at nitrate reductase protein level

Preformed cyanobacterial nitrate reductase, as experimented with *A. nidulans* (Herrero *et al.*, 1984), is subjected to environmental changes prevailing inside the cells. These factors strongly influenced the enzyme activity by modifying and/or degrading the protein. Apparently this enzyme was decayed in a biphasic manner. First, oxidative modification of the active centres caused massive reversible inactivation. This followed a more rapid and irreversible proteolytic degradation of the enzyme, mainly the modified one. Actively photosynthesizing cells can generate superoxide anion (O_2) and H_2O_2 as a consequence of excess reductant load at PS I centre, which in turn would reduce molecular oxygen to these toxic radicals. Such radicals and H_2O_2 can directly interact with nitrate reductase protein, leading to its inactivation. Infact, an inactivation of *N. muscorum* nitrate reductase was achieved following H_2O_2 treatment (Bagchi *et al.*, 1987b). Purified NR from *Plectonema boryanum* lost its activity following xanthine/xanthine oxidase treatment causing production of superoxide (Mikami & Ida, 1986). Presently, a great variety of enzymes, including cyanobacterial nitrogenase, have been found to respond to the oxygen radicals in similar pattern. Proteolytic degradation of modified nitrate reductase could be partially protected by external nitrogenous compounds such as NO_3 (Herrero *et al.*, 1984). It has been reported that superoxide ion is involved in $NADH_2$-mediated inactivation of green algal and higher plant enzymes which include NR (Mikami & Ida, 1986).

Virus-induced regulation of nitrate assimilation

Although several cyanobacteria are known to serve as hosts responsible for host specific cyanophage multiplication, limited host-virus systems have been studied for the likely changes in the host nitrate assimilatory pathway. Both with *N. muscorum* /N-1 (Bagchi *et al.*, 1987b) and *P. uncinatum* /LPP-1 (Bisen *et al.*, 1986; Bagchi & Kaloya, 1987) as test systems, it was found that infection caused a massive increase in the nitrate utilization capacity, in general, and enhancement of molybdenum-cofactor activity and thereby NR activity in particular. Further, the enzyme from infected host managed to escape the H_2O_2 caused oxidative inactivation normally observed with the uninfected counterparts (Bagchi & Kaloya, 1987; Bagchi *et al.*, 1987b). These adjustments were necessary to meet the high demand of nitrogenous compounds required for virus multiplication.

Molybdenum-regulation of nitrate reductase and implication of other molybdo-enzymes

Molybdenum is an essential component of NR and tungsten a competitive inhibitor, replaces molybdenum from the active centres of various molybdoenzymes, including NR (Hewitt, 1975; Losada, 1975/76). It is now almost established that molybdenum binds to a low molecular weight polypeptide to form a molybdenum-cofactor which also contain pterin like compounds (Bagchi et al., 1987a). Various molybdoenzymes, excluding nitrogenase which contains an iron-molybdenum-cofactor (Ugalde *et al.*, 1985), share a common molybdenum-cofactor (Ketchum & Swarin, 1973; Lee *et al.*, 1974; Hewitt, 1975). It is nevertheless, certain that molybdenum-cofactor and iron-molybdenum-cofactor share a common route of synthesis at some step after molybdenum-intake (Ugalde *et al.*, 1985). It, therefore, would not be surprising if nitrogenase and NR share some common element of molybdenum processing. Singh *et al.* (1978a) first hypothesized this in *N. muscorum* and was subsequently confirmed by several experiments (Bagchi & Singh, 1984; Bagchi *et al.*, 1985). *N. muscorum* grown in the absence of combined nitrogen source showed nitrogenase reaction confined to the heterocysts and nonoperative NR localized in the vegetative cells (Bagchi & Singh, 1984). Molybdenum-cofactor activity of N_2 cultures was much lower than the NO_3 cultures (Bagchi *et al.*, 1985b) lacking active nitrogenase. This observation suggests a competition for molybdenum-components between NR and nitrogenase. Further, supporting this view, Kumar *et al.* (1985) detected Mo-cofactor activity in the isolated heterocysts showing nitrogenase activity but lacking NR activity. Conditions that favoured expression of *N. muscorum* NR (N_2 and NO_3 medium) led to an excessive synthesis of the apoprotein moiety whereas the molybdenum-cofactor synthesis become limiting, which decided the level of net cellular NR protein (Bagchi *et al.*, 1985b). Therefore, molybdenum control of NR is at the level of net cofactor synthesis and of the proportion that diverts for the processing of other molybdoenzymes.

Genetic and molecular aspects of nitrate reduction

The genetic approach has been used to isolate nitrate reductase-less mutants and mutants altered in regulation of this enzyme. Investigation on physiological properties of these mutants and on the wild-type-like strains derived from complementation or transformation with wild-type genome fragments unravel the genetic organization of nitrate assimilation system.

Spontaneous mutants of *N. muscorum* scored on chlorate failed to assimilate nitrate due to defect in a) nitrate uptake, b) nitrate reductase or c) both (Bagchi & Singh, 1984). In some mutants, loss of nitrate reductase was accompanied with a loss of nitrogenase (Singh *et al.*, 1977). With a non-N_2 fixer, *P. uncinatum* chlorate-resistant nitrate reductase was experienced in a resistant mutant which appears to be a regulatory mutation (Bagchi *et al.*, 1992).

Alternatively, mutants were obtained by chemical MNNG and transposon (Tn 901) mutagenesis and scored for poor growth on nitrate. The genetic analysis of cyanobacterial nitrate assimilation was greatly hampered by the lack of suitable gene transfer system. Progressive development of gene cloning system in unicellular cyanobacterium *A. nidulans* allowed elaborate genetic study. A cosmid gene bank of strain R2 (PCC 7942) was constructed in shuttle cosmid pUC 29 and was used to transform nitrate reductase-less mutants. The *nar*+ characteristic was chosen among transformants. Using this technique, at least three genes for nitrate reductase were identified and cloned (Kuhlemeier *et al.*, 1984a,b). Two more genes responsible for nitrate assimilation have been identified by transformation work on additional and varied classes of defective mutants (Madueno *et al.*, 1988). Unfortunately, none of these genes correspond to the *nar* genes of heterotrophic micro-organisms; distinctly responsible for apoenzyme synthesis genes, regulatory genes or genes for molybdenum-cofactor synthesis (Marzulf, 1981).

CONCLUSIONS

A significant feature among all cyanobacterial NR is its tight association with photosynthetically active thylakoid membranes and dependence upon ferredoxin as the sole natural electron donor (Hattori & Myers, 1967; Manzano *et al.*, 1976; Ortega *et al.*, 1976). Cases like iron starvation caused replacement of ferredoxin by flavodoxin as an electron donor (Flores *et al.*, 1983a). Ferredoxin reduced by illuminated thylakoid membranes (Manzano *et al.*, 1976; Ortega *et al.*, 1976, 1977) or by illuminated 5-diazariboflavin in the presence of suitable electron donor like EDTA (Candau *et al.*, 1980) or by low concentrations of sodium dithionite (Flores *et al.*, 1983a), can donate electron to NR. High concentrations of dithionite was non-effective (Hattori &

Myers, 1967; Manzano *et al.*, 1976) presumably due to the establishment of a stable and inactive complex between reduced ferredoxin and NR.

Ferredoxin can effectively be replaced by other artificial electron donors such as reduced FAD and FMN (Hattori, 1970) and dithionite reduced methylviologen (Manzano *et al.*, 1976; Ortega *et al.*, 1977). The latter is a routine practice for the cell-free assay of cyanobacterial NR. Nitrate reduction associated with anoxygenic photosynthesis has been achieved with chlorophyll containing particles plus ferredoxin when suitable electron donor (eg. DCPIP, ascorbate or H_2) to photosystem I was supplied (Flores *et al.*, 1983a).

NiR from an unicellular cyanobacterium, *Anabaena nidulans* (Guerrero *et al.*, 1974; Manzano *et al.*, 1976; Manzano, 1977) and filamentous *Anabaena* 7119 (Mendez & Vega, 1981, Mendez *et al.*, 1981) has been partially purified and characterized. The enzyme from both cyanobacteria exhibited a Km value for nitrite in the range of 70 to 100 µM and possessed a single polypeptide chain with a molecular weight of about 52,000 daltons containing iron presumably as siroheam. Like NR, NiR of cyanobacterial origin physiologically receives electron from ferredoxin, a typical plant like character (Hattori & Myers, 1967; Manzano *et al.*, 1976; Ortega *et al.*, 1976). Under iron starvation, flavodoxin takes the responsibility of ferredoxin (Bothe, 1977; Manzano, 1977). Ferredoxin also serves as an artificial electron donor for NiR once reduced by illuminated photosynthetic preparations (Manzano *et al.*, 1976; Ortega *et al.*, 1976; Mendez *et al.*, 1981) or by illuminated 5-diazariboflavin in presence of a suitable electron donor (Candau *et al.*, 1980). Dithionite reduced ferredoxin can also donate electrons to NiR (Manzano *et al.*, 1976; Vega *et al.*, 1980; Mendez *et al.*, 1981), and in the cell-free system dithionite reduced methylviologen serves as the reductant source.

Non-standard abbreviations

DCMU, Dichlorophenyl dimethyl urea; DCPIP, Dichlorophenol indophenol; GOGAT, Glutamate synthase; GS, Glutamine synthetase; LPP, *Lyngbya Plectonema Phormidium*; MSX, Methionine sulfoximine, Nir, Nitrite reductase; Nif, Nitrogenase; NR, Nitrate reductase; MNNG, Nitrosoguanidine; PS, Photosystem; W-R, Tungsten resistant mutant lacking active molybdenum cofactor.

References

Arizmendi, J.M., Fresnedo, O., Martinez Bilbao, M., Alana, A. & Serra, J.L. (1987). Inorganic nitrogen assimilation in the non-N_2-fixing cyanobacterium *Phormidium laminosum*. II Effect of the nitrogen source on the nitrite reductase levels. Physiol. Plant. **70**, 703-707.

Avissar, J.Y. (1985). Induction of nitrate assimilation in the cyanobacterium *Anabaena variabilis*. Physiol. Plant. **63**, 105-108.

Azuara, M.P. & Aparicio, (1984). Effects of light quality, CO_2 tension and NO_3 concentration on the inorganic nitrogen metabolism of *Chlamydomonas reinhardii*. Photosynth. Res. **5**, 97-103.

Bagchi, S.N. & Singh, H.N. (1984). Genetic control of nitrate reduction in cyanobacterium *Nostoc muscorum*. Molec. Gen. Genet. **109**, 82-84.

Bagchi, S.N., Rai, U.N., Rai, A.N. & Singh H.N. (1985a). Nitrate metabolism in the cyanobacterium *Anabaena cycadeae*: Regulation of nitrate uptake and reductase by ammonia. Physiol. Plant. **63**, 322-326.

Bagchi, S.N., Rai, A.N. & Singh, H.N. (1985b). Regulation of nitrate reductase in cyanobacteria. Repression - derepression control of nitrate reductase apoprotein in the cyanobacterium *Nostoc muscorum*. Biochem. Biophys. Acta. **838**, 370-373.

Bagchi, S.N. & Kaloya, P. (1987). Cyanophage LPP-1 induced changes in the synthesis and stability of *Phormidium uncinatum* nitrate reductase. Proc. Indian Natl. Sci. Acad. B **53**, 461-464.

Bagchi, S.N., Sherman, T.D. & Funkhouser, A.E. (1987a). Biochemical characterization of molybdenum-cofactor in a cyanobacterium, *Nostoc muscorum*. Plant Cell Physiol. **28**, 1411-1419.

Bagchi, S.N., Kaloya, P. & Bisen, P.S. (1987b). Effect of cyanophage N-1 infection on the synthesis and stability of *Nostoc muscorum* nitrate reductase. Curr. Microbiol. **15**, 61-65.

Bagchi, S.N., Palod, A. & Chauhan, V.S. (1989). Photosynthetic control of nitrate metabolism in *Phormidium uncinatum*, a cynanobacterium. Curr. Microbiol. **19**, 183-188.

Bagchi, S.N., Chauhan, V.S. & Palod, A. (1990). Heterotrophy and nitrate metabolism in a cyanobacterium *Phormidium uncinatum*. Curr. Microbiol. **21**, 53-57.

Bagchi, S.N., Palod, A. & Chauhan, V.S. (1992). Sustained nitrate metabolism by a chlorate-resistant mutant of the cyanobacterium, *Phormidium uncinatum*. J. Plant Physiol. **139**, 764-766.

Beevers, L. & Hageman, R.H. (1969). Nitrate reduction in higher Plants. Ann. Rev. Plant Physiol. **20**, 495-522.

Bergman, B. (1984). Photorespiratory ammonium release by the cyanobacterium *Anabaena cylindrica* in the presence of methionine sulfoximine. Arch. Microbiol. **137**, 21-25.

Bisen, P.S., Bagchi, S.N. & Audhlia, S. (1986). Nitrate reductase activity of a cyanobacterium *Phormidium uncinatum* after cyanophage LPP-1 infection. FEMS Microbiol. Lett. **33**, 69-72.

Bothe, H. (1977). In Encyclopedia of Plant Physiology, Vol. 5, Edited by A. Trebst and M. Avron. Springer-Verlag, Berlin. pp 217-221.

Candau, P. (1979). Ph.D. Thesis, University of Sevilla, Sevilla.

Candau, P., Manzano, C. & Losada, M. (1976). Bioconversion of light energy into chemical energy through reduction with water of nitrate to ammonia. Nature (London) **262**, 715-717.

Candau, P., Manzano, C., Guerrero, M.G. & Losada, M. (1980). Ferredoxin-dependent enzymatic reduction of nitrate with a deazaflavin photosystem. Photobiochem. Photobiophys. **1**, 167-174.

Cook, R.J. & Anthony, C. (1978). Regulation of ammonia transport in *Aspergillus nidulans*. J. Gen. Microbiol. **124**, 275-286.

Fernandez, E. & Cardenase, J. (1981). *In vitro* complementation of assimilatory NAD(P)H-nitrate reductase from mutants of *Chlamydomonas reinhardii*. Biochim. Biophys. Acta. **657**, 1-10.

Flores, E. (1982). Ph.D. Thesis, University of Sevilla, Sevilla.

Flores, E., Guerrero, M.G. & Losada, M. (1980). Short-term ammonium inhibition of nitrate utilization by *Anacystis nidulans* and other cyanobacteria. Arch. Microbiol. **128**, 137-144.

Flores, E., Ramos, J.L., Herrero, A. & Guerrero, M.G. (1983a). Nitrate assimilation by cyanobacteria. Photosynthetic prokaryotes: Cell differentiation and function, Edited by G.C. Popageorgious and L. Packer. Elsevier Science Publication, New York, pp 363-387.

Flores, E., Guerrero, M.G. & Losada, M. (1983b). Photosynthetic nature of nitrate uptake and reduction in the cyanobacterium *Anacystis nidulans*. Biochim. Biophys. Acta. **722**, 408-416.

Flores, E., Ramero, J.M., Guerrero, M.G. & Losada, M. (1983c). Regulatory interaction of photosynthetic nitrate uptake and carbon dioxide fixation in the cyanobacterium *Anacystis nidulans*. Biochim. Biophys. Acta. **725**, 529-532.

Foy, R.H. & Smith, R.V. (1980). The role of carbohydrate accumulation in the growth of planktonic *Oscillatoria* species. Br. Phycol. J. **15**, 139-150.

Guerrero, M.G., Manzano, C. & Losada, M. (1974). Nitrite photoreduction by a cell-free preparation of *Anacystis nidulans*. Plant Sci. Lett. **3**, 689-699.

Guerrero, M.G., Vega, J.M. & Losada, M. (1981). The assimilatory nitrate-reducing system and its regulation. Ann. Rev. Plant Physiol. **43**, 169-204.

Hattori, A. (1962). Adaptive formation of nitrate reducing system in *Anabaena cylindrica*. Plant Cell Physiol. **3**, 371-377.

Hattori, A. (1970). Solubilization of nitrate reductase from blue green alga *Anabaena cylindrica*. Plant Cell Physiol. **11**, 975-978.

Hattori, A. & Myers, J. (1967). Reduction of nitrate and nitrite by subcellular preparations of *Anabaena cylindrica*. II. Reduction of nitrate to nitrite. Plant Cell Physiol. **8**, 327-337.

Herrero, A., Flores, E. & Guerrero, M.G. (1981). Regulation of nitrate reductase levels in the cyanobacteria *Anacystis nidulans, Anabaena* sp. strain 7119, and *Nostoc* sp. strain 6719. J. Bacteriol. **145**, 175-180.

Herrero, A., Flores, E. & Guerrero, M.G. (1984). Regulation of nitrate reductase level in *Anacystis nidulans*. Activity decay under nitrogen stress. Arch. Biochem. Biophys. **234**, 454-459.

Herrero, A., Flores, E. & Guerrero, M.G. (1985). Regulation of nitrate reductase cellular levels in the cyanobacteria *Anabaena variabilis* and *Synechocystis* sp. FEMS Microbiol. Lett. **26**, 21-25.

Herrero, A. & Guerrero, M.G. (1986). Regulation of nitrite reductase in the cyanobacterium *Anacystis nidulans*. J. Gen. Microbiol. **132**, 2463-2468.

Hewitt, E.J. (1975). Assimilatory nitrate-nitrite reduction. Ann. Rev. Plant Physiol. **26**, 73-100.

Hewitt, E.J., Hucklesby, D.P. & Notton, B.A. (1976). Nitrate metabolism. In Plant Biochemistry, Edited by J. Bonner and J.E. Varner. New York Academic Press. pp 633-681.

Ida, S. & Mikami, B, (1983). Purification and characterization of assimilatory nitrate reductase from cyanobacterium *Plectonema boryanum*. Plant Cell. Physiol. **24**, 649-658.

Ingram, L.O., Calder, J.A., Van Baalen, C., Plucker, F.E. & Parker, P.L. (1973). Role of reduced exogenous (algae): Photoheterotrophic growth of a 'heterotrophic' blue-green bacteria. J. Bacteriol. **114**, 695-700.

Ketchum, P.A. & Swarin, R.S. (1973). In vitro formation of assimilatory nitrate reductase: presence of the constitutive component in bacteria. Biochem. Biophys. Res. Commun. **52**, 1450-1456.

Kleiner, D. & Castroph, H. (1982). Inhibition of ammonium (Methylammonium) transport in *Klebsiella pneumoniae* by glutamine and glutamate analogue. FEBS Lett. **146**, 201-203.

Kramer, E. & Schmidt, A. (1989). Nitrite accumulation by *Synechococcus* 6301 as a consequence of carbon - or sulfur - deficiency. FEMS Microbiol. Lett. **59**, 191-196.

Kuhlemeier, C., Logtenberg, T., Stoorvogel, W., Van Heugten, H.A.A., Borrais, W.E. & Van Arkel, G.A. (1984a). Cloning of two nitrate reductase genes from the cyanobacterium *Anacystis nidulans*. J. Bacteriol. **159**, 36-41.

Kuhlemeier, C., Teeuwsen, V.J.P., Janssen, M.J.T. & Van Arkel, G.A. (1984b). Cloning of a third nitrate reductase gene from the cyanobacterium *Anacystis nidulans* R2 using a shuttle gene library. Gene **31**, 109-116.

Kumar, A.P., Rai, A.N. & Singh, H.N. (1985). Nitrate reductase activity in isolated heterocysts of the cyanobacterium *Nostoc muscorum*. FEBS Lett. **179**, 125-128.

Lee, K.Y., Pan, S.S., Erickson, R. & Nason, A. (1974). Involvement of molybdenum and iron in the *in vitro* assembly of assimilatory nitrate reductase utilizing *Neurospora* mutant *nit*-1. J. Biol. Chem. **249**, 3941-3952.

Losada, M. (1975/76). Metalloenzymes of the nitrate reducing system. J. Mol. Catal. **1**, 245-264.

Madueno, F., Borrais, W.E., Van Arkel, G.A. & Guerrero, M.G. (1988). Isolation and characterization of *Anacystis nidulans* R2 mutants affected in nitrate assimilation: Establishment of two new mutant types. Molec. Gen. Genet. **213**, 223-228.

Manzano, C. (1977). Ph.D. Thesis, University of Sevilla, Sevilla.

Manzano, C., Candau, P., Gomez-Moreno, C., Relimpio, A.M. & Losada, M. (1976). Ferredoxin dependent photosynthetic reduction of nitrate and nitrite by particles of *Anacystis nidulans*. Mol. Cell. Biochem. **10**, 161-169.

Martin-Nieto, J., Herrero, A. & Flores, E. (1989). Regulation of nitrite reductases in dinitrogen fixing cyanobacteria and Nif mutants. Arch. Microbiol. **151**, 475-478.

Marzulf, G.A. (1981). Regulation of nitrogen metabolism and gene expression in fungi. Microbiol. Rev. **45**, 437-461.

Mendez, J.M., Herrero, A. & Vega, J.M. (1981). Characterization and catalytic properties of nitrite reductase from *Anabaena* sp. 7119. Z. Pflanzenphysiol. **103**, 305-315.

Mendez, J.M. & Vega, J.M. (1981). Purification and molecular properties of nitrite reductase from *Anabaena* sp. 7119. Physiol. Plant. **52**, 7-14.

Mikami, B. & Ida, S. (1986). Purification of nitrate reductase by methylviologen-bound CM sephadex C-50 from a cyanobacterium, *Plectonema boryanum*. Agricul. Biol. Chem. **47**, 1653-1654.

Miller, J.B. & Amy, N.K. (1983). Molybdenum cofactor in chlorate resistant and nitrate reductase deficient insertion mutants of *Escherichia coli*. J. Bacteriol. **155**, 793-801.

Muller, A.J. & Grafe, R. (1978). Isolation and characterization of cell lines of *Nicotiana tabacum* lacking nitrate reductase. Molec. Gen. Genet. **161**, 67-76.

Ohmori, M. & Hattori, A. (1970). Induction of nitrate and nitrite reductases in *Anabaena cylindrica*. Plant Cell Physiol. **11**, 873-878.

Ortega, T., Castillo, F. & Cardenas, J. (1976). Photolysis of water coupled to nitrate reduction by *Nostoc muscorum* subcellular particles. Biochem. Biophys. Res. Commun. **71**, 885-891.

Ortega, T., Rivas, J., Cardanas, J. & Losada, M. (1977). Metabolic interconversion of Ferredoxin-nitrate reductase and NADP - reductase of *Nostoc muscorum*. Biochem. Biophys. Res. Commun. **78**, 185-193.

Palod, A., Chauhan, V.S. & Bagchi, S.N. (1990). Regulation of nitrate reduction in a cyanobacterium *Phormidium uncinatum*: Distinctive modes of ammonium-repression of nitrate and nitrite reductases. FEMS Microbiol. Lett. **68**, 285-288.

Ramos, J.L., Guerrero, M.G. & Losada, M. (1982a). Optimization of conditions for photoproduction of ammonia from nitrate by *Anacystis nidulans*. Appl. Environ. Microbiol. **44**, 1013-1019.

Ramos, J.L., Guerrero, M.G. & Losada, M. (1982b). Sustained photoproduction of ammonia from nitrate by *Anacystis nidulans*. Appl. Environ. Microbiol. **44**, 1020-1025.

Romero, M.J. & Lara, C. (1987). Photosynthetic assimilation of NO_3 by intact cells of the cyanobacterium *Anacystis nidulans*. Plant Physiol. **83**, 208-212.

Romero, M.J., Lara, C. & Guerrero, M.G. (1985). Dependence of nitrate utilization upon active CO fixation in *Anacystis nidulans*: a regulatory aspect of the interaction between photosynthetic carbon and nitrogen metabolism. Arch. Biochem. Biophys. **237**, 346-401.

Serrano, A., Guerrero, M.G. & Losada, M.M. (1981). Nitrate and nitrite as '*in vivo*' quenchers of chlorophyll fluorescence in blue-green algae. Photosynth. Res. **2**, 175-183.

Serrano, A., Rivas, J. & Losada, M. (1982). Changes in fluorescence spectra by nitrate and nitrite in a blue-green alga. Photobiochem. Photobiophys. **4**, 257-264.

Singh, H.N., Rai, U.N., Rao, V.V. & Bagchi, S.N. (1983). Evidence for ammonia as an inhibitor of heterocyst and nitrogenase formation in the cyanobacterium *Anabaena cycadeae*. Biochem. Biophys. Res. commun. **11**, 180-187.

Singh, H.N., Sonie, K.C. & Singh, H.R. (1977). Nitrate regulation of heterocyst differentiation and nitrogen fixation in a chlorate resistant mutant of blue-green alga, *Nostoc muscorum*. Mut. Res. **42**, 447-452.

Singh, H.N., Vaishampayan, A. & Singh, R.K. (1978a). Evidence for the involvement of a genetic determinant controlling functional, specificity of group VI B elements in the metabolism of N and NO in the blue-green alga *Nostoc muscorum*. Biochem. Biophys. Res. Commun. **81**, 67-74.

Singh, H.N., Vaishampayan, A. & Sonie, K.C. (1978b). Mutation from molybdenum dependent growth to tungsten dependent growth and further evidence for a genetic determinant common to nitrogenase and nitrate reductase in blue-green alga *Nostoc muscorum*. Mut. Res. **50**, 427-443.

Srivastava, H.S. (1980). Regulation of nitrate reductase activity in higher plants. Phytochemistry **19**, 725-733.

Stevens, S.E. Jr. & Van Baalen, C. (1974). Control of nitrate reductase in blue-green alga. The effect of inhibitors, blue light and ammonia. Arch. Biochem. Biophys. **161**, 146-152.

Stewart, W.D.P. & Rowell, P. (1975). Effects of L methionine-DL-sulfoximine on the assimilation of newly fixed NH acetylene reduction and heterocyst in *Anabaena cylindrica*. Biochem. Biophys. Res. Commun. **65**, 846-856.

Tischner, R. & Schmidt, A. (1984). Light mediated regulation of nitrate assimilation in *Synechococcus leopoliensis*. Arch. Microbiol. **137**, 151-154.

Ugalde, A.R., Imperial, J., Shah, K.V. & Brill, J.W. (1985). Biosynthesis of the Iron-molybdenum cofactor and the molybdenum cofactor in *Klebsiella pneumoniae*: Effect of sulfur source. J. Bateriol. 1081-1087.

Vega, J.M., Cardenas, J. & Losada, M. (1980). Ferredoxin-nitrite reductase. Methods Enzymol. **69**, 255-270.

15

Biotechnology — biovillages and a better biofuture

M.S. Swaminathan
Centre for Research on Sustainable Agricultural and Rural Development
Madras, India

ABSTRACT

Biotechnology applications in the genetic improvement of crops and farm animals as well as in the production of sero-diagnostics and vaccines have a comparative advantage in the area of cost and environmental safety. Genetic resistance to biotic and abiotic stresses helps in eliminating the need for applying market purchased chemicals. The challenge lies in packaging the new technologies in such a manner that there is a significant impact on the productivity and profitability of major farming systems. The new technologies before they are popularised in rural areas must be subjected to an impact analysis based on considerations of ecology, economics, employment, and equity.

INTRODUCTION

The quantitative and qualitative dimensions of the economic and livelihood security challenges facing developing countries and of the ecological challenges facing all countries on our planet are formidable. Some of these are:

- World population will probably have reached between 8-9 billion by the year 2025.

- Over 65% of the world population will live in the coastal zone area, e.g. within 60 km from the coast, by the year 2000.

- Resources utilization for food production is approaching carrying-capacity on a local and regional scale. Soil degradation and ground water depletion are extensive.

- Limiting factors for energy utilization will be others than accessibility, e.g. zero increase of or negative development of greenhouse gas levels in the atmosphere.

- Yet, no production of fusion energy available. Fission energy accessibility still limited.

- Mineral and biomass for food, fibers, and chemical products, as well as water resources decline in qualitative and quantitative availability.

- Regional polarization (North-South, East-West) regarding technology know-how and education will increase. Strong infrastructure and economic and social constraints for the transfer of available scientific and technical techniques to the market will be felt.

- Local and regional ecological disturbances and devastations will lead to social instability and migrations. Such effects are expected to reach global scale by way of changes in biogeochemical flows, e.g. global climate change.

Thus, we have to increase the productivity and production of the wide range of agricultural commodities under conditions of shrinking land resources and diminution in both the biological potential of the soil and in the biological wealth. Any strategy for agricultural research should, therefore, take into consideration the following requirements:

1. Technology must be related to what we may call land and water-saving crop husbandry; in other words, enhancement of biological productivity per unit of land, water, and time on an ecologically sustainable basis. Technologies have to be land-saving since we have no option except to produce more and more food from less and less land.

2. We need grain-saving animal husbandry. Our animal nutrition patterns have to be different from those of Western Europe or USSR. This will imply greater reliance on non-traditional feeds, enriched cellulosic wastes, high-yielding fodder legumes and grasses, and locally available biomass.

3. Ecological security, that is, the protection of long-term productivity of land, water, flora and fauna, and atmosphere must be a primary goal of all R & D efforts.

4. Technologies must promote, not endanger, the livelihood security of landless labour families and those without assets, particularly women. The displacement of people from their traditional jobs with no alternative avenues of employment in sight, if the principle that "job destruction and job creation must be concurrent events," is not built into the technology development and dissemination system, a fact known too well, will prove counterproductive.

5. We have to learn to operate under an unfavourable international trade environment because of the problems caused by protectionism, on the one hand, and heavy subsidies, on the other, which the industrialized countries are adopting. In other words, we have to develop our own competitiveness both in terms of negotiations in GATT and in terms of efficiency of farming and quality of produce.

6. A careful identification of priorities in R & D investment based on comparative advantages is absolutely essential because of our resource constraints. The poorer the sources, the more important it is to have correct sense of priorities.

7. Finally, patents and intellectual property rights could lead to "an orphan remaining orphan" situation. In other words, problems which are not commercially attractive might

remain unattended to, even if they are of great importance to the livelihood security of the poor. Substitutes for natural products may threaten agricultural income and trade in some cases. Technology alert systems should analyze the social and trade implications of such technologies.

The above challenges can be faced only by a new paradigm of research and development which integrates principles of ecology, economics, and equity. To convert such a paradigm into an operational reality, the establishment of "Biovillages", to begin with in China and India have been recommended (Swaminathan, 1991). A Biovillage helps to achieve desirable blends of modern biological technologies with traditional technologies.

Biovillages

The most desirable technologies are often the result of blending traditional and frontier technologies. For this purpose, the concept of biovillage is a good one. Each Biovillage or a cluster of Biovillages are serviced by a Bio-Centre which provides key services. The aim is to promote decentralised production supported by a few key centralised services. The improved water management that will deal with sanitation, hygiene, and availability of drinking water, forms an important aspect of this concept.

The technology options are in the area of production of crops and animal products. Here, the intervention can be at the levels of both material and management. For example, although cashew varieties yielding nearly 40 kg of nuts per tree per year are available within the country, farmers still grow varieties which hardly give about 2 kg of nuts. In the area of management, there is considerable scope for introducing bio-pesticides, bio-fertilizers and bio-energy sources. The other area relates to poor post-harvest operations and marketing. Here, there is considerable opportunity for the preparation of value added products from the agricultural biomass. For this purpose, biorefineries can be established.

The greatest challenge lies in bringing the benefits of new technologies to rural women and landless labourers. Women often lack opportunities for skilled jobs. This is where biotechnology applications could provide additional jobs of skilled nature, as for example, in tissue culture, integrated nutrient supply, and integrated pest management. Where water is available aquaculture and poultry farming can be undertaken.

Challenges Ahead

It is now reasonably clear that inspite of the deceleration being witnessed in world's population growth, the population may well rise at least to somewhere in the range of 10 to 15 billion or two or three times the current population level during the next century. The burden of this growth will fall largely on the poorer nations. According to the World Development Report of the World Bank published in 1991, the population of low income countries would rise from 2.9 billion in 1989 to 5.2 billion in 2025. In sub-saharan Africa, population would treble in the next

40 years. As a result, rich-poor divide which is already very high will get wider still (see UNDP Human Development Report, 1992).

Thanks to advances in genetics and breeding, much of the progress made during the last three decades in increasing crop production has been due to yield improvement. How long can we sustain this pathway of production advance? China, for example, has set an annual foodgrain production target of 500 million tons by the year 2000, as compared to the present production level of 400 million tons per year. At the same time, the National Report prepared by China in August 1991 for the U.N. Conference on Environment and Development states, "the ecosystem of China's agriculture is faced with pressures greater than those of any other country. The problems of excessive demands on the agricultural ecosystem could hardly be alleviated in the short run in view of the sharp contradiction between population increase and annual decrease in arable land, and the problems of soil erosion, desertification of the grasslands and natural disasters like floods and droughts". The increased use of mineral fertilizer has been responsible for about half of the 100 million ton increase in grain production between 1978 and 1988, as well as for much of the increase in commercial and high value crops. Nitrate pollution of ground-water is already an environmental problem and additional doses of chemical fertilizer may aggravate this problem. The productivity of soils has also begun to decline in some farming areas because of the use of continuous cropping instead of crop rotations that sustain nutrient levels and limit pests and diseases. Such a situation leads to the extension of farming to marginal lands. According to a World Bank Environmental Strategy paper for China (1992; Mimeographed), "probably the most visible environmental problem that agriculture has created to China is soil and water loss due to the cultivation of semi-arid or sloping lands. Some such lands have long been farmed, but large areas have been deforested or brought under cultivation since 1949".

Similarly, an estimate states that over 50 million tons of additional food grains will have to be produced annually from existing or less land by the year 2000 to provide food for a billion in India.

Ruttan (1991) has catalogued some of the major constraints on sustainable growth in agricultural production into the 21st century. These relate to technological, resource, environmental and health concerns that condition the capacity of the agricultural sector to respond to the demands that population and income growth will place on this sector, particularly in developing countries.

Swaminathan (1991) has proposed a new paradigm of agricultural research and development based on concurrent attention to issues of ecological sustainability, economic viability and social equity. He had pointed out that achieving such a paradigm shift would need intensive research on blending traditional and frontier technologies. It will also need new measurement and monitoring tools and a global grid of genetic resources and enhancement centres. These challenges will impose a heavy responsibility in the areas of scientific ingenuity and innovation and international collaboration.

The onward march of the green revolution, i.e., yield enhancing and land-saving technologies, is an ecological and economic imperative in countries like India and China. However, the next phase of the green revolution should really be "green", i.e., based on environmentally benign technologies and public policies. The "greening" of the revolution and spreading a new environment-friendly yield revolution to more geographical areas and farming systems should be a major item on the technological, organisational, and public policy agenda for the next phase in the evolution of our progress from Nature to Crop Production. The World Resources Report (1992-93) has cited alarming figures on global soil degradation. Although as much carbon is fixed on the sea as on land, marine fish catch has already reached the sustainable level in many parts of the world. Thus, soil has to be the saviour for providing our daily bread. Deforestation, and habitat destruction are further compounding the problem, since they are leading to species extinction at a time when genetic engineering has made all plants and living organisms valuable. Issues relating to Plant Breeders' Rights and gene patenting have also created new difficulties in international cooperation in saving and sharing plant genetic resources.

Mendel helped to keep the Malthusian prediction of population growth outstripping food production at bay during this century. Will it be possible to maintain this position in the coming century? Human induced problems of soil degradation, water depletion and pollution, loss of biological diversity, global warming and ozone layer depletion can be solved only through concerted political, public and scientific action. Otherwise, we may witness the emergence of the Malthusian era in the 21st century. Dr. Norman Borlaug has rightly warned that "there is no time to relax" in our quest to make hunger a problem of the past.

Biotechnology for a Better Common Future

The environmentally sound and safe application of biotechnology has the potential for addressing issues such as health care, levels of food security, the efficiency of industrial development process, the purity of water supplies and environmental problems arising from water pollution, pesticides residues, and soil degradation. There is a need to accelerate the development and application of biotechnologies, particularly in developing countries.

For this purpose, major efforts are required to build up institutional capacities at national and local levels, especially in terms of research and training, raising public awareness, the development of human resources and information facilities, matched by appropriate levels of financial support and backed by equivalent development and support for traditional methods and techniques as practiced by local and indigenous communities.

The call for a major effort in the biotechnological transformation of the quality of life of rural urban poor in developing countries needs for its realisation new patterns of international cooperation backed by appropriate institutional structures. It calls for action at the local national, regional and global levels. It may be worthwhile to review some of the major components of a "Biotechnology for a better common future" programme.

A number of centres have initiated active agricultural biotechnology programmes. The ICGEB for example is playing an active part in the international rice biotechnology programme. The specialised expertise of ICGEB will be of much benefit to the International Agricultural Research Centres (IARC's).

At the regional level, several networks already exist, some of them with support from regional development banks. These networks should be designed to forge a trilateral partnership among academic and research institutions, private and public sector industry and user groups.

At the national level, different countries have different mechanisms at Governmental level to stimulate and sustain biotechnology research and development. A standing Committee could deal with Biotechnology, Biodiversity and Sustainable Development. Pioneer projects like the Biovillage programme supported by the Asian Development Bank in China and India need urgent support (Swaminathan, 1991).

There is also a need for specialized genetic resources centres for assisting genetic engineers. Establishing two specialised gene pools for use in recombinant DNA experiments is essential. One can deal with the assembly of candidate genes which can confer tolerance to sea water intrusion. Another gene pool can be an assembly of genotypes having bearing on sustainability factors, such as:

- Germplasm of nitrogen fixing tree species and shrubs, stem-nodulating legumes like *Sesbania rostrata*, as well as of *Azolla* and blue green algae.

- Plant species of importance in pest control including tree species, annual plants, fungi and bacteria which can serve as repellents of pests and those which control soil nematodes and weeds.

- Germplasm of species which can enhance the efficiency of fertilizer use including Neem (*Azadirachta indica*) and other trees species whose seed cakes have a nitrification inhibition capacity.

- Species which can help to prevent/reduce soil erosion and protect local food security, often referred to as ecological or economic key species, such as vettiver grass (*Vetiveria zizancides*), *Chenopodium* species, and grain and leaf Amaranths.

- Tree species and shrubs of value in agroforestry and alley farming practices and in restoring the fertility of degraded and wastelands, and

- Plant species of value in veterinary and human medicine.

Such specialised genetic garden for sustainable agriculture need to be established. In addition to such specialised genetic gardens, we also need genetic enhancement centres where novel genetic combinations can be developed for distribution among field level plant and animal breeders. ICGEB in association with IBPGR and other IARC's could develop a global grid of specialised genetic resources and enhancement centres.

There is also an urgency for national, regional, and global efforts in promoting basic research on topics of relevance to the application of biotechnology.

Dissemination of Information and Awareness Generation

There is need for special information centres at the regional level for giving credible information to financial institutions on the cost, risk and return aspects of new biotechnological innovations. UNIDO can help to establish such a network in collaboration with FAO, UNESCO, UNEP, WHO and UNCTS, as well as the Economic Commissions for Africa, Latin America and the Caribbean and Western and Southern Asia. Such Regional Information Centres for the commercialization of biotechnology should have a network of sources of credible information.

Another area where there is need for public awareness and action is the protection of habitats rich in biological diversity. The pressure of population as well as some of the pathways of development chosen have resulted in a gradual destruction of major ecosystems like hill, wetland and coastal ecosystems and to the loss of habitats rich in fauna and flora. Because of the diversity of soil, climate and growing conditions, India is rich in its endowments of plant species. Habitat destruction and the extension of agriculture to forest areas promote species extinction. But for the green revolution, more forest land will have to be diverted for the cultivation of annual crops. For example, India produced 12 million tons of wheat from 14 million hectares in 1964, whereas in 1990, 55 million tons of wheat were produced from 24 million hectares. Hereafter, the additional food requirements will have to come from higher productivity per crop and higher intensity of cropping. Higher yield has to come from higher biomass production and better harvest index. Such progress will be possible only with the help of biotechnology.

Thus, sustainable advances in biological productivity are essential for meeting the food, fuel, fodder and other needs of our growing population. Biological diversity is essential for achieving sustainable gains in productivity per units of land, water, energy, and time. We can neither sustain national food security systems nor face the challenge of climate change if we fail to conserve and utilize in a sustainable manner the global genetic wealth in flora, fauna and microorganisms through the tools of biotechnology.

Even protecting the protected areas is becoming a major challenge. The number of species entered in Red Data Books is growing. There is a need for special international programme on Biotechnology for preserving biological diversity in order to save endangered species of plants and animals through cell and tissue culture and cryopreservation techniques.

CONCLUSIONS

Thus the Third World farmers and small scale manufacturers need the support of biotechnology most. They are well endowed in biological diversity and can derive immense benefit from molecular biology. The rural economy is largely biomass based in many developing countries and it will be sad if the fruits of biotechnological research do not reach them.

The foregoing account of the wide range of research and development activities currently underway would serve to provide a glimpse of the excitement, hope, and fear that biotechnology research has generated. The hope is particularly great in the area of overcoming some of the ecological ills associated with earlier technologies. Bioremediation involving the use of microorganisms to get rid of toxic chemicals, will find increasing application. Rapid afforestation of degraded forests can be undertaken with the help of micropropagation of plus trees of suitable species. Biomass refineries will help produce energy and value-added products. Thus, there is every hope that even a 10 billion population can be supported within the carrying capacity of supporting ecosystems.

Even the brief review presented in this paper on the wide spectrum of problems in agriculture, industry, medicine, and energy now being studied with the help of the tools of molecular biology and biotechnology would suffice to convey a sense of hope and optimism in relation to the potential for improving the quality of life for humankind during the remaining years of this century. It is equally obvious that biotechnology is one area of research which has attracted widespread interest among both private industry and the academic world. The implication of modern biotechnology research for human welfare are far-reaching. The very power of the new tools of genetic engineering make the adoption of codes of ethics and safety imperative. Earlier in this paper methods of imparting a pro-poor bias in the choice of research problems have been described. Biotechnology also helps to increase the carrying capacity of supporting ecosystems through new opportunities for employment generated by the efficient use of biomass and by higher crop and animal productivity.

References

Belcher, B. & Geoffery Hawtin (1991). A patent of life ownership of plant and animal research, IDRC, Ottawa, Canada, 40.

Dubashi, P.R. (1992). Drought and development. Econ. Polt. Wkly. **27**, 27-36.

Ruttan, V.W. (1991). Constraints in sustainable growth in agricultural production: into the 21st century. Outlook on Agriculture, C.A.B. International **20**, 225-234.

Sen, Amartya (1981). Poverty and Famines: an Essay on Entitlements and Famines. Clarendon Press, Oxford.

Swaminathan, M.S. (Ed) (1991). Reaching the Un-reached: Biotechnology in Agriculture - A Dialogue, McMillan India Ltd., Madras.

World Resources Report (1992-93).

16

Biodiversity and biotechnology

Habib Haque[1] and Ashwani K. Srivastava[2]

[1]*Department of Biochemistry,* [2]*Faculty of Basic Sciences and Humanities, Rajendra Agricultural University, Pusa 848 125, Bihar, Inida*

ABSTRACT

At present, the biosphere and consequently the biodiversity is under great stress due to manifold reasons, the prime factor being the human agency. As a rule, however, the biosphere keeps itself regularly transforming through a series of subtle changes in the homeostatistical balance, the bioitems as well manipulating their contents. Yet the recent 'shocks' have been severe and alarming, needing immediate and special attention. The proliferation in biotechnology has alarmed certain concerns that through it the biodiversity could be easily disturbed. The recent evidences, however, show that the species diversity would keep shrinking in an irreversible manner. The need, therefore, is emerging that through a nonconventional manner attempts are to be made to preserve, and 'freeze', the biodiversity. We review certain aspects of the complex problem and show that the biodiversity can be rescued through the judicious application of biotechnology so as to select and preserve the genetic content of the living species.

INTRODUCTION

The Concept of Biodiversity

Biodiversity stands for 'biological diversity', a self-explanatory term; and it represents the whole range of ecosystems, living organisms, and genetic material that humans co-inhabit the Earth with (Wilson, 1986). It includes not only wild-life (wild plants and animals) and natural habitats, but also domesticated species (crops, livestocks, etc.), and genetic materials.

The Geometry of Biodiversity Conservation

All natural resources are distributed in space, so are the human communities and all other living communal systems. Human communities appropriate natural resources for their day-to-day requirements. Appropriation is mediated by the social order through such social mechanisms that regulate the nature and pattern of appropriation of natural resources. In traditional agricultural communities, all the energy needs were met endogenously. Similarly the conditions of social production in these communities were dictated by natural limitations. The social space of such agricultural communities, consisting of the exploiting and the exploited classes, are dependable on the sustainable use of natural resources since this was crucial to the reproduction of the social order. The socio-economic order existed in a symbiotic relationship with its material base, one serving to reproduce the other. The geometry of the biodiversity conservation is a complex concept which is based on the objective and subjective existence of a local socio-ecological community in an ecological space.

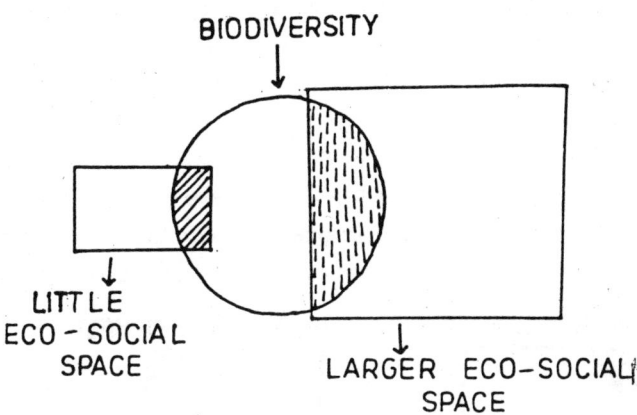

Fig. 16.1: The Relationship of Little and Larger Eco-social Spaces with the Biodiversity
The stippled areas suggest the utilization and preservation of the Biodiversity by the respective Space. See the text for detail.

The ideas expressed above can now be quantified as described in Fig. 16.1. We have presumed a Little Eco-social Space (LiES), which we believe is a peasant eco-niche. The LiES interacts with the Biodiversity in a manner which is specifically its own and is represented in the figure as a hatched area. It could be observed that the rest of the Biodiversity 'area' remains independent of interaction with the LiES and, therefore, could behave in any way it likes to. However, its contraction or expansion, if takes place, would necessarily impinge upon their relationships. Furthermore, apart from the LiES, there is a Larger Eco-social Space (LaES), which encompasses the Earth, and could be made up of a number of eco-social spaces, and this too interacts with the Biodiversity. This interaction is shown in the figure in a differently stippled manner. Again the same rider should hold: interaction of the Biodiversity with the LaES would leave the remaining biodiversity 'area' free to indulge in any kind of behaviour. To simplify the idea, we have left out the interactions which take place between the LiES and the LaES.

Thus, under specific circumstances, it is not essential that preservation and maintenance of biodiversity *in toto*, is a *sine qua non* for the life processes to keep going and the biosphere to keep functioning. If the system is under stress it is not necessarily due to the adverse effects, human or otherwise, on the biodiversity; the reasons are many and quite diverse.

The geometry of biodiversity conservation is related to the dynamics of biosphere transformations in terms of eco-social entities in a manner that could be visualized through a set of changes which could be depicted in the way as shown in Fig. 16.2, in which the eco-social entities are shown by differently hatched portions in the biospherical rectangles. Here, two 'initial' biosphere set-ups can be visualized as Biosphere Set-up I and Biosphere Set-up II. These could come in contact with each other (as hinted in the Fig. 16.1) and interact with each other by crossing their respective boundries to give rise to, in a dynamic manner (where entities are under specific biological stresses, including that of evolution and diversification), a system which is recognized as the Added Biosphere Set-up. This should ever remain in a transient stage, for the interactions now assume serious proportions. Bioentities enter the processes of contradictions and regressions, as well as expansions and development, which results in the transient set-up not remaining in an apparent static state but keeping its momentum going, transforming itself into a newer system depicted as the Transformed Biosphere Set-up. However, the newly-emerged system keeps itself involved in the process of continuous transformations. The 'end' never arrives; and the Transformed Biosphere Set-up can keep themselves further transforming. If again, the biodiversity involves the eco-social entities as their constituting elements,

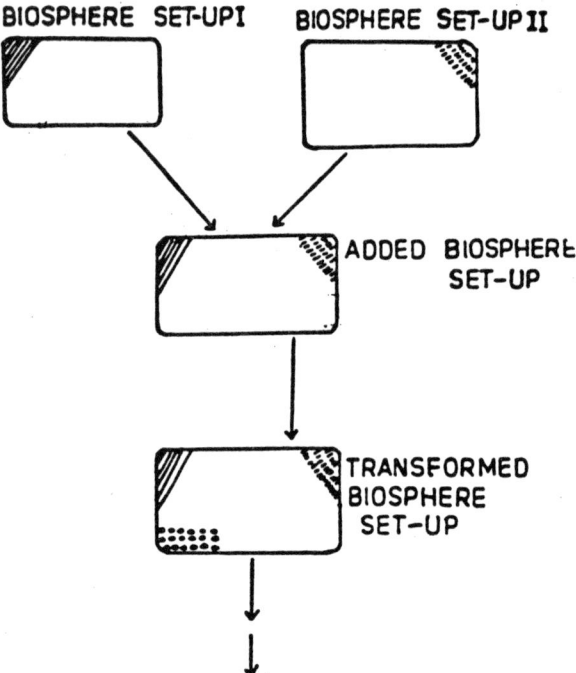

Fig. 16.2: The Dynamics of Biosphere Transformations

Biosphere Set-up I interacts with the Biosphere Set-up II to add up to a dynamically-transformed Added Biosphere Set-up, where not only interactions between the eco-social entities, showed by differently-hatched portions in the biospherical rectangles, take place but a situation develops, in a dynamical sense, that a new set-up arises, which is the Transformed Biosphere Set-up. But whatever its inner contradictions and regressions, the set-up does not remain static but keep its momentum and hence keep changing.

and as these do get changed, it could be visualized that a 'frozen' biodiversity is not what biosphere regards as the basis of their uniqueness and of their development: it remains in a dynamic situation, choosing, rejecting, improving, and evolving.

In a narrow view, science and technology are conventionally accepted as what scientists and technologists produce; and development is accepted as what science and technology produce. In a wider context, where science is viewed as 'ways of knowing' and technology as 'ways of doing', all societies, in all their diversity, have had science and technology systems on which their distinct and diverse developments have been based. Technologies, or system of technologies, bridge the gap between natures' resources and human needs. Systems of knowledge and culture provide the framework for the perception and utilisation of natural resources. The technological approach portrays the evolution of technology as self-determined and views social sacrifice as a necessity.

Agricultural research and technology, to take an example, is primarily a means of developing useful plants. Perhaps the most important barrier in utilising useful plants is the nature of the seed, which reproduces itself and multiplies. The seed thus possesses a dual character that links both ends of the process of crop production: it is both means of production and, as grain, the product. In planting each year's crop, the farmer also reproduces a necessary part of their means of production. The 'improvement' of the seed, as brought about by technology, is a change in the role of agricultural products and the role of ecological process, that is not going to affect the system adversely.

Biotechnology

It is an integration of several disciplines, and provides economical and efficient solutions to the rising problems of energy cost, pollution and depletion of world's renewable resources. The multidisciplinary nature of biotechnology can be gauged through the Fig. 16.3, which shows how biological and chemical sciences and biological engineering can come together to benefit the mankind.

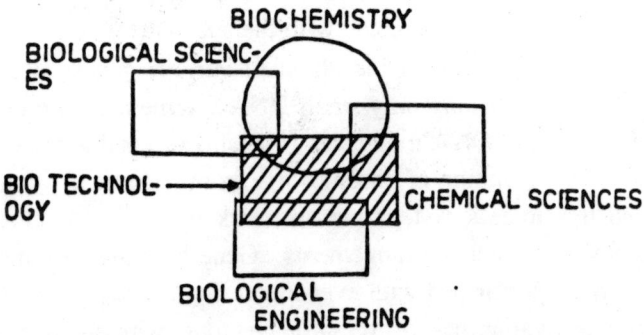

Fig. 16.3: The Multidisciplinary Nature of Biotechnology

Biochemistry is emphasized only because the biochemical techniques were first utilized to initiate the development in biotechnology more frequently. It needs to be noted that microbiological and molecular-biological techniques too were similarly utilized.

Older biotechniques, using biological organisms and fermentation processes, had a distinct empirical, rather than scientific foundation up to the second half of the last century in Europe. The scientific understanding of how these processes worked started developing after Pasteur's discovery that fermentation involves microorganisms (Sharp, 1985; Yanachinsky, 1985). The scientific and technological endeavours which started soon after, led to the use of bioprocess for the production of, e.g., citric acid, and prompted an increase in chemical engineering aspects of industrial microbiology especially in the field of cell culture and sterilization (OTAF, 1985; Sharp, 1985). These bioprocesses encouraged the manufacturing of solvents and related low molecular weight compounds (Whelan & Blacks, 1982; Daly, 1985).

Such developments in fermentation technology, in pharmaceutics and diagnostics kept up the development of biotechniques until biotechnology blossomed into an overall system of bioprocesses.

The advent of contemporary biotechnology is usually traced back to 1973, when Chang and Cohen at the University of Stanford and Boyer and Helling at the University of California in San Francisco developed the recombinant (r) DNA technique; and 1975, when Milstein and Kohler in Cambridge employed hybridoma technology to produce monoclonal antibodies. To these dual innovations in the biotechniques involved in protein engineering, reverse genetics and of polymerase chain reaction (PCR), which utilizes the techniques of homologous recombination, antisense RNA utilization, and DNA primer extension needs to be added.

Genetic engineering has generated a new technological paradigm that has created a new body of research, patents and processes. However, the emergence of biotechnology as a new technical paradigm has itself been made possible by a process of relatively slow and cumulative accumulation of scientific and technical knowledge in very different areas, which jointly has produced a new outlook and a new set of principles and techniques.

In the last decade or so, developmental biology has undergone a dramatic change. This change is primarily due to the ability to isolate and characterize wild-type and mutant alleles of genes that are expressed at specific developmental stages with major impetus from DNA sequencing (Prober et al., 1987), and the ability to generate transgenic organisms after introducing various mutant versions of the wild-type gene, with the subsequent study of their phenotypes.

Such studies allow for a defination of the effect of various genes. Recombinant DNA can now break out of its limits by defining only small DNA fragments. The developments in pulse field gel electrophoresis of large DNA molecules now allows resolution of DNA containing up to 10 mega-basepairs. Thus, isolation of chromosome in pure form is possible. A combination of various blotting and hybridization steps permit quick insight into chromosomal structural abnormalities, oncogenes, and other rearrangements. Gene location, chromosome mapping, and linkage analysis can now be performed with ever-increasing speed. There has been an unparalleled advancement in the availability of versatile vectors, which clone and deliver discrete genomic portions. The available vectors can be used to clone very large fragments (ca 250 kilobases) of DNA into yeast. These fragments are stable in yeast and replicate as linear molecules, and have telomeres at their termini. By adding autonomously replicating sequences, this stability is further enhanced (Burke *et al.*, 1987).

These experiments allow analysis and purification of chromosomes from various organisms. Very large fragments of these chromosomes can be cloned and propagated and then transformed into homologous organisms. This will, e.g., speed up mapping and sequencing of entire genomes, which will not only allow analysis of genetic diseases in humans, for example, but may permit manipulation of the events involved in the development of a fertilized egg.

The predictive power of rDNA technology allows logical planning towards the solution of complex biological problems. This has permitted the determination of gene structure through sequencing that provides clues to the function and evolution of many eukaryotic genes. DNA sequence analysis provides insights into the regulatory sites in the DNA and RNA molecules. The capability to modify precisely the genetic code through site-specific mutagenesis, using strategies based on synthetic oligonucleotide chemistry, may be useful to generate precise mutations so that effects of a single amino acid substitutions can be evaluated. Genomes of many phages, plasmids, viruses, human mitochondria and tobacco chloroplasts, amongst many others, have already been completely sequenced (Moores, 1987). Many more genomes, including those of humans, will be almost, if not completely, sequenced by the end of the century, thus allowing the scientists an access to the blueprint of life stored in their computers (Smith *et al.*, 1987).

The practice of introducing mutant genes into organisms and then studying the phenotypes is the reverse of classical genetics. Using homologous recombination, precise replacement of the resident gene is now possible in almost all the eukaryotes. In higher organisms, transformed DNA integrates randomly and is not specific to the homologous target sites. But with the possibility of direct selection for homologous recombination, it is possible that targeted events can be differentiated from random insertions.

Reverse genetics is a powerful technique for studying developmental biology and has potential application in gene therapy. It can also be used for deleting undesirable DNA sequences, replacing defective sequences, and for protein engineering.

Antisense RNA

Antisense RNA, being complementary to messenger RNA, inhibits gene expression by forming a double-stranded structure with the corresponding message. Antisense RNA regulated gene expression in phages and bacteria, can be used for preventing gene expression in plants and animals. Antisense cDNA library can be prepared to transform organisms that can be screened for desired phenotypes. Antisense is most effective when selective reduction of tissue-specific expression at certain stages of development, regulated by a specific promoter, is required. It can be used to construct viral resistance, and suppress activity of undesirable genes. It has been used to suppress synthesis of petal-specific chalcone synthase, an enzyme required for pigment synthesis, thus generating transgenic plants with altered flower colour. It has now become possible to use antisense RNA technology to alter the phenotype of a mature plants.

Polymerase Chain Reaction

Polymerase Chain Reaction, PCR (Erlich, 1989; Kumar, 1989; White *et al.*, 1989) is an analytical tool and is based on the principle of DNA primer extension and geometric progression. A DNA polymerase catalyzes template-directed synthesis of DNA initiated by a primer hybridized to the template. PCR involves simultaneous synthesis of two complementary DNA strands by the extension of two primers annealed to complementary strands at opposite ends of the template. Because of the unidirectional nature of DNA synthesis, the location of the two primers delimits the region amplified and each newly synthesized DNA strand contains the sequence for the annealing of the primer.

Since each cycle of PCR involves thermal denaturation of DNA, annealing of primers and primer-initiated synthesis (Fig. 16.4A) at every cycle, the amount of DNA is doubled, therefore doubling the number of templates for subsequential cycles of PCR. The initial round results in the synthesis of DNA that has a fixed 5' and a variable 3' end. Later, both ends of the synthesized DNA is fixed, because they are either derived from or determined by the sequences of the primers. This should result in the synthesis of a uniform product of defined length.

Thus, PCR can be utilized in many ways to understand the basic genetic structure. Two examples can be diagrammatically shown. Fig. 16.4B demonstrates how sequences amplified by PCR can be modified at their termini by the addition of noncomplementary extensions to the 5' ends of the primers. These include restriction sites, promoters and binding sites for proteins. Fig. 16.4C shows that unbalanced or asymmetric priming permits the generation of a molar excess of one of the two strands in the template. This is achieved by limiting the amount of one of the primers in the reaction. The single-stranded products can be used for sequencing and for diagnostic applications.

CONCLUSIONS

Subtle changes in the behaviour of the planet Earth keep the biosphere developing and evolving. Since the last few decades becasue of revolutionary changes in the human habitat, the homeostatistical balance which the biosphere maintains has also been undergoing changes. The climatic and other defects sent an alarm that the biodiversity is in great peril (Wilson, 1986; Barnes, 1989).

Biodiversity is being reduced in a devastating manner. The vanishing species, indeed all the species, are the unique and irreplaceable products of millions of years of evolution; and, like every species, ours is intimately dependent on others for its survival. What is, therefore, required is a novel manner of their preservation; and it seems now abundantly clear that biotechnology can assist in preserving their genetic elements. Such biopreservation can keep the genetic information safe for ages. And, furthermore, the species can be 'revived'.

Biotechnology is gradually bringing about a fundamental change in our understanding and utilization of biosphere (Krimsky, 1982; OECD, 1985; OTAF, 1985; Malik, 1989; Hacking,

ONE CYCLE OF P·CR

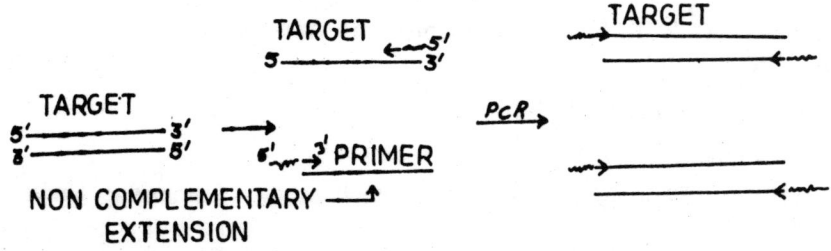

DENATURATION   ANNEALING   PRIMER EXTENSION

B GENERATION OF MODIFIED ENDS BY PCR

TARGET

TARGET

NON COMPLEMENTARY — ↑
EXTENSION

PRIMER

PCR →

C GENERATION OF SINGLE-STRANDED DNA BY ASYMMETRIC PRIMING IN PCR

SINGLE STRANDS

PCR →

DOUBLE STRANDS

Fig. 16.4: The Polymerase Chain Reaction Technique

(A) One Cycle of PCR: Three steps in the amplification of DNA by PCR. Double-stranded DNA is denatured by heating and cooled in the presence of an excess of oligonucleotide primers. The hybridized primers are extended by thermo-stable DNA polymerase to duplicate the original template. Each newly synthesized DNA molecule incorporates a primer at its 5' end.

(B) Generation of modified ends by PCR: Sequences amplified by PCR can be modified at their termini by the addition of noncomplementary extensions to the 5' ends of the primers. These include restriction sites, promoters, and binding sites for proteins.

(C) Generation of single-stranded DNA by asymmetric priming permits the generation of molar excess of one of two strands in the template. This is achieved by limiting the amount of one of primers in the reaction. The single-stranded product can be used for sequencing and for diagnostic applications.

1986; Kenney, 1986; UNESCO, 1988; Jacobsen & Jolly, 1989; Orsenigo, 1989; Cordle & Young, 1990; Crawford, 1990; Finn, 1990; Fox, 1990; Strange, 1990; Vasil, 1990a). The impact of biotechnological ideas and techniques on, e.g., cereals (Davies, 1990; Vasil, 1990b), as a result of the formation of transgenic maize plants, and those having defence genes (Doerner *et al.*, 1990), on preservation of food (Hardy, 1990), its utilization in clinical systems and in animal breeding, have been, and is going to be, tremendous and far-reaching. We are on the threshold, and as a matter of fact passing through the initial stages of a biological revolution.

References

Barnes, R.D. (1989). Amer. Zool. **29**, 1075-1081.

Burke, D.T., Carle, G.F. & Olson, M.W. (1987). Science **236**, 806-812.

Cordle, M.K. & Young, A.L. (1990). Toxicol. Environ. Chem. **28**, 25-36.

Crawford, R.L. (1990). Trends Biotech. **8**, 170.

Daly, P. (1985). The Biotechnology Business, Frances Printer, London.

Davies, J. (1990). Trends Biotech. **8**, 198-204.

Doerner, P.W., Sterner, P., Schmid, J., Dixon, R.A. & Lamb, C.J. (1990). Biotech. **8**, 833-840.

Erlich, H.A. (1989). PCR Technology Principles and Applications for DNA Amplification, Stockton Press, New York.

Finn, R.K. (1990). Food Biotech. **4**, 1-16.

Fox, J.L. (1990). Biotech. **8**, 392-394.

Hacking, A.J. (1986). Economic Aspects of Biotechnology, Cambridge University Press, Cambridge.

Hardy, R.W.F. (1990). Biotech. **8**, 615-617.

Jocobson, G.K. & Jolly, S.O. (1989). Biotechnology: Gene Technology, Vol 7B, VCH Publishers, Goettingen.

Kenney, M. (1986). Biotechnology: The University Industrial Complex, Yale University Press, New Haven.

Krimsky, M. (1982). Genetic Alchemy: The Social History of Recombinant DNA Controversy, MIT Press, Boston.

Kumar, R. (1989). Technique **1**, 133-152.

Malik, V.M. (1989). Adv. Appl. Microbiol. **34**, 263-306.

Moores, J.C. (1987). Anal. Biochem. **163**, 1-9.

OECD (1985). Biotechnology and Patent Protection: An Industrial Review, OECD, Paris.

OTAF (1985). Patent Profile: Genetic Engineering, US Govt. Printing Office, Washington DC.

Orsenigo, L. (1989). The Emergence of Biotechnology, Printer Publ. Ltd., London.

Prober, J., Trainor, G.L., Dam, R.J., Hobbs, F.W., Robertson, C.W., Zagarsky, P.J., Cocuzza, A.J., Jensen, M.A. & Baumeister, K. (1987). Science **230**, 593-596.

Sharp, M. (1985). The New Biotechnology. European Governments in Search of a Strategy, University of Sussex, Sussex.

Smith, C.L., Econome, J.G., Schutt, A., Kleo, S. & Cantor, C.R. (1987). Science **236**, 1448-1453.

Strange, C. (1990). Bioscience **40**, 5-9.

UNESCO (1988). Biotechnologies and Development, UNESCO, Paris.

Vasil, I.K. (1990a). Biotech. **8**, 296-302.

Vasil, I.K. (1990b). Biotech. **8**, 797-798.

Whelan, W.J. & Blacks, S. (1982). From Genetic Engineering to Biotechnology: the Critical Transition, Wiley, New York.

White, T.J., Arnheim, N. & Erlich, H.A. (1989). Trends Genet. **5**, 185-189.

Wilson, E.O. (1986). Biodiversity, National Academy Press, Washington, DC.

Yanachinsky, S. (1985). Setting Genes to Work, Viking, London.